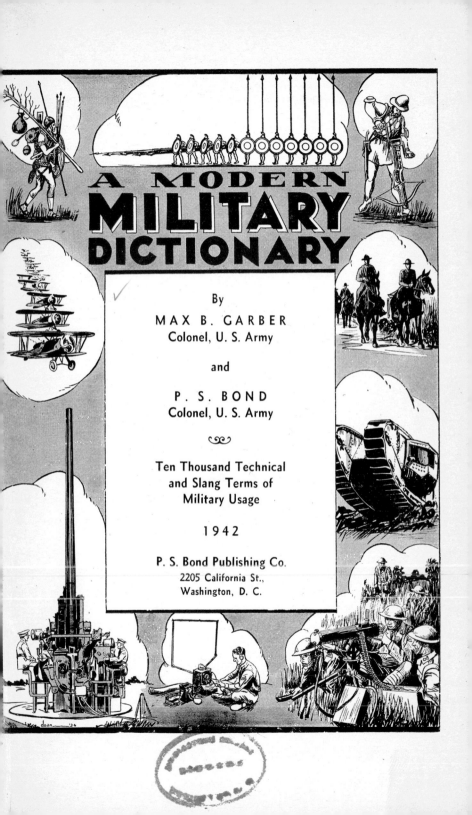

A MODERN
MILITARY
DICTIONARY

By

MAX B. GARBER
Colonel, U. S. Army

and

P. S. BOND
Colonel, U. S. Army

Ten Thousand Technical
and Slang Terms of
Military Usage

1942

P. S. Bond Publishing Co.
2205 California St.,
Washington, D. C.

Price, $2.00

PREFACE TO THE FIRST EDITION

There is need for a modern and comprehensive military dictionary. Every trade and profession has its characteristic language, and this is true of the military profession and its many branches. The World War and the period following it have witnessed a great and complex development in the military art. New and improved machines and weapons, the application of all known science to the conduct of war, and new tactical methods, have brought into use many new terms whose meanings are now becoming crystallized. Effective cooperation in every phase of military effort requires that members of each branch should be acquainted with the tactics and technique of all branches, and familiar with the terminology applicable thereto. In brief, military men must understand and speak the language of their profession. A standard work of reference promotes uniformity, saves time and effort, and often embarrassment.

This work was begun in 1922 when I was an instructor at the Infantry School, and has been prosecuted more or less continuously since that time. During its preparation I have had access to the libraries at the Infantry School, the Command and General Staff School and the Army War College, the Library of Congress, and many private libraries. In the search for terms and their meanings, most of the important military works that have appeared since colonial days, several hundred in number, have been critically examined, as well as publications of the War Department and the Service Schools of the Army, and military periodicals. From these sources, ancient and modern, about 10,000 terms have been selected. While emphasis is placed on current military terms of the United States Army, there have been included also important terms from other languages, and ancient terms, many now obsolete. These latter will be helpful in the study of foreign military writings or references thereto, ancient works on the military art, and military history.

The selection of terms to be included has naturally rested on my judgment. Definitions which appeared to be accurate and clear for the layman as well as the expert, have been accepted as found. When this was not the case, they have been rewritten to make them clear, but every effort has been made to retain unchanged the accepted military meaning of the term. Cross references have been sparingly used: thus all terms involving the common term "fire," will be found under that heading. Very condensed technical and historical articles have been included when this seemed desirable.

To the military man this work will serve as a practical working manual of the language of the military profession. To others it will be useful as a reference book of military terminology.

In the preparation of this text I have had the assistance of many persons, too numerous to be mentioned herein. I desire to express my appreciation of the valuable help which has been freely given me. Without it this work could not have been accomplished. as it is too great a task for the unaided effort of any one person. No book can be perfect; suggestions for the improvement of this one will be welcomed.

1936
Washington, D. C.

Max B. Garber,
Colonel, U. S. Army

PREFACE TO THE SECOND EDITION

Developments of the military art in the past few years have made this revision of A MODERN MILITARY DICTIONARY advisable. In making the revision the book has been brought up to date in every respect, and has been "streamlined" by the elimination of many ancient and foreign terms. Slang terms, of the old Army as well as the new, have been included in an Addenda. In its present form the authors believe the book will be of great value and interest to the average general reader, as well as a necessity to the military man.

The author welcomes as collaborator, Colonel P. S. Bond, who is probably the Army's best known writer on technical military subjects.

Washington, D. C.
1942

Max B. Garber,
Colonel, U. S. Army

A

Abatage. A piece is said to be in abatage when the wheels rest on the brake shoes. A demolition by high explosives.

Abatis. An obstacle consisting of trees felled or placed with their tops to the front, often interlaced with barbed wire. A live abatis is one consisting of saplings bent to the ground but not cut, so that the leaves do not wither.

Abnormal Shot. A shot whose point of impact is more than six probable errors from the center of impact.

About Face. A facing to the rear in the School of the Soldier, executed by turning to the right.

Abreast. Said of a line of men or units, side by side. Equally advanced. On the same front.

Absence Without Leave. Absence from post or duty without permission from proper authority and when there is no intention of deserting. Abbreviated, A. W. O. L.

Absent. A deficiency of any given number of officers or enlisted men. Not present.

Absentee. A soldier absent from any roll call or duty, whether authorized or unauthorized.

Absolute Deviation. The shortest distance between the center of the target and the point or center of impact or burst.

Absolute Error. The shortest distance between the center of impact and the point of impact.

Abstract. A prescribed form to accompany returns, consolidating the contents of several vouchers or other documents. A summary or epitome.

Abuses and Disorders. Minor offenses committed more by reason of a lack of discipline than of deliberate intent to do wrong.

Abutment. The block at the rear of a firearm which receives the rearward thrust of discharge. The support of a bridge at the shore end, including a connection to the approach.

Abutment Bay. The span of a ponton bridge at the shore end, usually supported by a trestle and an abutment sill.

Academic Board. The heads of the departments of instruction at the military or naval academy, when acting as a collective body.

Accelerated Velocity. Variable velocity in which the space covered from instant to instant is constantly increasing.

Accelerator. A cannon in which several successive charges are fired to give an increased velocity to the projectile within the bore. A part of the recoil system of a machine gun which accelerates the motion of the bolt.

Accelerograph. The apparatus for measuring the succession of pressures at a given point in the powder chamber by a powder charge acting on a projectile.

Accessories. Articles used in connection with a weapon or machine, but which are not an integral part thereof; as a magazine filler, oiler, etc.

Accessory Defenses. Artificial obstacles of any kind.

Accidental Errors. Errors which are nonsystematic in nature and which, in artillery fire, cause the dispersion of shots about the center of impact.

Accompanying Artillery. Artillery consisting of single guns, platoons, batteries, and sometimes battalions of light artillery, temporarily assigned to and under the direct command of the infantry battalion, regimental, or brigade commanders, to provide close fire support to the unit with which it is operating. Artillery attached to and physically accompanying a subordinate unit of attack troops (infantry or cavalry).

Accompanying Battery. A battery of light artillery attached to attacking infantry, usually a regiment, for close fire support of the latter unit in the attack.

Accompanying Gun. A gun attached to smaller infantry units, generally assault battalions, for use as an emergency weapon in close support.

Accompanying Tanks. Tanks which operate in small groups in close support of small infantry units (usually battalions), either preceding or following them.

Accompanying Weapons. Weapons, other than infantry weapons, attached to infantry units, or to troops acting as infantry, to assist in an attack. Weapons other than those pertaining to cavalry, which may be attached to a cavalry command.

Accountability. The condition of being required to account for the amount and condition of money or property.

Accountable Officer. An officer who has charge of property carried on a stock record, or of funds, of which he is required to render periodic returns or reports as to the status of the property or funds. Property may be issued to other officers or to organizations on memorandum receipts, whereupon such officers or the commanders of the organizations become responsible for such property.

Accounts. Systematic records of receipts and expenditures of funds or property.

Accounts Current. Running or continued accounts between disbursing officers and the accounting officers of the government.

Accouter. To furnish or equip with uniform and equipment.

Accouterments. Articles, other than arms, carried by the soldier outside his clothing.

Accuracy. The mechanical ability of a firearm, when clamped in position, to place a group of shots in a small area on a target. The accuracy of the weapon, as distinct from that of the man firing it.

Accuracy Life. The number of rounds, using the full service charge, which may be fired from a cannon with acceptable accuracy, and beyond which, due to erosion, the dispersion is excessive. For the purpose of fixing ammunition allowances, accuracy of life is estimated.

Accusation. The charges and specifications preferred against an accused.

Accused. A person who is arraigned before a military court for trial.

Accuser. An officer who prepares or signs charges against any person for misconduct in the military service.

Ace. A military aviator who has brought down a prescribed number of enemy aircraft (usually five) under prescribed conditions.

Acoustic Corrections. The corrections applied to azimuth and elevation to correct for sound lag. Sound-lag corrections.

Acoustic Corrector. An instrument for determining and applying the angular corrections to azimuth and elevation, as obtained by the sound locator, to compensate for sound lag. It also applies certain arbitrary corrections. A sound-lag corrector.

Acquit. To release or set free from an obligation, accusation, or suspicion. To perform a duty or obligation.

Acquittal. A formal finding of "Not Guilty" of all charges and specifications before a court-martial.

Acrobatics. Evolutions voluntarily performed with an aircraft other than those required for normal flight.

Acting. A prefix to a title which indicates that the person so styled is filling a position temporarily.

Action. An engagement or battle, usually one on a small scale. Mechanical functioning. Unlimbering of a gun preparatory to opening fire, or the command for same.

Active Defense. A class of defense which resorts to the active measures of counterattack, as well as passive resistance; and seeks to seize the initiative by the aggressive use of large reserves.

Active Duty. Serving on full pay on the active list. Performance of active military duties.

Active List. Officers and soldiers eligible for full duty, as distinguished from those retired from the service.

Active Service. Engaged in operations against the enemy or in campaign. Service on full pay on the active list, in contradistinction to the retired list.

Acts of Hostility. Acts of a diplomatic, commercial, or military nature, involving or preceding a state of war between two or more nations.

Actual Range. The horizontal distance from a gun or directing point of a rifleman to a target.

Adaptation to the Ground. The process of adapting the intrenchments to be constructed, the distribution of troops, and available weapons to the terrain to be occupied and defended.

Adapter. A device used in adjusting or fitting different parts of any equipment or apparatus together. A bushing that fits in the nose or base of a shell to adapt the shell to fit the fuze, sometimes having a booster charge attached.

Adjust. To correct the elevation and deflection of a gun, or the burst of its projectile, usually by observation of its fire, so as to bring the center of impact within effective distance of the desired point. To make to fit comfortably or properly.

Adjusted Elevation. An elevation corrected by observation of fire so as to place the center of impact on or very near to the target.

Adjusted Range. The range obtained by actual firing, which places the center of impact on or near the target. The range setting corresponding to the adjusted range.

Adjusted Range Correction. The range correction obtained or proven by actual firing which places the center of impact (or burst) on or near the target.

Adjusting a Traverse. When the plot of a closed traverse fails to close on itself or on a point whose location is accurately known, the errors of closure, both horizontal and vertical, are distributed proportionally so as to force closure.

Adjusting Point. The particular point or part of the objective on which fire is adjusted.

Adjustment. The process of correcting the elevation and deflection of a gun and the height of burst until the projectile bursts or falls at a desired point. Fire whose object is to determine, from the observed position of the shot and the target, the firing data necessary to produce the desired effect. See Fire, Adjustment of.

Adjustment Corrections. Corrections to firing data based on observation of fire.

Adjutant. A staff officer who assists his commanding officer in the performance of administrative duties.

Adjutant General's Department. The department of records, correspondence, and orders in the War Department, including the personnel thereof. The head of the department is the Adjutant General.

Adjutant's Call. The signal for organizations or guard details to assemble on the parade ground preparatory to formation for parade, guard mounting or other ceremony.

Administration. The management and direction of the affairs, especially the routine affairs of an organization, station or command. It comprises the care of troops and property, supply, discipline, records and reports, as distinct from training, maneuver and combat. The management and conduct of affairs connected with military organization.

Administrative and Supply Group. That portion of a headquarters which is concerned with administration and supply. In combat it forms the rear echelon of the headquarters. See also Command Group.

Administrative and Technical Staff. See Special Staff.

Administrative Control. Control in administrative matters without tactical command.

Administrative Map. A map on which information pertaining to administrative matters is graphically recorded, as supply, traffic, etc.

Administrative Order. An order covering administrative details, such as traffic, supply, and evacuation, when the instructions are too voluminous to be included in paragraph 4 of the field order, and at other times when necessary to publish administrative instructions to the command; usually issued by divisions and higher units.

Administrative Order, Paragraphs of. See paragraphs of an administrative order.

Administrative Unit. An organization whose headquarters exercises administrative functions, especially those pertaining to personnel and supply; it may be and usually is also a tactical unit, as a company, regiment, division.

Adrian Barrack. A type of portable barrack used extensively in the World War.

Advance. The progress of a command toward the enemy when within the range of small arms aimed fire. To move forward. To make progress in the direction of the enemy.

Advance by Bounds. An advance controlled by the designation of certain objectives for successive movements, usually from one terrain feature to the next, or from cover to cover.

Advance by Echelon. An advance of a unit by the successive movements of its component units. See also Leapfrog.

Advance by Rushes. An advance in which the various elements progress, usually alternately or successively, by a series of short rushes, the advance of the rushing element being covered by the fire of other elements.

Advance Cavalry. That part of the cavalry of the advance guard which precedes the support.

Advance Depot. A general or branch depot, designated as such by the commander of the theater of operations, located in the advance section of the communications zone, for the storage of supplies of balanced stocks forwarded from the rear or procured in the advance section for issue.

Advance Guard. A detachment which precedes and affords protection for the main body on the march.

Advance Guard Artillery, Cavalry, etc. Artillery, cavalry, etc., attached to an advance guard.

Advance Message Center. A message center established during combat to serve as a relay point for communications between superior and subordinate units; usually at the head of the axis of signal communications.

Advance On. A term used in an order when doubt exists as to whether the unit will reach the destination mentioned.

Advance Party. A detachment that is sent out by, and moves ahead of the support of the advance guard and forms the reconnoitering element of the support.

Advance Section. The forward or most advanced subdivision of the communications zone. An advanced base of supply for the combatant troops.

Advance To. A term used in an order when it is reasonably certain that the destination designated will be reached.

Advanced. Any portion of a force which is in front of the remainder. Posted or located in front of another. Promoted in rank.

Advanced Command Post. The location of the commander or a small staff, or both, other than at the command post or rear echelon of the unit.

Advanced Gun Drill. Machine gun drill to include the duties of all members of the squad, on varied terrain.

Advanced Position. A position occupied by troops, in attack or defense, which is in front of that held by the main force.

Advanced Post. A position well to the front, occupied for the purposes of security or observation. An observation or listening post placed well to the front.

Advanced Training. Higher training for individuals or organizations who have received basic training.

Advanced Works. Works of fortification placed in advance of the general line, but within supporting distance. Works exterior to the covered way, but connected to the main work.

Advancement. Honor, promotion, or preferment. Approach to a sentinel when so ordered, following a challenge.

Advancement, Order of. The priority in which persons approaching a sentinel's post at night are advanced for recognition; e. g., commanding officer, officer of the day, officer of the guard, etc.

Advancing the Attack. Pushing forward an attack by means of fire and movement. A phase of the attack extending from the time of opening fire to the assault. See Fire Fight.

Aerial. Of or pertaining to aircraft. Existing or performed in the air. Designed for, or operating from aircraft. The elevated conductor portion of a radio antenna.

Aerial Acrobatics. The stunting of aviators, looping, spinning, rolling, etc.

Aerial Barrage. A barrier of barrage balloons raised against aerial attack.

Aerial Bomb. See Air Bomb.

Aerial Bomber. A rating given to graduates of an aerial bombing school who have demonstrated their fitness for such rating.

Aerial Camera. A camera used in an aircraft for making aerial photographs. It may include a single lens, or a central lens and two or more inclined peripheral lenses (multiple-lens camera).

Aerial Flight. A journey in an aircraft; it begins when an aircraft takes off from a point of rest or support and terminates when it next comes to a complete stop at a point of support.

Aerial Gunner. A rating given to graduates of an aerial gunnery school, who have demonstrated their fitness for such rating.

Aerial Gunnery. The science of delivering gunfire from an aircraft.

Aerial Navigation. See Air Navigation.

Aerial Navigation Maps or Charts. Maps or charts prepared especially for the use of air navigators.

Aerial Observation. Observation from airplanes, balloons, or other aircraft.

Aerial Photograph. See Air Photograph.

Aerial Reconnaissance. Reconnaissance made from or by aircraft.

Aerial Sickness or Airsickness. A sickness affecting persons flying in aircraft, due to speed of flight and rapidity in changing altitudes, similar to seasickness.

Aerial Sound Ranging. The process of locating aircraft by means of the sounds emitted.

Aerial Spotting. Spotting or sensing of artillery fire by aerial observers.

Aerial Torpedo. Any inclosed charge of explosive designed to be propelled through the air either by its own motive power or by gravity. The explosive shell thrown by a trench mortar, constructed so as to fall with its point down. A torpedo discharged from an airplane.

Aerodrome. See Airdrome.

Aerodynamics. That branch of dynamics that treats of the motion of air and other gaseous fluids and of the forces acting on solids in motion relative to such fluids.

Aerofoil. See Airfoil.

Aerogun. An antiaircraft gun.

Aeromechanic. A mechanic expert in the making and repair of aircraft.

Aeromechanics. The science of the equilibrium and motion of air and gas, including aerodynamics and aerostatics.

Aeromotor. A light and powerful motor for use in aircraft.

Aeronaut. An aerial navigator; specifically, the pilot of an aerostat. A balloonist or aviator.

Aeronautical Charts. Maps upon which information pertaining to air navigation is shown.

Aeronautics. The science and art of rising from the ground and flying through the air by means of an aircraft. The science and art of flight. It includes aviation, or the operation of heavier-than-air machines, and aerostation, or the operation of lighter-than-air craft. As far back as written records extend we find references to man's desire to ascend into the air and to fly like a bird. As early as the 15th century A. D., clumsy attempts were made to fly by means of wings attached to the human body, and later by machines of various kinds. These attempts were uniformly unsuccessful and frequently disastrous. Aerostation first achieved real progress. In 1783 the Montgolfier brothers of Paris sent up balloons filled with hot air. This was followed in the same year by successful ascents by man. In 1795 two passengers crossed the English Channel in a hydrogen balloon. Efforts were made to devise some means for steering and propelling aerostats. By 1882 a dirigible balloon capable of a speed of about 6 miles per hour was devised. Beginning about 1900 many dirigibles appeared, especially in Germany and France. As early as 1905 a speed of nearly 30 miles per hour was attained. This was gradually increased to about 80 miles per hour. In 1929 the first round-the-world flight was accomplished in a little over 21 days.

The idea of personal flight with flapping wings was dropped, and the development of the heavier-than-air machine comprised numerous attempts with gliders, and efforts at propulsion with steam and gas engines. The first truly successful power-driven machine was constructed by the Wright brothers. It was tested at Kitty Hawk, N. C., in December, 1903, which date marks the beginning of modern aviation. Several successful flights were made, in which the machine rose from the ground by its own power. In one of these flights it was in the air for 59 seconds and traveled a distance of 852 feet. Following this successful demonstration of the possibility of mechanical flight, progress in aviation was rapid. Within two years the Wright brothers achieved a continuous flight of over 24 miles at an average speed of about 38 miles per hour. The World War gave a tremendous stimulus to aviation. Within two decades after its close the continents and oceans had been spanned, the globe girdled and both poles reached by airplane. The airplane has attained a speed of over 450 miles per hour, and has demonstrated its ability to remain indefinitely in the air. Many commercial, mail and passenger lines have been established all over the world.

Aeronautics have played a constantly expanding role in warfare. A military balloon was used as early as 1794, and Napoleon's army included a corps of balloonists. During the American Civil War balloons were used as observation posts. In the Franco-Prussian War balloons were employed to send dispatches and carriers out of Paris during its siege. At the outbreak of the World War (1914) both France and Germany were equipped with several hundred serviceable aircraft. As the war progressed thousands of improved planes were built and used by all the combatants. At present an air force is an essential part of the armed forces of all nations, and plays a dominant part in warfare over both land and sea.

Aerophotography. The art of photographing from or by means of aircraft. Aerial photography.

Aeroplane. An airplane.

Aeroscope. An instrument which indicates the direction and velocity of the wind and the density of the air.

Aerostat. A generic term for aircraft whose support is chiefly due to buoyancy derived from aerostatic forces. The immersed body consists of one or more containers filled with a gas that is lighter than air. A balloon or airship.

Aerostation. The art of operating aerostats.

Aerostier. A man serving with a balloon organization.

A-frame. A wooden frame, in the shape of an A, placed at intervals in a trench to support revetments and trench boards.

Agencies of Liaison. Agencies by which liaison may be maintained. The important agencies are: (1) All means of signal communication; (2) Individuals or bodies of troops employed to maintain connection between adjacent units, which are designated as connecting files and connecting groups, respectively; and (3) Officers or enlisted men sent by one organization to another to secure teamplay, who are designated as liaison officers and liaison agents.

Agencies of Signal Communication. The personnel and equipment necessary to operate message centers, signal intelligence, signal supply, and messenger, pigeon, radio, visual, sound and wire communication.

Agent. A officer or enlisted man attached to another command to secure and transmit information to or from his organic commander. One who represents another. A deputy.

Agent Officer. Any officer properly authorized to disburse public funds as the agent of an accountable disbursing officer. A Class A agent officer is one detailed by a commanding officer to make specified payments, as a company commander. A Class B agent officer is one detailed as finance officer at a post, camp, or other command, and empowered to make disbursements for general purposes.

Agent, Scout and Reconnaissance. An individual sent out to reconnoiter. A messenger.

Aggression. An unprovoked attack. An unwarranted invasion or encroachment upon the rights or territory of another. An act of hostility.

Aggressive War. A war carried on in the territory of the enemy, implying attack upon him. Offensive warfare.

Agraffe. A coupling pin used by field artillery. The clasp of a cuirass.

Aid. Assistance. Help. To assist. To support. An aide-de-camp.

Aid Station. An establishment of the Medical Corps provided for the emergency treatment, sorting and further disposition of casualties in combat. The first station on the route of evacuation, to which the wounded are brought on litters. Aid stations are usually established for each battalion in combat by the battalion medical detachment.

Aide-de-camp. A member of the personal staff of a general officer, in the relation of a confidential attendant.

Aids, The. The means employed in controlling a mount, consisting of the reins, and the legs, weight and voice of the rider.

Aiguillette. A decoration consisting of gold or gilt cords and loops worn on the right shoulder.

Aileron. A hinged or pivoted portion of an airplane wing, the primary function of which is to impress a rolling moment on the airplane in order to control its motion; it is usually part of the trailing edge of a wing. A shoulder-caponier or flank of a small earthwork.

Ailette. A small, square shield worn on the shoulder; the prototype of the epaulet.

Aim. The alignment of the bore of a firearm, or the shaft of a missile, so that the projectile will strike at a desired point; accomplished by eye, by the use of a sight, quadrant, elevation or azimuth scale, etc., or by any combination of methods. To establish by any means, an alignment of a missile weapon so that the missile will strike a desired point.

Aiming. The process of establishing a correct aim. The terms aiming, laying and pointing are often used interchangeably.

Aiming Circle. An instrument for measuring horizontal and vertical angles and for general topographic work in connection with artillery or machine gun fire.

Aiming Device. An attachment for a rifle, which permits an instructor to see the reflection of the sights and the object aimed at and thus determine errors in aiming.

Aiming Disc. A small disc, having a bullseye and pinhole in the center, and provided with a handle; used for instruction in sighting and aiming.

Aiming Drill. Exercises designed to teach the proper method of pointing and aiming firearms.

Aiming Line. A line along which the line of sight is directed when the piece is laid for direction.

Aiming Lights. Lights placed on aiming stakes for use in night firing.

Aiming Mechanism. The devices by which a gun is laid for elevation and direction.

Aiming Off. The procedure of aiming to the right or left of, or in advance of a target, to allow for the effects of wind when the sight has no wind gauge, or for the movement of the target. See also Lead.

Aiming Point. The point or object upon which the sights are aligned when laying the piece for direction.

Aiming Point Offset. The horizontal angle whose vertex is at the aiming point, measured clockwise from the piece to the observation post.

Aiming Post, Stake. A post or stake used as an auxiliary aiming point for direction and sometimes for elevation.

Aiming Rule. A device used to furnish the equivalent of an aiming point at infinity, consisting of a level bar on which slides a sleeve carrying a panoramic sight.

Aiming Stand. A device employed in teaching the proper method of aiming a rifle, such as a tripod, or a box, with means for supporting the piece.

Air Alert Method. The employment of pursuit aviation in antiaircraft defense by maintaining it on an alert status in the air above, or in the immediate vicinity of the defended objective.

Air Area. The area within which a tactical commander is responsible for air reconnaissance and observation. The area assigned for the operations of an aviation unit.

Air Attack. The attack of ground objectives by aircraft.

Air Base. A base of operations for aircraft, and for their storage, repair, and maintenance.

Air Beacon. A beacon on an established air route. See Beacon.

Air Bomb. A high explosive, chemical or incendiary bomb, designed to be dropped from an aircraft.

Air Brake. A movable surface or airfoil, operated by the pilot, used in landing to check forward motion by air resistance. A vehicle brake operated by compressed air.

Air Burst. The burst of a projectile in air, before impact.

Air Cylinder. A pneumatic buffer designed to absorb or check the recoil of guns and to store energy (compressed air) for the return of the piece to its firing position.

Air Defense Command. An organization for the coordination of all measures of defense against enemy air operations.

Air District. A territorial command for administrative and tactical supervision over aviation activities.

Air Field. See Landing Field.

Air Fleet. A large group or assemblage of military aircraft under a single command. The collective military aircraft of a nation.

Air Force. The organization which is charged with combining the capabilities and directing the operations of its component units so as to obtain the necessary continuity of air operations for the period of time essential to the success of the mission assigned to its designated theater of operations.

Air Gun or Pistol. A weapon which projects missiles a short distance by the force of compressed air; commonly used in target practice.

Air Hole or Pocket. A local region in the atmosphere which, because of peculiar meteorological conditions, affords little or no support for the sustaining surfaces of an aircraft. A partial vacuum in the air.

Air Lock. Two gas-tight blankets and a small interval separating them, in the entrance to a shelter. Only one blanket may be open at any time while persons are passing in or out.

Air Log. An instrument for recording the linear travel of an aircraft, relative to the air. One form consists of a windmill with revolution counter.

Air Mechanic. A rating of qualified enlisted men in the air service.

Air Mine or Torpedo. A shell containing a relatively large charge of explosive, projected a relatively short distance by a mine thrower or trench mortar. A torpedo discharged from an airplane.

Air Navigation. The art of determining the geographical position, and maintaining desired direction, of an aircraft relative to the earth's surface by means of pilotage, dead reckoning, celestial observation, or radio aids. Analogous to marine navigation.

Air Officer. The chief of aviation of a corps or army. The officer in charge of aircraft in any command.

Air Photograph. A photograph taken from an aircraft. See Oblique and Vertical Photographs; Mosaic.

Air Pocket. An air hole.

Air Power. The materiel and personnel, potential as well as actual, upon which a nation bases its plans for the employment of aircraft in war. Fighting strength in the air, analogous to sea power or military power.

Air Raid. A raid into enemy territory made by one or more military aircraft for the purpose of destroying or damaging an enemy force, materiel, communications, or harassing a civilian population by dropping bombs.

Air Reconnaissance and Observation. The gaining of information from the air through visual and photographic means.

Air Resistance. The resistance of the air to a projectile or an aircraft in flight.

Air Scouts, Air Guards. Ground personnel detailed to furnish warning of the approach of hostile aircraft.

Air Service. The service of a nation charged with the operation of aircraft in war. An organization operating aircraft, usually over other than prescribed courses and without regular schedules.

Air Speed. The speed of an aircraft with reference to the air through which it is flying.

Air Speed Meter. An instrument which indicates the speed of an aircraft, relative to the air.

Air Station. The general term applied to any stopping place for aircraft of any type and that is provided with facilities for their reception and servicing.

Air Superiority. Superiority over enemy aviation sufficient to permit air or ground operations in any desired locality without effective hostile air opposition. See Air Supremacy.

Air Supremacy. The ability to use observation and bombardment aviation in cooperation with the army or the navy, or independently, while denying the same ability to the enemy; a situation that may be only temporary.

Air Theater of Operations. A theater of operations that has been designated to contain an air force, but not a ground force.

Air Torpedo. See Aerial Torpedo.

8

Air-borne Infantry. Infantry which is moved to its place of employment by transport planes.

Air-borne Troops. A general term which includes parachute troops and other troops transported and landed by or from aircraft.

Aircraft. Any weight-carrying device, designed to be supported by the air, either by buoyancy or by dynamic action. Any form of craft designed for the navigation of the air. They are classed generally as aerostats (lighter than air) and airplanes (heavier than air).

Aircraft Artillery. Artillery designed for and used in aircraft.

Aircraft Defense. The defense against hostile ground attack provided by aircraft assigned to an area or sector.

Aircraft Machine Gun. A machine gun especially designed for use in aircraft.

Aircraft Observer. A passenger in or occupant of an aircraft, charged with duties of observation, as of troop dispositions, enemy activities, effects of artillery fire, etc., or who assists in maintaining liaison between elements of the combatant forces.

Aircraft Pilot. One who controls and navigates an aircraft in flight. For long journeys a separate navigator may be provided. A rating or qualification given upon satisfactory completion of a proper course of training.

Aircraft Plotter. A device for plotting tracks, headings, position lines, or bearings on a chart.

Aircraft Warning Service. A warning system for the primary purpose of detecting the presence of and courses followed by hostile aircraft and of distributing this information.

Airdrome. A landing field provided with facilities for the shelter, maintenance, and operation of aircraft, and the shelter and care of aviation personnel.

Airfoil. Any surface, such as an airplane wing, aileron, or rudder, designed to obtain reaction from the air through which it moves. A control surface.

Air-landing Troops. Troops moved by aircraft that disembark after the aircraft reaches ground.

Airman. One skilled in the operation of aircraft and in matters pertaining to aviation.

Airmanship. Knowledge and skill in the art of aerial navigation.

Air-photo Transfer. A transfer of fire in which the relation between the check point and the target is determined from an air photo.

Airplane. A mechanically propelled, fixed wing aircraft, heavier than air, supported in flight by the dynamic reaction of the air against its wings. Airplanes are classified in many ways: according to the number of wings, position of propellers, number of engines, type of landing gear, etc., and according to the purpose for which they are employed. See Aviation, Classes of.

Airplane Area. The area dependent on our aircraft for defense against enemy aircraft.

Airplane Bomb. A bomb filled with persistent vesicants or lacrimators, used in situations where greater persistence is desired, or when flying conditions prevent the use of tanks. A projectile dropped from an aircraft, whose flight through the air depends upon the speed of the craft and force of gravity.

Airplane Controls. The means provided to keep an airplane balanced, and to direct its flight.

Airplane Courier. A messenger using an airplane for transportation.

Airplane Flight. An elementary organization of aviation composed of a small number of airplanes of the same type.

Airplane Group. Two or more, usually four, squadrons, all of the same type of aviation.

Airplane Observer. See Aircraft Observer.

Airplane Pilot. A rating given to reserve pilots upon satisfactory completion of training in an Air Service Flying School.

Airplane, Pusher. An airplane with the propeller or propellers aft, or in rear of the main supporting surfaces.

Airplane Squadron. The basic tactical heavier-than-air unit, corresponding to the company. It consists of two or more flights, all of the same type of aviation.

Airplane, Tandem. An airplane with two or more sets of wings of substantially the same area (not including tail unit), placed one in front of the other and on about the same level.

Airplane Tank. A special tank mounted on the under side of an airplane for dispersing liquid chemicals into the air.

Airplane, Tractor. An airplane with the propeller or propellers forward of the main supporting surfaces. This type is now almost universally employed.

Airplane Wing. Two or more groups, which may be of different types of aviation.

Airport. An area, either of land or water, which is adapted for the landing and taking off of aircraft and which provides facilities for their shelter, supply, and repair. A place used regularly for receiving or discharging passengers or cargo by air.

Airship. An aerostat provided with a propelling system and with means of controlling the direction of motion; when its power plant is not operating, it acts like a free balloon. Any type of aircraft deriving its support mainly from the lifting effort of lighter-than-air gases and which has also some means of propulsion and control. A dirigible balloon.

Airship, Nonrigid. An airship whose form is maintained only by the internal pressure in the envelope.

Airship, Rigid. An airship whose form is maintained by a rigid framework.

Airship, Semirigid. An airship whose form is maintained by means of a rigid or jointed keel in conjunction with internal pressure in the gas containers and ballonets.

Airship Shed. See Dock.

Airship Station. A complete assembly of sheds, masts, gas plants, shops, landing fields, and other equipment required to operate airships and supply their needs. A base from which airships are operated.

Airship Wing. See Balloon Wing.

Airway. An air route along which aids to air navigation, such as landing fields, beacons, radio direction-finding facilities, intermediate fields, etc., are maintained.

Airworthiness. The quality of an aircraft denoting its fitness and safety for operation in the air under normal flying conditions. Analogous to seaworthiness of a ship.

Alarm. A noise or signal which denotes the presence of the enemy or other danger. A sudden apprehension cf being attacked.

Alarm Gun. Formerly, three guns were placed in advance of a camp, 100 paces from the artillery posts, ready to be fired, as an alarm in case of a sudden attack by the enemy.

Alarm Post. A place designated for men to repair in case of sudden alarm.

Alert. An alarm from a real or threatened attack. A warning signal for a guard. A period of time during which troops stand to in response to an alarm. An alarm against the presence or suspected presence of gas or other chemicals. A position in which the gas mask is carried when in the zone of possible gas attack in order to insure its rapid adjustment to the face. Vigilant. Watchful.

Alidade. A ruler or straightedge, provided with sights or a telescope, used to establish direction lines in plane table surveying and military sketching. The movable arm of a graduated instrument carrying sights or a telescope, by which an angle is measured from a base line.

Alien Enemy. One who owes allegiance to a hostile government.

Align. To form in line. To dress troops on a line. To bring the sights of a weapon in line with a target. To lay out the ground plan of.

Alignment. A straight line upon which several elements are formed, or are to be formed. The dressing of several elements upon a straight line. The course of a road or railroad, especially with reference to the amount of curvature.

All Around Fire Trench. A trench, not necessarily closed, which permits fire in any direction.

All Around Traverse. Traversing gear which permits the fire of a gun to be turned in any direction.

Allocation of Facilities. The assignment, by the Assistant Secretary of

War, of facilities to a supply arm or service for its particular use for procurement planning.

Allotment. A definite portion of the pay of an officer, warrant officer, Army nurse, enlisted man, contract surgeon, or permanent civilian employee of the War Department (when on duty outside the continental limits of the United States), which is voluntarily authorized to be paid to another person, or institution, in a manner prescribed by the Secretary of War. An allowance or assignment of troops, supplies or funds.

Allowance of Quarters. The prescribed number of rooms allowed to certain members of the military service, according to their rank or grade.

Allowances. Money or an equivalent in addition to prescribed rates of pay. Amounts of various supplies or accommodations prescribed for individuals or organizations as a matter of routine. Monetary payments in lieu of supplies in kind. Commutation.

All's Well. A call which is passed once each hour around the cordon of sentinels of an interior guard. It signifies that each sentinel is present and that all is well on his post.

Ally. A state or sovereign leagued with another by treaty, agreement, or common action.

Alternate Firing Position. A position to which a weapon is moved when its primary position become untenable, and from which its primary fire mission can be executed, at least in large part.

Altimeter. An instrument which registers or determines altitude, as of an airplane in flight, or the height of an aerial target.

Altimetric Roof. The figure showing the relation between the base line and the altitude of an aerial target in the two-station (horizontal base) system of altitude determination.

Altiscope. An instrument which enables the observer to see over objects intervening between himself and the object he desires to see.

Altitude. Elevation. The true height above sea level or other datum. The calibrated altitude corrected for air temperature and barometric pressure; it is always true unless otherwise designated. The height of an aerial target above the level of the directing point.

Altitude, Critical. The maximum altitude at which a supercharger can maintain a pressure in an engine equal to that existing during normal operation at sea level.

Altitude Determination. Determining the height of an aerial target; two systems are employed: the two-station system, in which simultaneous observations are made from the two ends of a horizontal base line; and the single-station system, in which a single self-contained height finder is employed.

Ambulance. A vehicle for the transportation of sick and wounded.

Ambulance Battalion. One of the battalions of a medical regiment, whose function is to furnish wheeled transportation for casualties from aid stations and dispensaries, or from litter bearers, to collecting stations and hospitals within the division area.

Ambulance Company. One of the companies of an ambulance battalion; there are two motorized and one horse-drawn company in the battalion.

Ambulance Shuttle. A relay system for the evacuation of casualties by ambulance.

Ambulance Station. A point for the administration and control of the movement of ambulance units.

Ambulance Train. A railway train designed for, or used as a hospital on wheels for the transportation of sick and wounded.

Ambulant. Walking. Able to walk.

Ambulant Battery. Heavy guns mounted on travelling carriages and moved as occasion may require. See Artillery, Mobile.

Ambuscade. A body of troops hidden for the purpose of taking an enemy by surprise. A position in ambush. To take up a position in ambush or to attack from ambush.

Ambuscade Defensive. A defensive which aims to lead an enemy into a trap or unfavorable position and then subject him to a surprise attack.

Ambush. A concealed place where troops lie hidden for the purpose of surprise attack. Troops posted in such a position. To attack from a hidden position. To prepare an ambuscade.

Amende Honorable. An apology for some injury done to another, or satisfaction given for an offense against the rules of honor or military etiquette.

American Eagle. See Bald Eagle.

American Flag. The flag adopted by the Continental Congress on June 14, 1777, consisting of 13 horizontal stripes of red and white alternating, to represent the original thirteen colonies, with a star for each state on a blue field in the upper corner next the staff. The Stars and Stripes. Old Glory.

American Legion. An association of the American veterans of the World War, organized in 1919.

American Legion Auxiliary. An association of American women who served with the armed forces during the World War.

Ammunition. Projectiles used with all classes of firearms, together with the requisite propellants, bursting charges, fuzes, cartridge cases, and means of ignition. The materials used in the discharge of firearms or other missile weapons. The first projectiles used were bolts, darts, stone shot, and langridge. Round metal shot was known at an early date but did not entirely supersede stone shot until the 17th century. Shell were used as early as 1376 A. D., but the lack of fuzes prevented their successful employment for several centuries. The original shell were known as grenades when spherical and bombs when oblong, the term shell being introduced later.

About the middle of the 15th century case shot was used at the siege of Constantinople, and incendiary shell was invented in 1460. In the 16th century hail shot, similar to case shot with the addition of a bursting charge, was used. In 1596 Sebastian Halle suggested the idea of the modern time and percussion fuzes, but the chemists could not furnish the necessary chemicals, and their development awaited a satisfactory percussion powder. General improvement in ammunition occurred during the 18th century. In 1784, General Shrapnel invented the shell which bears his name. The 19th century witnessed enormous improvements in war material generally. In 1807 Forsythe patented a percussion priming lock and about ten years later Hawkins invented a percussion cap. Percussion fuzes were introduced about 1842, time fuzes in 1864, and combination fuzes in 1867. With the introduction of breechloading rifled guns, elongated projectiles provided with the necessary means of rotation came into use. In 1865 steel shot were introduced, but due to their cost, were superseded by the Palliser chilled iron shot in 1867. In 1887 forged steel armor-piercing projectiles were introduced.

In small arms, a paper cartridge for muskets was used as early as 1596. Gustavus Adolphus made important improvements in ammunition. He first provided separate receptacles for each powder charge, and subsequently combined the powder with the projectile in the paper wrapper, which, until about the middle of the 19th century, formed the principal ammunition for small arms. The metallic cartridge case, on which the success of the breechloading principle depended, was curiously slow in gaining recognition. It was invented relatively early in its four distinct forms of pin fire, center fire, rim fire and needle. The pin fire type emerged gradually out of the Demondion cartridge of 1831, which embodied a tube detonator in its base, through a capped, or rather, nippled cartridge invented by Lepage about 1840, to the true pin-fire case in which a cap imbedded in the charge was struck by a wire pin, invented by Houiller, a Frenchman, in 1847. The needle gun cartridge dates from 1830, but it was not until 1840 that a breechloading needle gun cartridge was patented.

The primed metallic cartridge case, invented in France, was first used by troops in the American Civil War. It contained all the components of the ammunition in an envelope which also formed a gas check, greatly simplifying the breech mechanism. Substantially the same form of cartridge (usually center fire) is used today. In 1886 a new phase in the development of firearms was ushered in by the invention of smokeless powder, which permitted conversion to smaller calibers, and increase of velocity and range. The development of small arms ammunition since that date has been chiefly in the improvement of existing types. Cannon of small calibers employ fixed ammunition similar in construction to small arms ammunition.

Ammunition Belt. A belt with loops or pockets for carrying ammunition.

Ammunition Box or Chest. A box or chest designed to contain and for the easy transportation of ammunition.

Ammunition Bread. A loaf of bread, weighing about 5 pounds, which was formerly issued to soldiers, the allowance for 4 days.

Ammunition Chest. A chest mounted on a caisson and limber for the storage and transportation of ammunition. Any ammunition box.

Ammunition Detail, Squad. A group of men who maintain the supply of ammunition for a gun in action.

Ammunition Distributing Point. A point where ammunition is distributed to the combat wagons or representatives of the using units.

Ammunition Dump. A location in the combat zone where ammunition is temporarily stored for issue as the necessity arises.

Ammunition Hoist. The apparatus used in fixed emplacements for hoisting ammunition from the magazine to the loading platform.

Ammunition Park. A place, usually within easy reach of the divisional trains, but at such distance in rear of the front line as to insure freedom of movement, where ammunition is piled for distribution.

Ammunition Recess. A recess, as in the front slope of a trench, or machine gun emplacement, for local storage of ammunition.

Ammunition Refilling Point. The place where ammunition trains are refilled.

Ammunition Supply. The amount of ammunition available. The measures taken to provide ammunition.

Ammunition Train. The train employed in transporting the divisional ammunition from the refilling point to the point of issue to the combat trains.

Ammunition Truck. A truck, usually 3-wheeled, for carrying projectiles and powder charges to and loading into a large gun.

Amnesty. An act by which two belligerent powers at variance agree to forget and bury in oblivion all that is past. A pardon to rebellious persons, usually with some exceptions. An act of forgiveness for past offenses.

Amphibian. An airplane designed to arise from and alight on either land or water. A tank capable of crossing streams.

Anaglyph. A map in two colors which exhibits relief of the terrain when viewed through two-colored spectacles (macyscope).

Anchor. Any device for holding an object to the ground, or to any fixed position. To fix firmly. To secure by an anchor, as a vessel or guy rope.

Anchor Picket. A short picket of wood or steel, used to anchor revetments or wire entanglements.

Anemograph. An instrument used in target practice for measuring and recording the direction and velocity of the wind.

Anemometer. An instrument for measuring the velocity of the wind.

Anemoscope. An instrument used to indicate the direction of the wind.

Aneroid Barometer. An instrument which indicates the elevation (altitude) of a locality by measuring the pressure of the superincumbent atmosphere. In forms suitable for the purpose the instrument is used by aviators and surveyors to indicate or record altitudes.

Angel-shot. A segmental chain shot, formed of parts of a sphere, which spread when discharged. Angle Shot.

Angle, Critical. The angle of attack at which the flow about an airfoil changes abruptly.

Angle, Dead. The angle between the line joining the salients of two adjacent bastions and the lines of defense.

Angle, Landing. The acute angle between the wing chord and the horizontal when an airplane is resting on level ground in its natural position. Also called ground angle.

Angle of Approach. The lesser horizontal angle between the vertical plane containing the path of an aircraft and the vertical plane containing the line of sighting of an antiaircraft gun.

Angle of Attack. The acute angle between a reference line in a body and the line of the relative wind direction projected on a plane containing the reference line and parallel to the plane of symmetry. The angle at which a wing surface meets the air, as in climbing or diving.

Angle of Clearance. The angular difference by which the quadrant elevation

13

necessary to strike the target exceeds the minimum quadrant elevation necessary to clear the mask.

Angle of Concentration or Convergence (Battery). The horizontal angle toward the directing gun through which the other flank gun turns from a line parallel to the directing gun in order to engage a target of less frontage than the battery. The gun angle of concentration is the horizontal angle through which any gun turns from a line parallel to the directing gun in order to engage a target of less frontage than the battery. See Battery Angle of Convergence.

Angle of Deflection. The horizontal angle between the line of aim and the axis of the bore when the gun is laid.

Angle of Departure. The vertical angle between the line of departure and the line of site. See also Angles, Quadrant.

Angle of Depression. The angle formed by the axis of the bore of a gun with the horizontal when the gun is aimed below the horizontal.

Angle of Distribution or Divergence (Battery). The horizontal angle away from the directing gun through which the other flank gun turns from a line parallel to the directing gun in order to engage a target of greater frontage than the battery. The gun angle of distribution is the horizontal angle through which a particular gun turns from a line parallel to the directing gun in order to engage a target of greater frontage than the battery. See Battery Angle of Distribution.

Angle of Elevation. The vertical angle between the line of elevation and the line of site. See also Quadrant Elevation.

Angle of Fall. The vertical angle between the line of fall and the line of site or slope, or base of the trajectory.

Angle of Impact. The acute initial angle between the line of fall and the plane tangent to the surface at the point of impact.

Angle of Incidence. The vertical angle between the surface of the ground or target and the tangent to the trajectory at the point of impact.

Angle of Jump. The angular elevation of the line of departure above the axis of the bore at the time when the piece was pointed. The vertical component of the jump.

Angle of Landing. The vertical angle, measured in the plane of symmetry, between the ground line and the longitudinal axis of an airplane, when it rests on the ground.

Angle of Opening. The apex angle of the cone of burst of a shrapnel or shell.

Angle of Position. The vertical angle between the line of position and the base of the trajectory or horizontal. The angle of site.

Angle of Reference. The horizontal angle through which a gun turns from a reference point to a target or its zero line.

Angle of Safety. The minimum permissible angular clearance, at the gun, of the trajectory above friendly troops.

Angle of Shift. In indirect laying of machine guns, the angle through which the base gun turns from its base line in order to fire on its part of any particular target.

Angle of Sight or Sighting. The horizontal angle between the line of sighting and the line of fire.

Angle of Site. The vertical angle between the line of site and the horizontal.
the angle of site is set off or measured.

Angle of Site Mechanism. The mechanism on a quadrant or sight by which zontal.

Angle of Slope. The vertical angle which the surface of the ground makes with the horizontal, usually called slope.

Angle of Spiral or Rifling. The angle between a tangent to the spiral of the rifling in a gun barrel and the axis of the bore. See Twist.

Angle of Splay. The divergence between the sides or cheeks of a loophole or embrasure.

Angle of Stabilizer Setting. The acute angle between the longitudinal axis of an airplane and the chord of the stabilizer; it is positive when the leading edge of the stabilizer is higher than the trailing edge.

Angle of Superelevation. In antiaircraft fire the vertical angle between the line of future position and the axis of the bore when the gun is ready to fire. The additional elevation necessary to allow for the curvature of the trajectory.

Angle of the Face. In the bastion trace, the angle formed by the face and the line of defense.

Angle of the Flank. In the bastion trace, the angle formed by the flank and the curtain.

Angle of the Line of Defense. In the bastion trace, the angle formed by the flank and the line of defense.

Angle of the Polygon. The angle formed by two adjacent exterior sides of a fortification.

Angle of Traverse. The angle at a gun position subtended by the frontage to be covered by the fire of the piece at that position.

Angle of Wing Setting. The acute angle between the plane of the wing chord and the longitudinal axis of an airplane; it is positive when the leading edge of the wing is higher than the trailing edge.

Angle, Salient. The projecting angle formed by the two faces of a bastion, or other salient work.

Angle Shot. A kind of chain shot. Angel-shot.

Angle, Terminal. The angle which a tangent to the trajectory forms with the horizontal plane at the level point.

Angle, Wind-fire. The horizontal angle, measured clockwise, from the plane of fire to the direction from which the ballistic wind is blowing.

Angles, Quadrant. Vertical angles measured or laid off by means of a quadrant. Quadrant angles are measured above or below the horizontal. Thus the angle of elevation is the vertical angle between the line of elevation and the line of position; the quadrant angle of elevation is the vertical angle between the line of elevation and the horizontal. Similarly for the angle of departure, angle of fall, etc. Some quadrants are provided with a mechanism on which the angle of site or position may be set off.

Angular Height. In antiaircraft fire, the vertical angle between the line joining the gun and the target, and the horizontal. The angular elevation of a point in space (in the air), measured at the gun.

Angular Travel Computer. A device used to determine the angular travel of the target during the time of flight of the projectile.

Angular Travel Error. An error in a predicting angle obtained by multiplying instantaneous angular velocity by time of flight, due to changes in angular velocity during the predicting interval.

Angular Travel Method. A method of prediction in antiaircraft fire based upon the angular speed of a target in azimuth and elevation, and the time of flight. See also Lateral and Vertical Deflection Angles.

Angular Unit Method. A method of adjusting antiaircraft fire in which range deviations in mils obtained by a distant observer are converted into altitude corrections in yards by use of a conversion chart, for application at the data computer.

Angular Velocity. The rate of change of direction to a moving target, as an aerial target, expressed in angular measure; the two components are the vertical angular velocity and the lateral angular velocity. The velocity at which a body revolves about a fixed point or axis.

Animal-drawn Transportation. Transport, usually by wheeled vehicles, in which the motive power is furnished by animals.

Annexes. Orders covering the details of execution of certain aspects of operations and employed to amplify the field orders of divisions and higher units. Annexes usually consist of the field orders of the auxiliary arms or services.

Ante-bellum. Before the war; generally used with reference to the United States Civil War.

Ante-mural. An outwork consisting of a high, strong, turreted wall, for the defense of the gate of a fortification.

Anthrax. A contagious fever of horses and mules accompanied by discoloring of the eyes, swellings and bloody discharges from nose and bowels.

Antiaircraft. Designed for or capable of use against aircraft.

Antiaircraft Artillery. Artillery designed and intended for use against aircraft.

Antiaircraft Artillery Area Defense. A thoroughly organized and coordinated antiaircraft artillery defense of a definite area, which is protected by the mutually supporting fires of antiaircraft artillery guns and machine guns.

Antiaircraft Artillery Defense. Defense provided by antiaircraft artillery against attack from the air.

Antiaircraft Artillery Gun Area. An area within which all localities are protected by the fire of antiaircraft artillery guns.

Antiaircraft Barrage. A barrage of fire designed to intercept or interfere with hostile aircraft in flight.

Antiaircraft Defense. All means used to limit the aerial activity of the enemy. Defense provided by the coordinated employment of air and ground forces against attack from the air; it also includes the passive means of defense.

Antiaircraft Gunner. A rated gunner of antiaircraft artillery.

Antiaircraft Guns. Guns designed to fire at high angles of elevation and to great heights in order to reach aircraft in flight. Antiaircraft guns or cannon vary in caliber from .50 inch to 5 inches. The smaller weapons are machine guns or automatic cannon for quick fire against planes at low or intermediate elevations. Planes now fly at elevations up to 40,000 feet, and the 4.7 inch antiaircraft gun has a vertical range greater than this.

Antiaircraft Machine Guns. Machine guns designed or mounted for use against aircraft.

Antiaircraft Volume of Dispersion. A rectangular paralleopiped in space with its center at the center of burst, its faces being parallel and perpendicular to the mean trajectory, and its dimensions being 8 probable errors in range, in lateral deflection and in vertical deflection. It is assumed that 100 per cent of all bursts will fall within this volume.

Anti-dim. A chemical preparation used to prevent dimming of the eyepieces of a gas mask by smearing over the inner surfaces thereof.

Antimechanized Defense. Means and measures employed to protect troops, installations and establishments against attack by mechanized, motorized, or armored units.

Antitank. Designed for or capable of use against tanks.

Antitank Ditch. A ditch placed as an obstacle for tanks.

Antitank Fort. An area specially prepared for defense against tanks.

Antitank Mines. Land mines buried in the ground or scattered on the surface and camouflaged, used to destroy enemy tanks.

Antitank Obstacle. Any obstacle designed to stop or hinder the progress of tanks.

Antitank Weapons. Weapons whose primary mission is employment against armored vehicles.

Anvil. A metal part against which a cap is exploded. A small pennon at the head of a lance. An old name for the handle or hilt of a sword.

Aparejo. A pack saddle of the type formerly used in the United States Army.

Apparent Center of Impact. The center of impact of a limited number of shots, commonly referred to as the center of impact. The true center of impact is that of an infinite number of shots.

Application of Fire. See Fire, Employment of.

Applicatory Method or System of Instruction. A system of military instruction in which principles are inculcated through their application to concrete problems, either on a map or in the field.

Appointing Authority. The officer having the power to convene a military court and to take action upon its proceedings.

Appointment. Office, rank, or employment.

Appointments. The military accouterments of officers and soldiers.

Approach. An approach trench. A route by which a place or position can be approached by an attacking force. The route leading to anything, as a bridge.

Approach March. A phase of the attack. The advance, usually in extended formations, from the point where hostile medium artillery fire may be expected or air attack is encountered, to the point of effective small-arms fire. It ordinarily begins with the development of companies and larger units and terminates with complete or partial deployment as skirmishers.

An approach march is said to be covered when it is made in rear of advanced forces sufficiently strong to afford security against ground attack. When such protection is lacking or inadequate the approach march is said to be uncovered.

Approach Trench. A trench serving to connect fire trenches from front to rear, or an intrenched area with the rear, generally perpendicular to the front.

Approaches. Trenches whose general direction is perpendicular to the front; they are for fire or communication purposes or both, depending on their location and intended use. The covered roads, or trenches, by which the besiegers approach a fortified place. See Parallels and Approaches.

Approximate Lateral Deflection Angle. An approximation of the principal lateral deflection angle obtained by multiplying an instantaneous lateral angular velocity by a time of flight.

Approximate Vertical Deflection Angle. An approximation of the principal vertical deflection angle obtained by multiplying an instantaneous vertical angular velocity by a time of flight.

Apron. An inclined pavement of brush, concrete, etc., protecting an exposed surface, as an embankment, river bank, etc. A sheet of lead used to cover the vent of a muzzle-loading cannon. An inclined surface in a barbed wire entanglement.

Aptitude Tests. Tests to determine whether a candidate has the kind of specific knowledge or abilities required for a given grade or rating.

Arbalest. A medieval crossbow made of steel, set in a shaft of soft wood, with a string, a trigger, and a device for bending the bow.

Arbitration. The hearing and determination of a cause at controversy between two persons or nations.

Archer. A long bowman. One who uses a bow and arrows.

Area Depot. A supply depot pertaining to or serving a particular area.

Area of Burst. The area covered by the burst of one or more projectiles (shells, bombs or grenades). See also Effective Area of Burst.

Area of Dispersion. A rectangular horizontal area, with its center at the center of impact, within which 100 per cent of a series of shots is assumed to fall. See Dispersion Diagram.

Area Sketch. A hasty sketch of an area of terrain showing the important features. See Outpost, Place, Position Sketches.

Arm. A weapon. A combatant branch of the army. To furnish or equip with weapons. To prepare for combat. To prepare the fuze of a projectile or grenade so that it will function.

Arm Chest. A chest used for storing arms.

Arm Rack. A rack or stand used for safe keeping of arms in barracks.

Arm and Hand Signals. Prescribed or prearranged signals made with the arms and hands.

Armament. The arms and equipment of troops. The war material of a nation. The weapons of a fortification, place, war vessel, tank, or body of troops. Cannon of various sizes and powers, including their carriages.

Armament Error. The divergence, stripped of all personal errors and adjustment corrections, of an impact (or burst) from the center of impact (or burst) of a series similarly stripped.

Armed. Equipped with weapons. Set for firing, as a fuze.

Armed Neutrality. The condition of a neutral power, in time of war, when it maintains an armed force to resist any act of aggression or any violation of its sovereignty on the part of belligerent nations between which it is neutral.

Armed Services. The army and the navy of a nation.

Arm-guards. Metal plates which were buckled over mail to protect the outer surface of the upper arms and the front surface of the lower arms.

Arming a Fuze. Releasing the safety device of a fuze so that it will be free to function. This may be done by hand, as in a grenade, or by inertia (set-back) or centrifugal force in the bore of the gun upon discharge.

Armistice. A temporary cessation, by mutual agreement, of hostilities between two armies in the field, or between nations at war. A general truce.

Armlet. A protecting sleeve of leather or metal worn on the forearm.

Armor. A defensive covering, especially of metal, worn by or applied to a soldier, animal, airplane, tank, vessel, machine or weapon, designed to afford protection against weapons of any kind. The use of personal armor dates back to the beginning of warfare, when leather and other tough fabrics and wood were used for the purpose. Metal was soon added; first in the form of small plates or sequins attached to a garment of heavy cloth (undoubtedly suggested by the scales of a fish); later as a fabric of interlocking metal rings (mail); and finally in the form of plate armor, articulated at the joints. In its earlier forms armor was applied only to the vital parts of the body, especially the head and chest, but eventually full suits of mail or plate, covering every part of the body, were developed. The horses of the knights were also armored on their vital parts. Chain mail has been traced back to 400 B. C., and was very generally worn up to 1300 A. D., when it was largely replaced by plate armor. The development of firearms caused a gradual disuse of armor, because of the impracticability of an armor heavy enough to afford protection against the new weapons. The steel helmet of the World War, and the cuirass still worn by some heavy cavalry, are the last remnants of the armor of the ancients. The principal parts of the full body armor were:

Helmet, armet or bas-inet	Shoulder pieces, pauldrons, passegardes	Loin and thigh guards, tassets, tuilles, cuishes,
Visor, beaver, chin-piece or metonniere	Arm-guards, pallettes, rerebraces, vambraces,	gardes-reines
Neck piece, gorget	elbow pieces	Knee pieces, genouilleres
Cuirass or breastplate	Mailed gloves or gaunt-lets	Greaves or jambes
Back plate, culet		Shoes, sollerets or pedieux
Girdle	Waist piece, braguette	

Many other names were applied to the various parts.

Armor Bearer. One who carries armor for another. An armiger. An esquire.

Armor Plate. Steel plate of various thicknesses, usually alloyed and heat-treated to increase its resisting power, used as armor to protect the vital parts of war vessels, tanks, airplanes and other machines, and in armored turrets. Armor plate may be face hardened to break up projectiles, or non-face hardened to resist perforation.

Armored Car. An armed and lightly armored motor vehicle, designed primarily for reconnaissance.

Armored Force. A new combat element in warfare. A combined force, including generally all arms and services, combining reconnaissance, assault and supporting elements. The assault element consists essentially of tanks, the supporting element of motorized (armored) infantry and artillery. The force as a whole is characterized by great mobility and striking power.

Armored Forts. Fortifications in which the guns are emplaced in armored turrets, or behind armored shields.

Armored Fronts. A system of covering the extended intervals between great systems of fortifications with the fire of small guns arranged in two or three lines.

Armored Plane. An airplane having its vital parts protected by armor.

Armored Tractor. A tractor protected by steel plates.

Armored Trains. Armored railway trains used as escorts for work or traffic trains, and for other purposes.

Armorer. One who is charged with the care and condition of arms and armor. The caretaker of an armory.

Armor-piercing Cap. A cap of soft steel placed over the point of an armor-piercing projectile to prevent the point from breaking or bending on impact with hard-faced armor.

Armor-piercing Projectile. A projectile intended for the attack of armor. A specially designed and heat-treated projectile, provided with windshield and armor-piercing cap, for penetration of armor before explosion.

Armory. A place for the safe keeping of arms. An arsenal. A government factory or plant for making arms. A building for the use of troops, including a drill hall.

Arms. Instruments and devices of different forms and natures used for attack and defense. Any weapons used in warfare. The combatant components of the Army; they are as follows: infantry, cavalry, field artillery, coast artillery, air corps, corps of engineers, signal corps, chemical warfare service, armored forces, and air forces. See also Weapons.

Arms of Precision. Rifled weapons of all kinds.

Arms of the Service. The different combatant branches composing a military establishment. See Arms.

Armstrong Gun. A built-up, breech-loading gun, constructed according to the method introduced by Sir William Armstrong about 1850.

Army. An organized and disciplined force of armed men raised or maintained for military purposes. A body of troops composed of a headquarters, certain auxiliary troops and trains called army troops, and two or more corps, with certain troops from the General Headquarters Reserve attached from time to time as their special services are needed. The fundamental strategic maneuver force and, as such, the main battle unit. The armed land forces of a nation, considered as a whole. See Field Army.

Army Administration. The organization and other means by which the various administrative duties are performed, especially as relates to the care and supply of troops, and preparation for war.

Army Air Forces. A general term, including all air units of the U. S. Army.

Army and Navy Munitions Board. An agency charged with the coordination of planning for the procurement of military requirements considered necessary in the event of a national emergency.

Army Area. The territorial area assigned an army for administrative and tactical purposes. That portion of the combat zone assigned to a single army. One of the major military geographical districts of the United States, including several corps areas, organized under a single commander for purposes of defense, and in which it is planned to raise an army in time of emergency.

Army Artillery. The artillery forming part of an army, exclusive of the organic artillery of any subdivision. In our service, it consists organically of only army artillery headquarters, an ammunition train and an antiaircraft brigade, but is augmented as necessary by the assignment of units from the General Headquarters artillery reserve.

Army Cavalry. Cavalry which is an organic part of, or attached to, an army.

Army Combat Area. That portion of the combat zone occupied by the front line corps. It includes the front line division areas and the corps service areas.

Army Corps. See Corps.

Army Depot. A branch depot, designated as such by the army commander, located in the army area, for the reception and temporary storage of supplies that the situation demands be kept closer at hand than is possible in the advance section of the communications zone, and for the storage of supplies requisitioned in the combat zone.

Army Extension Courses. Military instruction, especially for National Guard and Reserve officers, conducted by correspondence.

Army Front. The advanced elements of an army. That part of an army nearest to and facing the enemy. The lateral extent of the area assigned to or occupied by an army in attack or defense.

Army Industrial College. A school for the training of officers in matters pertaining to industrial procurement and mobilization.

Army List and Directory. A periodical publication issued by the War Department, containing the names and addresses of all commissioned officers, their relative rank, and other information of military interest.

Army Mobilization Plan. A plan which prescribes the organization of the zone of the interior, the mobilization of manpower, the actual units or troops to be employed in the theater or theaters of operation at the outbreak of war, and the methods for concentrating these troops.

Army Nurse Corps. A corps of nurses forming part of the medical department, under the direction of the surgeon general, for the care of patients in military hospitals.

Army of Observation. A part of a besieging army, or a force acting in cooperation with a besieging army, whose function is to watch for and oppose any hostile operations designed to raise the siege or in any way assist the beleaguered garrison. An army in an advanced position whose function is to watch the enemy and be prepared to oppose his operations.

Army of Occupation. An army established in hostile territory to hold and control such territory, including the maintenance of law and order amongst the civil inhabitants (law of hostile occupation).

Army of the United States. The Army of the United States had its beginning when Washington assumed command of the Colonial Militia around Boston, on July 3, 1775. During the Revolution we employed a total of about 395,000 troops. At the close of that war the Continental Army was reduced to a detachment of 80 men. The separate colonies, and subsequently the states, continued to maintain, or to improvise as needed, a rather indifferent militia. An Act of Congress of 1792 (Militia Act) established the principle that every able-bodied male citizen of military age was liable for military service. But little was done to render the militia a really effective force for more than 100 years. At various times the Regular Army was reorganized, increased and decreased. At the outbreak of the War of 1812 it numbered some 6,700, scattered all over the country. For that war we employed 530,000 troops, mostly state militia for short terms of enlistment. After the war a regular army of 18,000 was authorized. But in 1821 its actual strength was about 6,000. At the outbreak of the Mexican War (1846) its actual strength was about 5,300, occupying about 100 stations. During the Mexican War we employed about 106,000 troops, mostly volunteers. At the close of this war the regular army was again reduced to 10,000, and at the beginning of the War between the States it numbered less than 16,000. In this, our greatest war, the federal government employed approximately 2,672,000 troops, the bulk of whom were state militia or volunteers, with officers locally elected or appointed. Following the war the regular army was reduced first to 37,000, and subsequently (1874) to 25,000, at which approximate figure it remained until the beginning of the Spanish War in 1898. For this war the regular army was expanded to some 47,000, and about 233,000 national volunteers were called to the colors. But of this great force barely 20,000, mostly regulars reached the scene of actual conflict. In 1901, for the first time after the close of a war, the regular army was increased. Its authorized strength was placed at approximately 100,000, but it never actually reached this figure. In 1903 and 1908 Acts of Congress extended the amount of federal control over and increased the allowances for the support of the state militia, which was then designated as the National Guard. In 1916, events having foreshadowed our entry into the World War, Congress passed the National Defense Act, which laid the foundations of our present military policy. It provided that our land forces should consist of a Regular Army of 175,000; a Volunteer Army; an officers' and enlisted men's Reserve Corps; and a National Guard. In May, 1917, shortly after our entry into the World War, Congress passed the Selective Service Act, providing for compulsory military service. During the World War we mobilized a total of about 4,057,000 officers and men, organized into a Regular Army, a National Guard and a National Army. Of this number about 2,086,000 went overseas, and about 1,390,000 were actually engaged in battle against the enemy. In 1920 Congress amended and expanded the National Defense Act of 1916. The amended act provided, in brief, for a Regular Army of 280,000 officers and men; a National Guard of 428,000 officers and men; and Organized Reserves of indeterminate strength. Limitations on the appropriations for the support of the military establishment have held the first two components far below the strengths authorized. See also Militia; National Defense Act; National Guard; Officers Reserve Corps; Reserve Officers Training Corps.

Army of the United States, Organization of. Consists of the Regular Army, The National Guard of the United States, The National Guard while in the service of the United States, and the Organized Reserves, including the Officers' Reserve Corps, and Enlisted Reserve Corps. It includes also all men called into service under a selective service act. As reorganized in 1942 for the duration of the war, the army consists of the War Department General Staff, and three main groups, each having its own commanding general, viz.: the Ground Forces, The Air Forces, and The Service of Supply. The Air Forces may be assigned to independent missions, but as a component of any task or expeditionary force they will be under the control of the commander of such force, who may be a ground officer, an air officer, or a naval officer.

Army Organization. See Organization.

Army Register. An official list of the officers of the army, published annually, giving the name, rank, branch, and brief military history of each officer of the army; with other data, such as pay tables, changes since last issue, etc.

Army Regulations. Abbreviated A.R. See Regulations.

Army Sector. Comprises, in defensive combat, the entire area, front and depth, assigned a field army.

Army Serial Number. See Serial Number.

Army Service Area. That portion of the combat zone in rear of the army combat area and extending back to the forward boundary of the communications zone.

Army Service Schools. Include general service schools and special service (branch) schools.

Army Service Troops. Troops assigned to army headquarters, and designed for service duty such as supply, administration, transportation, police, guard, labor, etc.

Army Signal Agencies. The army signal communications system and the special army signal service.

Army Signal Communications System. A communications system includes a message center, through which all communications pass, and the following means of communication: (1) Messengers (foot, horse, motorcycle, pigeon, dog, airplane, etc.); (2) Wire systems (telephone and telegraph); (3) Radio; (4) Pyrotechnics; (5) Panels (for signal to airplanes); (6) Visual systems (lamps, flags, etc.). See also Wire Systems.

Army Strategical Plan. A plan whose purpose is to furnish an estimate of the land forces required, the times and places for their initial concentration, so as to carry out the mission assigned the army by the joint plan.

Army Transportation. All means of transportation employed by an army in the field.

Army Transport Service. A branch of the quartermaster corps whose function is the transporting of troops and supplies by water.

Army Troops. Troops which are part of an army, but not a part of its component corps.

Army Wagon. A particular type of wagon for the transportation of ammunition, supplies, and foot soldiers. An escort wagon.

Army War College. A general service school located at Fort Humphreys, District of Columbia, for the education of officers in higher command and staff duties.

Arquebus. Same as Harquebus.

Arrest. Temporary restriction to limits, usually living quarters, imposed by moral restraint. To appehend and detain. To take into custody. See also Confinement.

Arrest in Quarters. Arrest involving restriction to quarters, imposed when the offense is not serious enough to warrant confinement in the guardhouse.

Arresting Gear. The gear incorporated in aircraft and in the landing area to facilitate landing in a limited space.

Arrow. A slender missile weapon shot by means of a bow. A work at a salient angle of a glacis.

Arsenal. A place for the manufacture or storage of arms and munitions of war. An armory.

Art of War. The military art. The conduct of war in its broadest sense, including the application of the recognized principles in the use of the available means. The art of war includes two great divisions, strategy and tactics, and also comprises logistics, engineering and communications.

Articles of War. The code of laws for the government of the Army and the Navy.

Arterial Road. One of the main lines of communication and supply for large units.

Artifice. A stratagem. A scheme used to deceive the enemy.

Artificer. A mechanic in a company or similar organization.

Artificial Horizon. A substitute for the natural horizon, such as the level of a liquid, pendulum or gyroscope, incorporated in a navigating instrument; a device that indicates the position of an aircraft with reference to the horizon. A basin of mercury, used to establish a horizontal mirror for sextant observations for elevation.

Artificial Obstacles. Obstacles prepared by human agency; they may be fixed, such as fixed wire entanglements, abatis, etc., or portable, such as chevaux-de-frise, certain types of wire entanglements, etc.

Artillerist. An artilleryman. One experienced in artillery and gunnery.

Artillery. Guns borne on wheels or mounted in fortifications and of calibers greater than those of small arms. Guns too heavy to be carried by hand. Engines and devices of any nature intended for the discharge of heavy missiles by the force of an explosive. That arm (or arms) of the service which mans and operates cannon.

Mechanical devices for throwing projectiles were produced early in the history of organized warfare and were employed throughout the ancient and medieval periods. The date of the first employment of cannon cannot be fixed, but there is good evidence that the Germans used cannon at Cividale in Italy in 1331. The principal use of artillery for the next three hundred years was to batter the walls of fortified places. The introduction of field artillery may be attributed to John Ziska who, about 1420 mounted several small guns in carts, called wagenburg. The most notable example of the use of artillery in the 15th century was at the siege of Constantinople in 1453, where the Turks used a large force of artillery, some of which was employed against a British squadron in 1807.

In the 16th century artillery played a prominent part, both in siege and field warfare. But with the development of infantry fire power the use of light field guns decreased, and it is due largely to this fact that artillery came to imply cumbrous and immobile guns. There was little maneuvering, the guns generally came into action in advance of the other troops and, due to their lack of mobility, were captured or recaptured with each changing phase of the battle. It was not until the appearance of Gustavus Adolphus, whose secret of success was vigor and mobility, that artillery began to assume its true role on the battlefield. Gustavus, about 1630, decreased the weight of his field artillery and introduced the use of grape shot. The 12-pounder was the heaviest field piece and the 9-pounder was the principal piece. Gustavus was the first to classify artillery as mobile and nonmobile, which classification has since been almost universally employed. Due to the elimination of the pike, infantry fire power ruled the battlefield during the greater part of the 18th century.

More than 300 years after the first employment of guns the men working them and the drivers were still civilians. Indeed, the guns were often the property of some master gunner whose services were for hire. From 1700 onward artillery gradually became an arm of the service, but the drivers were not actually soldiers until 1793 at the earliest.

In France uniformity of organization was attempted by Valliere in 1732, but to Gribeauval, who became inspector-general of artillery in 1776, is due the credit of having simplified artillery and introduced improvements in its equipment. He built up a complete system both of personnel and materiel, creating a distinct materiel for field, siege, garrison, and coast artillery. Uniform construction was adopted for all materiel, and the parts made interchangeable as far as practicable. During the long wars of the French Revolution and Empire the field artillery gradually became as we know it today. The guns were at last organized into batteries, each complete in itself, and there was steady improvement in maneuvering ability. With the deterioration of Napoleon's infantry, due to its use without regard to losses, artillery became the deciding factor in battle at this time.

From this time onward the history of artillery is the history of its technical improvement and tactical effectiveness. The use of artillery on the battlefield has gradually increased and in the World War thousands of guns were often massed for important operations. See also Ballistic Machines; Gun.

Artillery Annex. An amplification of the artillery sub-paragraph of paragraph 3 of a field order.

Artillery, Antiaircraft. Artillery designed for use against aircraft. Artillery whose primary mission is the ground defense of troops and important facilities against activities of enemy aircraft; it may be fixed or mobile, and it includes guns and machine guns.

Artillery, Classification of. (1) According to arm, field artillery and coast artillery, including antiaircraft; (2) According to nature of mount, fixed and mobile; (3) According to type of weapon, guns, howitzers, mortars, and trench artillery; (4) According to weight, light, medium, and heavy; (5) According to tactical assignment, division, corps, army and G. H. Q.; (6) According to means of transportation, horse, horse-drawn, pack, portee, railway, self-propelled, tractor-drawn and truck-drawn. An old classification was: mountain, field, siege, and seacoast.

Artillery Counterpreparation. See Counterpreparation.

Artillery District. A seacoast artillery area united for administrative purposes under a single commander.

Artillery, Division, Corps, Etc. See Artillery, Classification of.

Artillery Duel. Counterbattery fire between the artillery of the opposing forces, which marks the opening of a battle.

Artillery, Field. Artillery, other that antiaircraft and trench artillery, which may accompany an army in the field.

Artillery Fires. See Fire.

Artillery, Fixed. Artillery which is permanently installed in fortifications on land and sea frontiers, or for the protection of important places.

Artillery, Garrison. Artillery, usually permanently emplaced, employed in permanent or semipermanent land or seacoast defenses.

Artillery, Heavy. The heavier guns and howitzers of the field artillery; specifically, the 155mm (6 inch) gun, 210mm (8 inch) howitzer, and all heavier pieces.

Artillery, Horse. Light artillery which is intended to accompany cavalry in the field. It differs from horse-drawn artillery only in that the cannoneers are individually mounted upon horses to increase its mobility.

Artillery, Horse-drawn. Light artillery which is intended to accompany infantry in the field; the type weapon is the 75mm gun. Any artillery drawn by animals.

Artillery Information Service. The personnel of army and corps artillery assigned to the duty of obtaining, collating, and distributing information of value to the artillery.

Artillery, Light. Mobile guns and howitzers of 105mm or smaller caliber; the type weapons are the 105mm howitzer and the 75mm gun.

Artillery Mass. A concentration of mobile artillery usually employed after the weak points of the enemy's lines have been determined. The artillery employed in a single operation considered as a whole.

Artillery, Medium. Guns and howitzers intermediate in size between the 105mm howitzer and the 155mm gun; the type weapon is the 155mm howitzer.

Artillery Mission (Aviation). A flight for the purpose of locating targets, adjusting artillery fire, reporting the effect of fire, or for the surveillance of scheduled artillery fire and enemy activity.

Artillery, Mobile. Artillery which is designed for movement from place to place. See Artillery, Classification of.

Artillery, Mountain. Light artillery designed for use in mountainous or difficult country, usually transported on the backs of animals. Pack artillery.

Artillery of Position. See Artillery, Fixed.

Artillery, Pack. Light weapons which may be disassembled for transportation on pack animals. It is especially suitable for operations in mountains or in jungles. The type weapon is the 75mm howitzer.

Artillery Park. A depot or park of guns, materiel, etc. The camp of one or more batteries.

Artillery, Portee. Artillery, usually of 75mm caliber, which is transported in trucks. It can be moved rapidly from place to place, but it becomes relatively immobile after being emplaced.

Artillery Position. A position selected for and occupied by an artillery fire unit for the delivery of fire. Artillery positions are generally in rear of and covered by the advanced infantry lines. Artillery positions are classified according to the method of laying and the amount of concealment or cover as direct laying, indirect laying, open, unconcealed, concealed, unprotected, protected. According to tactical use they are classified as positions, alternate positions, supplemental positions, dummy positions, positions in observation, positions in readiness.

Artillery Preparation. Artillery fire delivered according to schedule, preparatory to an attack, for the systematic destruction or neutralization of enemy defenses, materiel, and personnel. See also Counterpreparation.

Artillery, Railway. Artillery designed to be transported on and fired from railways, although some types may be transported on railways and fired from prepared positions. It includes guns, howitzers, and mortars of large caliber.

Artillery Reserves. Artillery units held in readiness, or directed to reserve their fire pending the development of contingencies during the course of an action. G.H.Q. reserves of artillery used as reinforcements when necessary. The true reserve of an artillery unit in combat is the ammunition available.

Artillery Ricochet. Effect deliberately obtained by ricochet with artillery of the 18th century. See Ricochet.

Artillery, Self-propelled. Artillery which combines carriage and motive power in a single vehicle.

Artillery, Siege. Artillery designed and used for the attack of fortified places, having either mobile or fixed mounts. Artillery of similar types is also used as garrison artillery. Artillery units armed with siege artillery.

Artillery Support. The fire of designated artillery units in support of the attack or defense of other arms. The artillery units so employed.

Artillery Supports. Troops of other arms with the mission of protecting artillery units against direct attack.

Artillery Tactics. The art and science of maneuvering artillery and employing its fire power in coordination with other combatant branches.

Artillery, Tractor-drawn, Truck-drawn. Artillery drawn or towed by tractors or trucks.

Artillery Train. A mobile artillery force, including guns, ammunition, etc., and transport. A siege train. The train of an artillery unit, as distinguished from its combat elements. An artillery ammunition train.

Artillery, Trench. Short-range, special-purpose weapons which are most suitable for use in stabilized situations or in mountainous country.

Artillery with an Army (Corps). A term used to designate all the artillery with an army (or corps), including both the organic artillery of the army and the organic artillery of its component units.

As You Were. A command used to cause men engaged in drill to resume the previous position when any movement has been improperly begun or executed; also to revoke a preparatory command.

Ascending Branch of the Trajectory. That portion of the trajectory between the muzzle of the gun and the summit of the trajectory.

Asphyxiating Gases. Gases having a suffocating and poisonous effect.

Asphyxiating Shell. A shell charged with a proportion of asphyxiating gas in addition to its bursting charge.

Assault. A furious but regulated effort of troops directed against an enemy's works or positions in order to carry them by a single or concentrated attack. The final rush to close combat with the enemy in order to determine the issue by hand-to-hand encounter; especially when delivered by mounted troops it is also called the charge. The last advance to the enemy's position without intermediate halt. The culmination of an attack. To deliver a concentrated attack from a short distance. To close with the enemy in hand-to-hand combat.

Assault Battalion. A battalion with all or part of its elements in the assault echelon. A front line battalion in the attack.

Assault Company. A company with all or part of its elements in the assault echelon in attack.

Assault Echelon. The unit, or units, of an organization which are assigned to initiate the attack. The most advanced line of combat elements in attack.

Assault Position. A position from which the assault units are to launch an attack. A line or lines of departure. A jump-off position.

Assault Wave. The leading wave of an attack.

Assemble. The grouping, in close order, of the elements of a command. To come together in one place or locality. To come together in proper order. To put together, as to assemble a weapon.

Assembly. Formation in ranks or column. The regular grouping, in close order, of the elements of a command. A change from extended order to close order. Ployment. The collection of the elements of a command in one or more places as a preliminary to any operation, ceremony or other duty. The signal or bugle call to fall in or assemble. A group of parts of any machine, as an automatic weapon, pertaining to a particular function of the machine, as the buffer group.

Assembly Area. A position, or an area, containing a number of positions in which troops are assembled in preparation for attack or other operation.

Assembly Point. A locality at which troops assemble in preparation for attack or other maneuver, ceremony or duty. A place designated for reassembly and reorganization in case troops have become disorganized or scattered as a result of a rapid advance, forced withdrawal or retreat. A rendezvous or rallying point for tanks.

Assembly Position. A position or an area in which a unit is to be grouped preparatory to further action. A place where attacking troops assemble and make dispositions for the attack.

Assembly Trenches. Trenches designated or prepared for the assembly of troops preparatory to an attack or night raid. Jump-off trenches.

Assign. To fill an organic vacancy in an organization or position by the assignment of an individual or unit thereto.

Assignment. The act of assigning an individual or unit to an organization or duty. The condition of being assigned. The duty to which assigned.

Assistant Chief of Staff. An executive assistant to a chief of staff who is in direct charge of one of the functional branches or departments into which the work devolving upon a general staff is divided.

At Ease. A condition of rest in drill or other formation in which the men preserve silence and remain in their proper places.

A-tent. The common name for a small tent which, from the front, resembles in shape the letter A.

Atmosphere Slide Rule. A slide rule with two slides for determining the atmospheric reference number, and the muzzle velocity to be expected.

Atmospheric Absorption. Absorption of radio waves due to imperfect electrical characteristics of the atmosphere.

Atmospheric Message. A statement by the meteorological service of atmospheric conditions for the meteorological datum plane expressed in a form convenient for the artillery. A meteorological message.

Atmospherics. Strays produced by atmospheric conditions; quite generally known in the United States as static.

Attach. To cause individuals or units to act with troops to which they do not organically belong.

Attached. Applied to a person or unit placed temporarily under the orders of a commander other than his or its regular commander. Temporarily assigned, without becoming an organic part of.

Attached Artillery (Cavalry, etc.). Artillery (cavalry, etc) detached from its organic organization and placed temporarily under the orders of another commander.

Attack. An offensive action of any kind. An advance upon the enemy with intent to drive him from his position. The act of attacking. To take the initiative.

Attack Aviation. See Aviation, Attack.

Attack Bomber. See Aviation, Classes of.

Attack Echelon. The leading echelon in attack.

Attack, Forms of. See Forms of Attack.

Attack Order. A field order announcing the plan of attack and the missions of the component units.

Attack, Phases of. The phases of a normal attack include: the advance in route column; deployment (including development and approach march); advancing the attack; assault; reorganization; and pursuit or organization of the ground.

Attack Unit. The battalion, whether operating alone or as part of a larger unit.

Attention. A cautionary command to assume a condition of readiness preparatory to some military maneuver or exercise. The state of being at attention. The position of a soldier.

Attention to Orders. The command or signal given by the adjutant to announce to the troops that he is about to publish orders.

Attrition. The act or process of gradually wearing away the enemy's strength and reducing his fighting power.

Authenticate. To make authentic or authoritative, as by signing a document, and sometimes fixing a seal thereto. To certify.

Authentication. The signing of an order by, or in the name of the proper commander or authority.

Authority. The right to command and enforce obedience. The right to perform certain acts or take certain measures.

Authorize. To sanction. To confer authority upon.

Authorized Abbreviation. An abbreviation whose use is authorized by the War Department.

Autofrettage. A method of manufacturing cannon by subjecting a single forging to interior hydraulic pressure; also called the cold-worked process.

Autogyro. See Gyroplane.

Automatic Anchor. An anchor which holds a submarine mine at a constant depth below the surface of the water.

Automatic Firearms. Guns which, after the first shot, automatically continue to fire until the ammunition in the magazine or feed belt is exhausted, or the pressure on the trigger is released. They are operated by the force of recoil or the pressure of the powder gases.

Automatic Mine. See Mine, Contact.

Automatic Pilot. An automatic control mechanism for keeping an aircraft in level flight and on a set course. Sometimes called gyro pilot, mechanical pilot, or robot pilot.

Automatic Pistol. See Pistol.

Automatic Pistol, Caliber .45. The pistol in use in the United States Army; it weighs 2 pounds 7 ounces, and the magazine holds 7 cartridges.

Automatic Primer. A primer used in grenades which explodes automatically after the expiration of a certain period of time.

Automatic Rifle. A light, individual weapon which fires small arms ammunition on the automatic principle. It is gas-operated, magazine-fed, air-cooled, and fired from the shoulder. There are many types of this weapon; that used in the United States Army is the Browning; it weighs 21 pounds, has a magazine capacity of 20 shots, and may be used for either automatic or semi-automatic fire. See also Machine Gun.

Automatic Safety. A device which automatically blocks the hammer so that the gun cannot fire until the safety is withdrawn.

Automatic Soldier. A device controlled by radio, for firing an automatic rifle.

Automatic Supply. Deliveries of specified kinds and quantities in accordance with a predetermined schedule.

Automatic Weapons. Beginning with the invention of firearms, efforts were made to develop weapons that would repeat the discharge and fire successive shots as rapidly as possible. Merely adding barrels was the simplest method of increasing fire power, but it was never satisfactory in handarms because of the bulk and weight. In 1807, the Reverend Alexander Forsythe patented the application of fulminate of mercury to the discharge of firearms and this opened the way to the development of modern firearms. The first multi-firing arm that had any real value as a military weapon was in the form of a single-barreled pistol or rifle, patented by Samuel Colt in 1836, which had a rotating chambered breech holding from 5 to 8 charges. The next development was the metallic, pin fire cartridge, which was invented by Houiller, a Frenchman in 1847. Repeating rifles were first developed and were used to some extent in the Civil War in America. The first real machine gun was the Gatling gun, first used in the Civil War. It consisted of a number of barrels, usually ten, each having its own lock, the whole revolving about a central axis, and operated by a crank. The mitrailleuse was adopted by the French

Army in 1869, and used with poor results in the Franco-Prussian War. The next development of note was the Maxim gun, the first truly automatic gun, which was adopted by the British Army in 1889. This gun introduced several new features; the action was wholly automatic, the operator merely holding his finger on the trigger. The barrel and breechblock were locked together as each cartridge was fired, and the force of recoil was utilized to perform the operations of extracting, feeding and firing the cartridges. This gun and developments of it were used up to and to some extent in the first World War. Another method of utilizing the expansion of gases resulting from discharge was invented by John M. Browning, who took gas from a port near the muzzle for the operation of the weapon. Guns of this type are known as gas operated machine guns. Since 1900 developments in automatic arms have been rapid. The general trend has been toward a light and a heavy type, the light being used as an automatic rifle, and the heavy type as a machine gun. See also Automatic Rifle, Pistol, Machine Gun.

Automatic Wing Slot. A wing device on an airplane which reduces the camber at low speeds.

Automobile Torpedo. A self-propelling torpedo.

Auxiliaries. Foreign or subsidiary troops furnished to a belligerent in consequence of a treaty of alliance, or for pecuniary consideration. In the old Roman armies troops composed of barbarians.

Auxiliary. Giving or furnishing aid or support, especially in a subordinate or secondary manner.

Auxiliary Aiming Mark or Point. A point selected for each gun to lay on, from round to round, as nearly as possible in the line of sight.

Auxiliary Arm. Any arm that assists the principal arm assigned a mission in combat.

Auxiliary Line of Sight or Sighting. The line of sighting to an auxiliary mark other than the target.

Auxiliary Target. A visible point whose range and azimuth or position relative to the target are known; used as an adjusting point for fire on other targets.

Auxiliary Troops. Troops furnished to a belligerent power by an ally or for pecuniary consideration. Troops designed and trained to assist combatant troops in duties not directly connected with battle.

Avant Garde, Poste. Advance guard, outpost. (Fr.)

Avenues of Approach. Favorable routes by which a place or position can be approached by an attacking force.

Aviation. The art of navigating the air. The operation of aircraft heavier-than-air.

Aviation, Classes of. Airplanes are designed to meet certain requirements, and are classified according to the principal functions they are intended to perform, as follows: (1) **Observation.** Observes and reports the dispositions and activities of all hostile forces. Is equipped for aerial photography, and is lightly armed for its own protection. Its radius of action is about 500 miles. (2) **Reconnaissance.** Engages in reconnaissance, usually at long range and for strategic purposes. Has elaborate photographic and radio equipment. It is classified further as light, medium, and heavy. Its radius of action is from 1,000 to 3,500 miles. (3) **Bombardment.** Organized primarily for the attack of personnel and materiel objectives on land and sea, especially by bombing, but also by gunfire. Classified further as light, medium, and heavy; long range, high altitude; attack and dive bombers. Bombardment aviation is often supported by pursuit aviation, or heavy bombers by light bombers. (4) **Pursuit.** Organized primarily for aerial combat with hostile aviation. Has great speed and power, and is armed with machine guns, and often with small cannon. It intercepts and pursues enemy aviation, protects other classes of friendly aviation, local areas and installations. (5) **Transport.** Organized primarily to carry personnel and cargo. (6) **Training and Special Purpose.** Organized, equipped, and trained especially for the training of flying personnel and for other special purposes not directly connected with air operations. Classes (3) Bombardment, and (4) Pursuit Aviation are collectively known as Combat Aviation.

Aviation, Combat. Bombardment and Pursuit Aviation.

Aviation Pay. See Flight Pay

Aviation, Strategical. Aviation whose principal aim will be to hamper the production of enemy industrial establishments within its reach.

Aviation, Tactical. Aviation whose principal aim will be to attack and destroy such enemy organizations and installations as are out of range of artillery.

Aviator. The pilot of an aircraft heavier-than-air. One who practices the art of flying in heavier-than-air aircraft.

Aviette. A heavier-than-air flying machine which flies by motive power furnished solely by the aviator.

Awkward Squad. A term, now in disfavor, meaning the most backward of the recruits, who were grouped for special instruction.

Axes of an Aircraft. Three fixed lines of reference, usually centroidal and mutually perpendicular. The horizontal axis in the plane of symmetry, usually parallel to the axis of the propeller, is the longitudinal axis; the axis perpendicular to this in the plane of symmetry is the vertical or normal axis, and the third axis perpendicular to the other two is the lateral axis.

Axial Observation or Spotting. Observation from a point on or very near the line of fire.

Axial Road. The main route of supply for a unit, designated as such by the proper commander; usually approximately perpendicular to the front.

Axis. A line about which or along which a number of posts or establishments are grouped. A meridian, central, guiding or control line. A line about which anything rotates.

Axis of a Gun. The central line of the bore.

Axis of Signal Communications. The route of movement of a command post, designated in orders which state the controlling points through which the post will pass. The line formed by the successive positions of the command post of a unit in combat.

Axis of Supply. The route of advance of supplies, designated in orders which state the controlling points through which the supplies will pass.

Axle Traverse. A method of traversing a gun in which the carriage slides or rolls along the axle.

Azimuth. The angle of direction to any point measured from a base direction line which may be the true meridian, the magnetic meridian, or a north and south grid line, used for measuring or expressing direction; it can be measured in degrees or mils. Azimuths are measured clockwise from the north point. The horizontal direction given to a gun in pointing.

Azimuth Adjustment Slide Rule. A circular slide rule for the rapid computation of azimuth correction, when the correction for some other elevation or zone is known.

Azimuth Circle. An instrument for measuring azimuths. The azimuth scale of a sight, racer of a gun carriage, etc.

Azimuth Deviation. The difference between the azimuths from the directing point of the battery to the center of the target and to the point of strike of the projectile.

Azimuth Difference. The difference between two azimuths of a point as read from two other points. Parallax.

Azimuth Difference Chart. See Difference Chart.

Azimuth, Grid, Magnetic, True. Azimuth measured from the grid, magnetic or true meridian.

Azimuth Instrument. An instrument for measuring or laying off azimuths.

Azimuth Scale. The graduated scale of azimuths on an instrument, or gun carriage or platform.

Azimuth Type Range Finder. A self-contained horizontal base range finder operating on the coincidence principle.

B

Back Azimuth. The azimuth plus or minus 180 degrees or 3200 mils. The opposite direction.

Back Sight. In traversing, a sighting from a located station to the station from which the location was made. In leveling, a sight on a point whose level has been previously established.

Back Step. A movement in the school of the soldier, the step (15 inches) being used for short distances only.

Backfire. A premature explosion in the cylinder of a gas or oil engine tending to drive the piston in a direction the reverse of its proper direction.

Backwards. Movements to the rear in drill. A retrograde movement of troops.

Badge. A medal, ribbon or the like, used as a mark of honor, office, or rank.

Baggage. Clothes, tents, and other like equipment of an organization.

Baggage Guard. A guard in charge of the baggage.

Baker Rifle. A crude flintlock rifle introduced into the British Army in 1800. It had a caliber of .625 inches.

Balance. That point of the rifle at which it will balance; i. e., the center of gravity. A condition of steady flight in which the resultant force and moment on the airplane are zero. A state of equilibrium. To maintain equilibrium by means of an aircraft's control surfaces, either automatically or manually.

Balanced Force. A force which includes the proper proportions of various classes of troops for the accomplishment of the mission assigned to it.

Balanced Profile. A profile of a trench, center line of a road, etc., in which the amount of excavation is equal to the amount of embankment.

Balanced Ration. A ration containing the proper proportions of food components.

Balanced Stocks. Accumulations of all articles of supply of the types and in the relative amounts that have been determined to be necessary for the needs of the number of troops to be supplied.

Bald Eagle. The American Eagle; the symbol of the power and majesty of the United States.

Balista or Ballista. A ballistic machine of the crossbow type, used in ancient and medieval warfare. It discharged great arrows or darts, and derived its projectile force from the recoil of tightly twisted cordage. See also Ballistic Machines; Catapult; Trebuchet.

Balk Carriers, Lashers. Men who carry, and who lash the balk or stringers of a ponton bridge.

Balks. The longitudinal stringers which support the chess, or flooring, of a ponton bridge.

Ball. Any spherical projectile larger than a small shot.

Ball Cartridge. A cartridge for small arms containing a projectile and a propelling charge.

Ballast. Weight carried to steady a balloon, ship, etc. Broken stone, or other material, laid on the subgrade of a railroad to facilitate lining and leveling of tracks, to distribute the load, hold the ties in place, and insure drainage. Road metal.

Ballast Ceiling. The maximum altitude to which an airship may ascend and return to the surface of the earth in equilibrium.

Ballistic Coefficient. A factor which is a measure of the relative ballistic efficiency of a projectile.

Ballistic Corrections. Corrections to firing data based on ballistic conditions as distinguished from those based on observation of fire.

Ballistic Density. A fictitious constant density of the atmosphere which would have the same total effect on the projectile as the varying densities actually encountered.

Ballistic Efficiency. The capability of a projectile to overcome the resistance of the air. It depends chiefly on the weight, area of cross-section and shape of the projectile.

Ballistic Machines. A general term used to describe a great variety of machines used to discharge heavy missiles, such as stones or large arrows and darts, by means of the torsion of cordage, springs, weights and levers. They became obsolete following the introduction of gunpowder and the use of cannon. We have no direct evidence as to when engines for throwing projectiles were first used. The first allusion to them is in the Bible where we read of engines "to shoot arrows and great stones" (2d Chron. 26-15). Dionysius of Syracuse had a siege train in his operations against the Carthaginians in 397 B. C. At the siege of Jerusalem, 70 A. D., catapults which threw 60 pound stones to a distance of 450 yards were used. From this time onward, until near the close of the 14th century, projectile-throwing engines were used.

The three types of projectile-throwing engines used in ancient and medieval warfare were the Balista, which discharged great arrows, and the Catapult and Trebuchet, which cast pieces of rock and great stones to a distance of 300 to 450 yards. The balista and catapult derived their projectile force from the recoil of tightly twisted cordage, while the trebuchet owed its power to twisted cordage or the fall of a heavy weight. The balista was used in the attack of personnel; the catapult was used to throw stones over the walls of besieged places; while the trebuchet, which threw a stone 5 or 6 times as heavy as that thrown by the catapult, was used to batter walls. The trebuchet was used as late as 1480. See Balista; Catapult; Trebuchet.

The following names were commonly, and often indiscriminately, applied to these weapons and their modifications:

Acquereaux	Cabule	Jauclide	Petrary
Balista	Calabra	Lide	Polibole
Beugle	Catapulta	Mangonel	Robinet
Blida	Engin	Martinet	Scorpion
Blude	Engin a Verge	Manganum	Springale
Bly	Espringale	Matafunda	Tormentum
Blyde	Fronda	Mategrifon	Trebuchet
Briche	Frondibale	Onager	Trepeid
Bricole	Fundibulum	Palitone	Tripantum

Ballistic Tables. Tables of standard ballistic data, such as ranges, times of flight, etc.

Ballistic Wind. A fictitious wind, constant in direction and velocity, which would have the same effect on the projectile during its flight as the true winds actually encountered, which vary in direction and velocity.

Ballistics. The science that deals with the motion of projectiles.

Ballonet. A gas-tight compartment of variable volume constructed of fabric, and placed within a balloon or airship. It is usually partially inflated with air to compensate for changes of volume in the gas contained in the envelope.

Balloon. A lighter-than-air type of aircraft. An aerostat which derives its buoyancy from a contained gas lighter than air. An aerostat without a propelling system.

Balloon Barrage. A barrier of captive balloons, with or without connecting cables or supported nets, against which airplanes may be expected to run.

Balloon, Barrage. A small captive balloon, used to support wires, or nets, which are intended as a protection against attacks by aircraft.

Balloon Bed. A mooring place on the ground for captive balloons.

Balloon, Captive. A balloon restrained from free flight by means of a cable attaching it to the earth.

Balloon, Ceiling. A small free balloon, whose rate of ascent is known, used to determine the ceiling.

Balloon Company. The basic lighter-than-air organization, which operates one balloon.

Balloon, Free. A balloon, usually spherical, whose ascent and descent may be controlled by releasing ballast or gas, and whose direction of flight is determined by the wind. A balloon that floats freely in the atmosphere with only static means of control.

Balloon Group. A headquarters, one service company, one photographic section, and two or more companies.

Balloon, Kite. An elongated form of captive balloon, fitted with lobes to keep it headed into the wind; it usually derives increased lift from the inclination of its axis to the wind. A sausage balloon.

Balloon, Observation. A captive balloon used to provide an elevated observation post.

Balloon, Pilot. A small balloon sent up to show the direction and speed of the wind.

Balloon, Propaganda. A small free balloon sent up without passengers but with a device by which papers or documents may be dropped at intervals.

Balloon Wing. A headquarters, one communications section, one airdrome company, and two or more balloon groups.

Bandage. A wrapping of cloth used to secure a dressing on a wound, or splints to a fracture. To apply a dressing and bandage to a wound.

Banded Mail. A type of armor which consisted of alternate rows of cotton, of leather and single chain mail.

Banderol. A small flag used to signal a saluting battery, or to mark the flanks of an organization at ceremonies. A marker flag.

Bandoleer. A belt or sling designed to be worn over the shoulder for the carrying of additional ammunition. An ancient contrivance used to facilitate the loading of muskets.

Bangalore Torpedo. A long metal tube charged with high explosive, used to cut breaches in wire entanglements.

Banjo Breech Mechanism. A seacoast breech mechanism, so-called because of its resemblance to a banjo when viewed from the rear.

Bank. The position of an airplane when its lateral axis is inclined to the horizontal. The lateral inclination of an airplane as it rounds a curve. Superelevation of the outside of a curve of a highway to compensate for centrifugal force. To incline an airplane laterally.

Bann. A proclamation issued at the head of a body of troops, or in the various quarters of an army, by sound of trumpet or beat of drums, either for observing military discipline, punishing a deserter, or the like.

Banner. A piece of cloth attached to a pole and usually bearing a device or emblem. The body of men who follow a banner.

Banquette or Banquette Tread. A raised way or platform of sufficient height above the terreplein to enable cannon or riflemen of medium height to fire over the parapet of a fortification.

Banquette Slope. The slope in rear of a banquette.

Baptism of Fire. A term signifying that soldiers have passed through their ordeal of fire in battle.

Bar Shot. A projectile consisting of two cannon balls, or half balls, united by a bar of iron.

Barbed Wire. A naked steel wire, often a twisted pair, armed at intervals with sharp points; used in making obstacles.

Barbed Wire Entanglement. An obstacle of barbed wire, usually fastened to upright pickets, intended to prevent or delay the advance of the enemy.

Barbette. A mound of earth or other platform upon which a gun can be mounted so as to fire over a parapet without an embrasure.

Barbette Battery. A number of guns mounted on a barbette or barbettes.

Barbette Carriage. A stationary carriage on which a gun is mounted to fire over a parapet.

Barbette Gun. A gun mounted on a barbette.

Bare Necessities. The minimum of personnel equipment or supplies required for a particular purpose. A principle which implies that in war all non-essentials must be eliminated; time, materiel, personnel, and transportation rigidly conserved; and that all structures must be sufficient for their purpose, but no more than sufficient.

Barometer. An instrument for measuring atmospheric pressure. See Aneroid Barometer.

Barong. A kind of bolo.

Barrack. A permanent building for the housing and accommodation of soldiers. Formerly a temporary hut for the shelter of troops. To live or lodge in barracks. To supply with barracks.

Barrack Bag. A bag issued to each soldier in which to stow articles of personal equipment.

Barrage. A barrier to the advance or retreat of enemy troops, or a protection to friendly troops, formed by the rapid and continuous fire of artillery or machine guns, or both, on a designated line or narrow belt. A dam or curtain of fire laid down in accordance with a prepared plan. A barrage may be either moving or fixed, and may be used offensively or defensively.

Barrages are classed as standing, rolling (jumping), normal, emergency (eventual or contingent), box. and antiaircraft barrages. See type of barrage in question, also Balloon Barrage.

Barrage Chart. A chart showing graphically the barrage plan, with pertinent data.

Barrage Line. The line on which the fire of a barrage is desired, or is to be placed.

Barrage, Machine Gun. A barrage of machine gun fire, either standing or rolling. The organized, controlled and coordinated fire of several machine guns, usually in batteries of two or more, laid on a line or series of lines. Either direct or indirect laying may be used, and a machine gun barrage may be combined with artillery fire.

Barrage Patrol. An airplane patrol whose mission is to keep enemy aircraft out of a designated portion of friendly territory.

Barrage Plan. The assignment of barrage missions to the batteries of a command when on the defensive, and the assignment of rolling barrage missions when on the offensive.

Barrel. The principal part of a firearm, its function being to concentrate the force of the gases generated by the explosion of the powder charge on the base of a projectile and give it proper initial velocity and direction. That part of the hilt of a sword adapted to be grasped by the hand.

Barrel Extension. The frame into which the barrel of a machine gun is screwed and which serves as a mounting for certain of the parts.

Barrel Floats. Casks or barrels utilized in the building of floating bridges when pontons are not available.

Barricade. A barrier or defense hastily constructed of trees, earth, or other available material. To construct a hasty obstacle to close a street, etc. To defend by a barricade.

Barrier. Any line of boundary or separation, natural or artificial, placed or serving as a limitation or obstruction. A fort or fortified place on the frontier of a country, commanding any line of approach. A kind of fence erected as an obstacle. Strong gates placed so as to defend the entrance of a passage into a fortified place. A barricade.

Barrier Forts. Small, isolated forts designed to bar passage of a defile, or to command the approaches to an important point or place.

Barrier Lights. Searchlights used in certain exceptional cases to detect the passage of vessels toward or through a channel or harbor entrance.

Barrier Line. A line, designated in the control of traffic, beyond which vehicles will not be permitted to pass except as provided in the priority schedule of the plan of circulation.

Barrier Tactics. The use of obstacles defended by fire to check a hostile invasion.

Barrier Works. Fortifications designed to close defiles, mountain passes, or stream crossings to the enemy, usually on the flanks of an army.

Barr-Stroud Range Finder. A self-contained horizontal base range finder. It consists of a tube about 30 inches long, containing a simple arrangement of prisms, and is reasonably accurate and practicable.

Base. The element on which a movement is regulated, or on which a formation is based. A place or area where the line or lines of communication originate, where stores are gathered, and where the business of supplying the field troops is organized and controlled. The lower part of the crust or wearing surface of a road. The smallest piece of ancient cannon, having a caliber of 1½ inches, weighing 200 pounds, and having a range of 700 paces. A skirt of mailed armor. To form or make a base.

Base Angle. The horizontal angle between the base line and the orienting line, or between the lines connecting the gun with the target and the initial aiming point.

Base Deflection. The angle read at the end of a horizontal base line in position finding. That deflection, for the base piece, which will direct the plane of fire of the piece on the base point. The deflection setting when the battery is on the base line.

Base Depot. A general or branch depot, designated as such by the commander of the theater of operations, located in the base section of the communications zone, for the reception and storage of supplies received from the zone of the interior or procured in the base section.

Base End Stations. The observation stations at the end of a base line.

Base Fuze. A fuze inserted in the base of a projectile.

Base Hospital. A large hospital located in the base section of the communications zone.

Base Line. The line on which the depots and means of supply of an army are established. The line on which the troops march. A line of known length and direction between two base end observation stations, the position of which with respect to a battery is known, used to determine ranges by means of triangulation. The line joining the base piece of a battery with the base point. The line on which the pieces of a battery are established, when in the firing position. A battery is on the base line when the base piece is laid on the base point and the other pieces are parallel to it.

Base of Operations. The line or area from which an army begins its forward movement and on or in which supply depots are organized.

Base of Supplies. The area from which an army draws its supplies.

Base of the Trajectory. The long chord or straight line joining the muzzle of the piece and the level point, or point on the descending branch of the trajectory which is at the same altitude as the muzzle.

Base Pay. The initial annual or monthly pay of a rank or grade under certain conditions, without increases for length of service.

Base Piece. The piece for which the initial firing data are computed. A gun which is the directing point for a battery.

Base Plate. A metal bearing plate, specifically one that sustains the recoil of a trench mortar.

Base Plug. A plug which closes the base of a projectile.

Base Point. A well defined point of the terrain which, as one of the points defining the base line, is used as an origin for direction or shifts, from the gun position. A point at which the base piece is pointed in the original orienting of the battery.

Base Printing Plant. A large, stationary plant for printing and map reproduction.

Base Ring. A metal ring in the concrete platform of an emplacement to support the weight of a fixed gun, howitzer or mortar, and on which the piece turns. In smooth bore guns, the ring which encircles the breech, connected with the body of the gun by a concave molding.

Base Section. The rear area or subdivision of the communications zone.

Base Unit, Base of Movement. See Base.

Basic Arm. The infantry is considered to be the basic arm, and upon its success depends the success of the army; all other branches are organized, equipped and trained to assist the infantry in its needs, functions, and methods of war.

Basic Communication. An original letter, report or the like, as distinguished from inclosures and indorsements.

Basic Data. The elements on which firing data are based; direction, height, width, and range of target. The uncorrected firing data determined from the map or firing chart.

Basic Principle of Training. To stimulate and develop the national individual characteristics of initiative, self-reliance, and tenacity of purpose, and so mold these characteri tics that they will at all times be responsive to the lawful directions of a superior.

Basic Procurement Plan. The plan which sets forth the general policies that will govern the mobilization of the procurement activities of the supply arms and services and their procurement planning to meet the requirements of the War Department mobilization plan.

Basic Training. Training in the essential fundamentals. Primary training.

Basket. The leather guard around the hilt of a fencing sword. The car suspended beneath a balloon for passengers, etc.

Basket-work. Interwoven withes and stakes, such as gabions, fascines, hurdles, wattling, wickerwork, etc., used in revetting. Brush-work.

Bastard. Not of standard or accepted design. Unusual. Inferior.

Bastile or Bastille. A tower or other elevated work designed for the defense of a fortified place. A small fortress. A movable tower or shelter for besiegers. One of a series of intrenched huts. A French term for a strong fortress defended by towers and bastions. To fortify with a bastile.

Bastion. One of the salients of the bastion trace, consisting of two faces, meeting at the salient angle, and two flanks, connecting the faces to the curtain between bastions. The angles between the faces and flanks are the shoulder angles, and those between the flanks and the curtain are the curtain angles. The bastion was first used and developed in Italy about 1450.

Bastion System. Fortifications based on the utilization of the bastion. A bastioned fort consists of a number of bastions, often 5, connected by curtain walls, the whole forming an enceinte.

Bastioned Trace, or Front. A system of fortification dating from the 16th century, in which the trace or magistral consists of a series of salients, called bastions, and reentrants, so arranged that all portions of the ditch could be swept by fire from the main parapet.

Bateau. A small, flat-bottomed boat, used as a ferry or float for a bridge.

Bateau-bridge. A ponton or floating bridge supported by bateaux.

Baton. A staff or truncheon borne as the symbol of office. The staff with which a leader beats time.

Battalion. The attack or tactical unit, consisting of a headquarters and a number of companies, or batteries. The organization next above a company.

Battalion Aid Station. A station established close to the front by battalion medical personnel, especially in defense, where the wounded are given emergency treatment and returned to duty or evacuated to the rear.

Battalion Commander's Detail or Battalion Detail. A portion of the headquarters battery which assists the battalion commander in his tactical duties.

Battalion Commander's Party. A portion of the battalion detail which normally accompanies the battalion commander on reconnaissance.

Battalion Defense or Defensive Area. See Defensive Areas.

Battalion Reserve Line. One of the lines of a defensive position, from 400 to 900 yards in rear of the line of resistance, occupied by the reserve companies of the front line battalions. In a prepared position it is marked by more or less continuous trenches.

Batter. A cannonade of heavy guns. Inclination from the vertical. The backward slope of a revetment or retaining wall. An inclined structural member, as a batter pile, the batter posts of a bridge, etc. To fire with heavy artillery on some point or place.

Batteries, Types of. Artillery organization in the United States service includes batteries of the following types: gun, howitzer, mortar, machine gun, searchlight, submarine mine, sound and flash ranging, headquarters, service, ammunition. Batteries armed with any type of cannon are distinguished as firing batteries.

Battering. Firing with heavy artillery on some enemy fortification or strong point with a view to demolishing it.

Battering Charge. The heaviest charge of powder used in a gun.

Battering Ram. A long, heavy beam used to breach the walls or force the gates of fortified places.

Battering Train. A train of artillery used exclusively for siege purposes. A siege train.

Battery. A combat and administrative unit consisting of a specified number of guns, usually four, other than small arms, and the personnel to man them. A number of guns grouped for the purpose of control of their fire. An emplacement used for a battery. An assault of artillery. A bombardment.

Battery Angle of Convergence or Divergence (Distribution). The angle through which the flank gun of a battery must turn from a parallel to the base piece in order to cover its part of a narrow or wide target, as the case may be. Used for converging or distributing machine gun fire by indirect laying.

Battery Angle of Parallax. The horizontal angle through which the flank gun of a battery must turn, from its original aiming point, so that its plane of fire will be parallel to that of the base gun. Used in indirect laying of machine guns.

Battery Box. A box to be filled with earth for making embankments or parapets, in the absence of gabions. A container for an electric battery.

Battery Chart. A chart on which are noted the firing data, targets, direction of fire, times of fire, and the traversing for each gun of the battery.

Battery Command. Two or more seacoast guns or mortars commanded directly by a single individual, together with all structures, equipment, personnel, etc., necessary for their emplacement, protection, and service.

Battery Commander's Detail or Battery Detail. A part of the headquarters of a battery which assists the battery commander in the conduct of fire.

Battery Commander's Party. A portion of the battery detail which normally accompanies the battery commander on reconnaissance.

Battery Commander's Telescope, Observation Instrument. A powerful telescope, made in several types, used by a battery commander in observation of fire. It measures both horizontal and vertical angles.

Battery Manning Table. A table showing the posts of the personnel of the battery, by name and duty.

Battery Salvo. The simultaneous or successive discharge of a single shot from each gun or mortar in a battery.

Battery and Store Wagon. A wagon which accompanies a field battery and transports equipment, spare parts and stores.

Battle. The combined tactical employment of all available moral, physical, and mechanical strength to obtain or gain physical ascendency over an enemy, and then crush his powers of resistance. A general contest between armed forces. An old term for an organization similar to the present company, but having a strength of about 450 men. An ancient mass battle formation. The main body. The most important part of the troops. A kind of halberd, formerly much used in northern Europe. To contend in battle. To assail in battle. To fight. To put in battle array. To equip or fortify with battlements.

Battle Area. The area within which a battle takes place. An area or zone organized in preparation for battle, including one or more battle positions. The area covered by the armament of a coast defense or fort command.

Battle Axe. An axe-shaped weapon formerly used. A halberd or bill.

Battle Chart. A chart used in fort, fire, or mine command stations, showing the water area covered by the armament of their respective commands with immediately adjacent land, locations of guns, observing stations, etc.

Battle Flag. A color or standard carried by troops in battle.

Battle Honors. Silk streamers or silver bands on the staff of the color or standard of any organization to indicate service in war, meritorious service in action for which the unit has been mentioned in War Department orders, or an award of a decoration by a foreign government, the latter being used only when decorations and service medals are prescribed by the commanding officer.

Battle Maps. Maps on which tactical features and the location of units are indicated.

Battle Position. The position on which the main effort of defense is, or is to be made, consisting of a system of mutually supporting sectors (areas) disposed laterally and in depth. The most important part of any defensive system and, in general, the most highly organized. See Defensive Position.

Battle Range. The range corresponding to the maximum danger space for any gun. The range within which most of the fire action of a battle occurs. The range within which any piece or battery normally delivers its fire.

Battle Reconnaissance. The continuous observation of all enemy forces engaged in battle with or in immediate contact with our own principal forces preliminary to and during battle, including artillery missions, infantry missions, and command missions. Reconnaissance carried out by troops during battle.

Battle Sight. The position of the rear sight of a rifle when the leaf is down; used for fire at close ranges; for the 1903 Springfield Rifle it corresponds to a range of 546 yards.

Battle Tactics. The tactics of combat.

Battle Zone. The territorial area within which troops are tactically disposed and the ground organized for combat. See also Combat Zone.

Battlefield. An area of the terrain on which a battle is fought or is to be fought. In includes any area in which actual conflict takes place.

Battlefield Illumination. Includes all devices, such as flares, rockets, star shells, and searchlights, used to illuminate targets or areas at night.

Battlement. A parapet with a crenellated fire crest, or provided with loopholes or embrasures.

Battle-plane. A fast, high-powered, military airplane, mounting a gun or guns, and designed especially for aerial combat. See Aviation, Combat.

Bay. One support and one span of a floating bridge. That part of the face of a truss or fuselage between adjacent bulkheads, struts, frames, etc.

Bayonet. A weapon of the dagger type made to fit on the muzzle of a rifle. The origin of bayonet is disputed, but it is generally credited to Bayonne, France, where a short dagger called bayonette was first made about 1490. It is known that a bayonet consisting of a steel dagger with a wooden haft which fitted into the muzzle of the musket was used at Ypres in 1647. The use of the plug bayonet prevented the firing of the musket, and a ring bayonet was introduced in 1678. About 1700 the socket bayonet was introduced and became, with the musket or other firearm, the typical weapon of infantry, replacing the pike. A bayonet which fastened to the musket by a spring clip was introduced into the British service in 1805. A triangular bayonet was used for some time, but when the magazine rifle was adopted it was generally replaced by the sword or knife bayonet. To stab with a bayonet.

Bayonet Assault. The last resort, either in attack or defense, employed when troops have closed and are engaged in hand-to-hand combat.

Bayonet Exercises. Exercises designed to instruct a soldier in the use of the bayonet as a combat weapon.

Bayonet Scabbard. A scabbard for carrying the bayonet.

Bayonet Stud. A bracket near the muzzle of a rifle barrel, to which the bayonet is attached.

Beach Party. A detachment detailed to assist in the disembarkation of troops and supplies.

Beachhead. A locality or area on a hostile shore which is garrisoned and organized to cover a landing of troops or supplies.

Beacon. A fixed or revolving light, or a conspicuous marker visible by day, or both, indicating a location or direction, generally as an aid in aerial or marine navigation. See also Radio Beacon.

Bearing. A method of expressing horizontal directions, especially those indicated by the compass, by measuring the angular divergence of a direction either east or west of the north or south point; e.g., N30°E is the same as azimuth 30°, S30°W is the same as azimuth 210°. True bearings are those measured from true north or south; magnetic bearings are those measured from the magnetic meridian. In the military service azimuths are preferred in indicating direction. That on which something rests, or in which a shaft or axle turns.

Beaten Zone. The pattern formed by the cone of fire when it strikes the ground.

Bedding Roll. A combination bedding and clothing roll of canvas for use in the field.

Beleaguer. To besiege. To surround with troops so as to preclude escape.

Bell or Swell of the Muzzle. A bell-shaped enlargement or flare at the muzzle of a cannon.

Bell Tent. A circular conical-shaped tent.

Belleville Springs. Steel disc springs used in the recoil systems of certain types of gun carriages.

Belligerent. A nation or state recognized by other sovereign powers as carrying on war. A person engaged in a recognized warfare. In a state of war.

Belligerency. The act or state of waging war. Warfare.

Bell-mouthed. Flaring at the muzzle, as a blunderbuss.

Belt. A band of leather or webbing, worn around the waist and used as a support for weapons, ammunition, etc.; as a cartridge belt, ammunition belt, pistol belt. A strip of fabric with loops to hold cartridges which are fed into the breech of an automatic weapon. A strip of terrain, usually parallel to the front.

Belt Filling Machine. A machine designed to fill the belt of an automatic gun with cartridges, at a much faster rate than by hand.

Belt Road. A lateral road. A road generally parallel to the front.

Belt, Safety. The belt or strap which secures the pilot or passenger to his seat in an aircraft.

Belt-fed. Supplied with cartridges from a feed belt, as certain automatic weapons.

Benet-Mercie Machine Gun. A machine gun, weighing about 22 pounds, used in the United States Army about 1912-1918.

Beneficiary. The relative of an officer or enlisted man who, under regulations prescribed by the War Department, is entitled to receive the gratuity of six months pay in case of the death of the officer or soldier.

Bengal Light or **Bengola.** A brilliant signal light.

Bent. A vertical bridge support of framed timbers or piles. One span of a bridge, including one support.

Berm. A horizontal offset in a slope, embankment, or excavation. A horizontal shelf inserted in the profile of a trench to prevent caving banks falling into the trench, or to provide a footing for riflemen.

Besiege. To invest a fortified place with the object of forcing a surrender.

Besieged. The garrison of an invested place. The state of siege.

Besiegers. The troops that lay siege to a place.

Bilateral Observation or Spotting. Observation of fire by two observers, one on each side of the line of fire.

Bill. An infantry weapon used as late as the 17th century, which consisted of a two-edged sickle-shaped knife or sword, weighing from 9 to 12 pounds and mounted on a handle about 4 feet long.

Billet. To assign or place in billets.

Billeting. The temporary quartering of troops in the homes of the inhabitants of an area or country.

Billeting Party. A party sent in advance to make arrangements for billeting a command.

Billets. Private or non-military buildings used as quarters for troops.

Biplane. An airplane having two main supporting surfaces or wings, one placed above the other.

Bipod. A two-legged support, as for the muzzle of a gun.

Bit. The mouthpiece of a horse's bridle. A snaffle or curb bit. A small drill, screwdriver or other tool for use with a brace. To fit with a bit.

Bit and Bridoon. The combination of a curb and snaffle bit, each with its own rein so that it may be used independently of the other.

Bit-wise. Said of a horse when he obeys the slightest pressure on either bar of the bit.

Bivouac. The arrangement of troops for resting on the ground without shelter. A night's encampment without shelter. The watch of an army by night when in danger of attack. To camp for the night in the open air.

Black and White Print. A print made by actinic processes on sensitized paper, which shows black lines on a white background. See also Blue Print.

Black Flag. A flag of black color displayed as a warning that no quarter will be given.

Black Powder. Gunpowder made with fully burnt charcoal.

Blacking Out. Loss of consciousness, or partial and temporary loss of consciousness by the crew of an airplane as a result of excessive speed or centrifugal force in maneuvering.

Blackout. To extinguish or conceal all the lights of a camp, garrison, or civil community, as a defense against night attack or threatened attack by hostile aviation. The act of taking such precaution.

Blade. One of the arms of a propeller. A sword. The cutting part of a weapon or tool.

Blank Cartridge. A cartridge which contains powder, but no shot or shell.

Blank File. A file which has no rear rank man. A front rank man without a rear rank man.

Blanket Roll. A roll consisting of the shelter tent half, blankets, and other prescribed articles of equipment, carried in the field.

Blanketing Smoke. See Smoke, Blanketing.

Blast Marks. Erosion of the ground in front of the muzzle of a gun, caused by the blast of firing.

Blasting Cap. A cap, charged with fulminate or other priming compound, with a booster of tetryl, ignited by safety and instantaneous fuze, to detonate a blasting or demolition charge of TNT.

Blasting Gelatin. A high explosive used in blasting.

Blasting Machine. A magneto exploder for setting off blasting charges.

Blasts. Small mines used in demolitions or intrenching.

Blending. The process of mixing powders of the same or different lots to obtain charges of uniform characteristics.

Blimp. A small non-rigid airship.

Blind. A place of concealment. An ambush. A slang term signifying a sentence to forfeiture of pay by a court-martial.

Blind Flying. See Instrument Flying.

Blind Landing. See Instrument Landing.

Blind Sap. A sap having a splinter-proof cover.

Blindage. A screen or covering for concealment or protection.

Blinded Batteries. Batteries, usually placed along the second parallel, whose guns are protected by armored parapets and bombproof blinds.

Blinding Hostile Observation Points. The placing of a smoke screen in front of or around enemy observation points to prevent or limit his observation.

Blitzkrieg. A sudden, violent, fast moving and highly coordinated offensive, usually by mechanized forces, employing the element of surprise, intended to destroy quickly the effective organized resistance of the enemy. A term of German origin, meaning literally "Lightning War."

Blob Stick. A bayonet training stick.

Block and Fall. A simple tackle rigging, as a whip or runner.

Block Carrier. A carrier to support the breechblock of a cannon when open. A breechblock carrier or tray.

Block, Tackle. A frame or shell supporting one or more rope sheaves or pulleys, used in combination with rope to form various tackle riggings. Blocks are designated as single, double, triple, according to the number of sheaves. A snatch block is one that may be opened at the side to permit insertion or removal of a rope.

Blockade. Cutting off intercourse between a belligerent and neutrals, by land or sea. Specifically, closing the seaports of an enemy by the use of war vessels. To subject to blockade.

Blockade Runner. A vessel engaged in running supplies through a blockade.

Blockading Army. An army, either independent or auxiliary to a besieging army, which is employed to invest a town or fortified place.

Blockhouse. A small, isolated, defensible building which gives shelter and protection against enemy fire, and from which fire may be delivered.

Blocking of Trenches. The placing of a barricade of some sort across a trench to prevent further advance of the enemy along the trench and to keep him beyond a point where he can make effective use of grenades.

Blouse. A military short coat.

Blowback. The reaction of the powder gases in a firearm directly to the rear against the bolt or breechblock. It is utilized as the force of operation in some types of automatic weapons.

Blue Print. A print made by actinic processes on sensitized paper, which shows white lines on a blue background. The original used is a tracing on translucent paper or cloth. A blue line print is made from an intermediate brown or Van Dyke negative, and shows blue lines on a white background.

Blues. The term used for the friendly side in the preparation and solution of map problems, exercises, etc. See also Reds.

Bluing. The blue colored finish of the metal parts of firearms. See also Browning.

Blunderbuss. An obsolete, short, muzzle-loading musket with a large bore and flaring muzzle, which made a loud noise and did great execution at short ranges.

Board of Officers. A number of officers convened by proper authority for the transaction of designated business.

Board of Survey. An officer appointed for the purpose of establishing facts or opinions by which questions of administrative responsibility may be settled.

Boat Telephone. A telephone used in a boat by temporary connection to the shore cable of a submarine mine group.

Boat-tailed Projectile. A projectile having a tapered base, which increases its range and accuracy.

Bobbing Target. A target which is fully exposed to the firer for a limited time; the edge of the target is toward the firer when the target is not exposed.

Body of a Field Order. That part which contains information and instructions for the command, and is arranged in numbered paragraphs.

Bofors. A Swedish arms firm. The Bofors gun is also made in the U.S.A. It has a vertical range of 16,200 feet, a rate of fire of about 130 rounds per minute, and is used against aircraft at low (dive bombing) and intermediate heights.

Bofors Breechblock. A threaded breechblock of ogival shape, affording a large bearing surface.

Bogie. A small wheeled truck used in the transportation of heavy guns, at the front end of a locomotive, etc., to facilitate the rounding of turns.

Bolo. A kind of large single-edged knife, commonly used by natives in the tropics as a weapon and as a tool.

Bolt. That part of a military rifle which pushes the cartridge into place and closes the breech. A pointed missile weapon fired from a crossbow or catapult. An elongated solid projectile for rifled cannon.

Bolt-lock. Part of the mechanism of a bolt type firearm which locks the bolt in position during firing.

Bomb. An explosive shell, such as those fired from mortars or dropped from aircraft. To drop bombs, as from aircraft. To bombard. To attack with bombs. Air bombs include high explosive, incendiary, gas and gas spray bombs. High explosive bombs are of five classes: (1) Fragmentation; (2) Demolition; (3) All purpose; (4) Land mines; (5) Delayed action. These bombs vary in weight from 30 to 4,000 pounds.

Bomb Lance. A lance with an explosive head.

Bomb Rack. A device used to carry bombs on an aircraft, and to drop them when desired.

Bomb Release Line. A perimeter surrounding a defended area, within which bombs released from aircraft may strike the defended area.

Bomb Screen. A screen of wire netting placed in front of or over trenches, or other open work, tanks, etc., for protection against bombs and grenades.

Bomb Sight. An aiming device used in dropping bombs from airplanes.

Bomb Thrower. A device for throwing bombs, as a howitzer or catapult.

Bomb Trap. A pit in a field work or at the bottom of the entrance to a shelter, to catch and confine the explosion of grenades.

Bombard. The earliest type of cannon. An ancient cannon which used gunpowder and was very short, thick, large bored, and fired a 300 pound projectile. The missile fired by a bombard. An engine for throwing large stones. To subject to shell fire.

Bombardment. An artillery preparation or fire upon a fortified place, position, or line. A sustained attack with shot, shell, etc.

Bombardment Aviation. See Aviation, Classes of.

Bomber. One skilled in the handling and projection of bombs. A bombing plane.

Bombing Plane. An airplane designed and equipped for bombing purposes.

Bombing Post. A pit from which hand or rifle grenades are thrown.

Bombing Squad. A squad of men specially trained for bombing purposes.

Bombproof. A structure of such strength and thickness that bombs or shells cannot penetrate it. Secure against bombs or shells.

Bombshell. A hollow metal globe, filled with explosive, and thrown from a mortar. A bomb.

Bonnet. A small two-faced work placed at a salient angle of the glacis of larger works. A part of the parapet elevated to screen another part from fire. A camouflaged hood placed upon a parapet for purposes of observation.

Bonus. Money, or other reward, given in addition to the agreed compensation for military service or as an inducement to enlist or reenlist.

Booby Trap. A grenade or other charge of explosive concealed in an abandoned trench, dugout or other place, with an arrangement whereby it may be unintentionally exploded by an unsuspecting person, as by moving some apparently innocent object, etc.

Boomerang. A curved wooden missile used by the natives of Australia, which circles about and returns to the point from which it was thrown. Any plan or scheme that reacts unfavorably upon its originator.

Booster. A moderately sensitive explosive which is detonated by a primer, fuze, or other detonator and which, in turn, is used to initiate the detonation of a less sensitive explosive; commonly used in high explosive shells.

Boot. A high-topped shoe. A scabbard for the rifle of a mounted man, attached to the saddle.

Boots and Saddles. A trumpet call as the first signal for mounted drill or other mounted formation.

Bore. The cylindrical cavity in a gun barrel. That portion of the interior of a barrel in front of the breechblock, including the powder chamber, slopes, and the rifled portion of the barrel, which is called the main bore. Caliber. The interior diameter of the cylinder of an engine.

Bore Sighting. Making the line of sighting parallel to the axis of the bore.

Bottle-necked Cartridge. A cartridge in which the powder chamber of the case is of larger diameter than the bullet.

Bottom Carriage. The lower assembly of the carriage of a large cannon. The chassis.

Bouchon. The firing mechanism used in hand grenades.

Bound. An advance in a single movement from one concealed position or firing point to another. A change of data to move the center of impact in direction or range. That part of the path or trajectory of a projectile comprised between two grazes. To make a bound or leap. To inclose. To fix the boundary of.

Boundaries. The lines defining the limits of a sector, zone, or other area.

Bounty. A premium given for enlistment or to induce men to enlist.

Bounty Jumper. One who enlists and deserts after receiving his bounty.

Bourrelet. The finished portion of a projectile just back of the ogive, which is larger than the remainder of the body and serves to center the projectile in the bore of the gun.

Bow. A weapon made of wood or other elastic material, with a cord connecting the two ends, by means of which, when drawn back and quickly released, an arrow is projected. The origin of the bow is lost in the mists of antiquity. Amongst the great peoples of antiquity, the Egyptians relied upon the bow as their principal weapon in war. Their bows were somewhat shorter than a man and their arrows were about 2½ feet in length. Nearly all the Asiatic nations relied heavily upon the bow, but the early Europeans were indifferent to it. But from the 4th to the 12th century, A. D., the bow was used extensively. A saddlebow. See also Crossbow; Long Bow.

Bowman. An archer. A soldier armed with a bow.

Bow-shot. The distance traveled by an arrow shot from a bow. The ordinary range of the English long bow was 300 to 400 yards.

Box Barrage. A standing barrage inclosing or partly inclosing an area to prevent the entrance or exit of hostile troops.

Box Trail. A gun trail having two side pieces or flanks and a top plate.

Boyaux. Winding or zigzag approach trenches. Short branches from a trench.

Bracket. The difference between two ranges or elevations, one giving a center of impact which is over, and the other a center of impact which is short of the target. The term is similarly applied to deflection, height of burst, etc. A support on a gun carriage for a sight, fuze-setter, or other device. The cheek of a mortar carriage. To fire so as to establish a bracket. To inclose a target in a bracket.

Bracket Adjustment. The process of determining a bracket, the depth of which depends on circumstances and the character of the target, for the purpose of searching it immediately with zone fire for effect.

Bracket Fuze Setter. A fuze setting instrument, so-called because it is mounted on a bracket.

Bracketing. A salvo or other similar series of fire is said to be bracketing when the number of rounds sensed to be short is equal to the number sensed to be over.

Bracketing Elevation. An elevation of a gun which gives an equal number of overs and shorts.

Bracketing Salvo. A salvo in which the number of splashes or impacts sensed short is equal to the number sensed over.

Branch. A subdivision of any activity. One of the primary subdivisions of the military forces of a nation, such as the infantry, cavalry, quartermaster corps, etc. A tributary of a stream.

Branch Depot. A supply establishment for the storage and distribution of supplies pertaining to a single supply branch, as an engineer depot. One of the sections of a general supply depot.

Brassard. A badge or band of distinctive color worn on the arm as a distinguishing mark, as for medical corps men in the field, umpires at maneuvers, etc. Armor for the arm.

Breach. A gap or opening forcibly made in a wall, fortification, or position. To create such a gap or opening.

Breach of Arrest. The act of an officer in leaving his place or limits of confinement without proper authority.

Breaching Batteries. Siege batteries employed for the special purpose of battering breaches in the walls of fortified places.

Break Camp. To strike tents, pack equipment, and prepare for a move.

Break Ranks. To fall out or leave ranks by order.

Break Step. To cease marching in cadence.

Break-through. An attack which penetrates through and beyond the depth of a defensive position or zone.

Break-through Guns. Machine guns which are sited to deliver direct fire on the enemy after he has reached the line of resistance of the battle position, and as he attempts to penetrate the defenses in rear.

Breast Reel. A reel for conductor wire, carried and operated by one man.

Breastplate. Armor for the protection of the chest or front of the wearer.

Breastwork. An earthwork which affords shelter for the defenders in a standing position, firing over the crest. A defensive work constructed entirely or largely above the surface of the ground, as in damp or rocky soil.

Breech. That part of a firearm in rear of the bore. The mechanism of a breechloading gun.

Breech Bushing. That part of the interior surface on which are formed the threaded and slotted sectors of the breech of a gun.

Breech, Face of. The rear plane of a cannon, perpendicular to the axis of the bore.

Breech Mechanism. The mechanism which opens and closes the breech of a gun and fires the charge. It includes the breechblock, obturating mechanism, firing mechanism, operating mechanism, and safety devices.

Breech Plate. The rear face of a cannon.

Breech Recess. The opening in a cannon which receives the breechblock.

Breech Reinforce. The part of a cannon in front of the breech and in rear of the trunnion band.

Breech Ring. A ring or bushing screwed to the jacket of a cannon, in which the breechblock engages, and which receives the longitudinal stress from the powder gases.

Breech Sight. The rear sight of a gun, usually mounted near the breech.

Breechblock. A movable piece, in the shape of a wedge, cylinder, or semi-spheroid, which closes the breech of a cannon and sustains the pressure of the powder gases. Common types of breechblock are the sliding wedge (including the drop block), interrupted or slotted screw, and eccentric.

Breechblock Carrier. A hinged support for a breechblock, especially in medium and smaller cannon, usually designed to permit the opening or closing of the breech by the continuous motion of an operating lever.

Breechblock Tray. A wide tray, hinged to the breech of a large cannon, which supports the breechblock when it is withdrawn, and permits it to be swung clear of the breech. The removal of the breechblock usually includes three motions; i.e., releasing the block by rotation, sliding it to the rear upon the tray and swinging the tray clear of the breech.

Breechloader. A firearm that is loaded at the breech.

Breechlock. A mechanism which closes and locks the breech opening of any breechloading firearm.

Bren Machine Gun. A light machine gun, of Czechoslovakian design, recently adopted in the British Army, replacing the Lewis Gun. It is gas operated and weighs, with bipod, 21 pounds, and with tripod, 25 pounds.

Brevet. A commission giving an officer higher honorary rank without an increase in pay or the right to exercise command in the brevet grade, except by special assignment. To confer rank by brevet.

Brevet Rank. Honorary rank conferred upon an officer above the rank he holds in his own branch.

Bricole. Harness used by men in dragging guns when animals are not available. A machine for projecting quarrels or darts.

Bridge Equipage. Bridge material especially for floating bridges, which accompanies large bodies of troops in the field.

Bridge Train. A field train carrying bridge equipage, which may be attached to an army or smaller organization. according to its needs.

Bridgehead. A position at the far edge of a bridge or ford or the outlet of a defile, occupied and defended by an advanced force to cover the passage of the bridge, ford or defile by the remainder of the command.

Bridoon. A snaffle bit and rein. A bridle equipped with a snaffle bit.

Brigade. An organization consisting of two or more regiments, or the equivalent thereof, and which is the proper command for a brigadier general. To form into a brigade. To place in a brigade. To attach infantry, cavalry or artillery units temporarily to a brigade or division other than that to which they belong.

Brigade, Regimental Detail. See Battalion Commander's Detail, which performs similar duties.

Brigadier General. An officer next in rank above a colonel and below a major general. The lowest grade of general officer. A brigade commander.

Broadaxe. A battle axe having a broad blade.

Broadsword. A sword with a broad blade for cutting only. A claymore.

Broken Line of Battle. A line of battle in which some of the elements are separated by the accidents of the terrain.

Brown Bess. A flintlock musket extensively used by both sides in the American Revolution, also in the wars of Frederick and Marlborough. It weighed 11½ lbs., was 4 feet and 11 inches long, caliber .75 inch, and fired a ball weighing 1½ ounces to a distance of 125 yards.

Brown Powder. Gunpowder made with underburnt charcoal.

Browning. A coating given to parts of small arms as a protection against the action of the atmosphere, and to prevent them from reflecting light. See also Bluing.

Browning Automatic Rifle. An air-cooled, gas-operated, magazine-fed automatic small arm, weighing 15½ pounds, invented by John M. Browning.

Browning Heavy Machine Gun. The machine gun now used in the United States Army. It is recoil-operated, belt-fed and water-cooled, and mounted on a tripod. The gun weighs 41 pounds and its tripod 51 pounds. It fires .30 caliber small arms ammunition at a practicable usable rate of 125 rounds per minute. There is also a .50 caliber model, mechanically similar, used to attack aircraft and unarmored or lightly armored vehicles.

Browning Light Machine Gun. A weapon mechanically similar to the heavy machine gun, except that it is air-cooled and somewhat lighter.

Browning Machine Rifle. A weapon mechanically similar to the automatic rifle, provided with a bipod rest and cooling fins.

Brun Spiral. A portable obstacle similar to the ribard cylinder, but less rigid and effective. A concertina.

Brush. A skirmish. A short, brisk encounter. A stick of carbon, strip of metal, or bundle of wires or plates, which bears upon the commutator or collecting ring of a dynamo (or motor) and carries current to or from the machine.

Brush Revetment. A revetment of brushwood materials, including fascines, gabions and hurdles.

Buckhorn Sight. A kind of rear sight which takes its name from the deep curved form of the notch in the rear sight.

Buckler. A shield or piece of defensive armor, usually having two handles, one held in the crook of the elbow, the other in the hand. It was often 4 feet long and covered the whole body.

Budget Officer. An officer who prepares departmental estimates of funds required for the coming fiscal year.

Budget System of the War Department. Provides for estimates and justifications for departmental, military and non-military activities of the War Department, and for the preparation of legislation involving financial matters relative to the War Department.

Buffer. A contrivance for checking the recoil of a gun. Any device for absorbing shock.

Bugle. A wind instrument, without keys, somewhat smaller than the trumpet.

Bugle Call or Signal. A command or signal blown on a bugle or trumpet. Bugle calls are used as the warning and assembly calls for all routine duties of a post, such as reveille, mess, etc., and as commands at drill, especially for artillery and cavalry.

Bugle Calls, List of. The routine bugle calls at a post or station include the following: reveille, mess, first call, assembly, drill, recall, guard mounting, adjutant's, first sergeant's, officers', fatigue, school, sick, stables, fire, water, to horse, boots and saddles, church, retreat, to the color, tattoo, to quarters, taps.

Bugler. A soldier detailed and trained to sound calls on the bugle.

Built-up Guns. Guns in which the principal parts are formed separately and then united by welding or screwing the parts together, or by shrinking one part over another.

Bullet. A missile, usually of lead, which is discharged from a rifle, pistol, or other small arm.

Bullet, Armor Piercing. A bullet designed to pierce light armor.

Bullet Ramp. An inclined surface at the entrance to the chamber which guides the cartridge into its firing position.

Bullet Splash. The particles of metal spattered by a bullet on impact against armor or other hard material.

Bullet, Tracer. See Tracer Bullet.

Bulletin. An official publication which contains matter that is informative or advisory in nature and of permanent duration.

Bullet-proof. Capable of resisting penetration by bullets.

Bullet-shell. An explosive bullet for small arms, having an inclosed copper tube containing the bursting charge.

Bullseye. The center of a target. A spot in the center of a target for aiming at. A shot which hits the center of the target. In slang, the point aimed at.

Bulwark. A solid wall raised for defense. A rampart, parapet, bastion, or outwork. That which secures, defends, or discourages attack. To secure by a fortification.

Bump. A sudden acceleration of an aircraft caused by a region of unstable atmosphere characterized by marked local vertical components in the air currents.

Buoy. A float of any kind on the surface of the water to mark the location of something beneath.

Bureau Chiefs. The advisors of the Secretary of War and the Chief of Staff on all matters pertaining to their respective branches or services. The heads of the bureaus of the War Department.

Bureau of War Risk Insurance. A federal bureau having charge of all matters relating to war risk insurance; now a part of the Veterans' Administration.

Burn-out Point. The point at which the composition in the base of a tracer projectile ceases to burn and be visible.

Burst. A number of shots delivered automatically between releases of the trigger of a machine gun. A relatively brief but intensive fire, or the period covered by such fire. The explosion of a projectile. To explode.

Burst Center. See Center of Burst.

Burst Effect. The distances, laterally and in the direction of range, within which the fragments of a shell or shrapnel are effective in producing casualties.

Burst Range. The distance in the plane of site from the muzzle to the point of burst.

Bursting Charge. Explosive placed in the cavity of a projectile and designed to explode with sufficient violence to rupture the shell and hurl the fragments with destructive effect. The principal explosives that are, or have been thus used, include TNT, dynamite, gun cotton, amatol, picric acid, picrates, explosive D, etc.

Burton. One of a variety of tackle riggings employing two or three blocks.

Bury the Tomahawk (Hatchet). In the language of the American Indian, to make peace.

Busby. A fur headdress worn by soldiers.

Bush Fighting. Fighting in bushes, trees, or thickets.

Bush Warfare. Irregular warfare in which one or both combatants are guerillas.

Bushing. A circular adapter, as for a shaft bearing.

Bushwhacker. A pretended noncombatant who secretly harasses a hostile force. A guerilla.

Butt. The embankment or other means employed on a target range to stop bullets in rear of the targets. The rear extremity of any small arm.

Butt Plate. A metal plate placed on the rear or butt end of a rifle.

Butt Stroke. A blow with the butt end of a rifle.

Butt Swing. A half-arm movement to bring the butt end of the rifle to the front with the barrel over the left shoulder.

Button. A spherical knob which forms the rear of the cascabel. A blunt tip for a foil, to prevent injury to an opponent in fencing.

Button Stick. A metal contrivance which slides behind button heads and protects the garment while the button is being cleaned.

Buttress. A supporting wall built in front of and at right angles to another and larger wall. To prop. To furnish with a buttress. See also Counterfort.

Butts. The parapet, pit, and back stop of a group of targets.

Buttstock. That part of the stock of a small arm in rear of the breech mechanism.

Buzzacott Oven. A cast-iron, telescopic oven, sometimes used by troops in the field.

Buzzer. An electro-magnetic device used in signalling. A buzzer telegraph.

Buzzerphone. An electrical instrument which may be used as a telegraph instrument or a telephone; messages sent by it by telegraph are practically impossible of interception.

By the Numbers. A command given at drill when it is intended that the exercise shall be executed slowly and carefully, one movement at a time, in obedience to the enunciation of prescribed numbers.

By-pass. A channel or passage through which oil passes from one end of a hydraulic cylinder to the other, as in the recoil of a gun.

C

Cadence. A uniform time and pace in marching. The number of steps soldiers march in a minute.

Cadet. A young man in training with a view to becoming an officer in the military service.

Cadre. A skeleton or nucleus of officers and enlisted men upon which a regiment or other organization is to be formed. The officers of a regiment forming the staff, or a list of such officers.

Caduceus. The wand of Mercury. The distinguishing insignia of the medical department.

Cage. An open-work steel support for rapid firing or other guns.

Caisson. A two-wheeled trailer vehicle, supporting an ammunition chest, specially designed for the transport of ammunition. The term is sometimes used to include a limber and caisson taken together. A water-tight box used in constructing foundations, and for other work under water, to float a sunken vessel, close the entrance to a drydock, etc.

Caliber. The diameter of the bore of a gun, usually expressed in inches, measured between the lands. The diameter of a projectile for a firearm.

Calibration. The determination of the corrections to be applied to each gun of a battery so that all guns will attain the same range. It is necessary because of variations in the wear on different pieces and to inherent differences in their construction. The determination of the corrections to be applied to the readings of any graduated instrument.

Calibration Corrections. The corrections applied to any piece to make it agree with the reference piece. Corrections which are applied on the guns as a result of calibration fire.

Calibration Point. A point in space at which calibration fire is conducted.

Calibration Trials. Tests to determine the performance of an aircraft.

Call. A signal sounded on a drum, bugle, trumpet, etc.

Call the Roll. To call off the list of names of men belonging to an organization in order to ascertain whether there are any absentees.

Call to Quarters. A call signifying that all men must repair to barracks.

Called to the Colors. The condition of a man on the reserve list, or a list subject to conscription, who has been ordered to report for service.

Calling the Hour. A call passed every hour around a cordon of sentinels at night, to verify the presence of each, thus: "No. 4, twelve o'clock, all's well."

Calling the Shot. Stating, from the observed alignment of the sights at the instant of discharge, where a bullet will strike the target.

Calling the Step. A method of insuring cadenced step by calling right, left, or 1, 2, 3, 4, as the feet should strike the ground.

Caltrops or **Caltraps.** See Crowsfeet.

Cam. An inclined or eccentric surface which imparts motion to another part bearing upon it, by sliding contact. To impart motion by means of sliding contacts; as in the operation of the valves of an engine by means of a camshaft.

Camber. A slight convexity of a member or part, as of an airplane wing. Convexity upward, as the deck of a ship or bridge. A slight inclination or tilt of the front wheels of an automotive vehicle which facilitates steering.

Camaraderie. The spirit of familiarity and good will that exists between comrades. Loyalty to one's associates. Comradeship.

Camel Corps. Troops mounted on camels. A slang term applied to infantry because of its overloaded appearance.

Camera Kite. A kite designed to carry a camera for aerial photography.

Camera Machine Gun. An aerial camera in the shape of a machine gun, which takes pictures when its trigger is pulled, recording the relation between the sight and the target; used for instruction in aerial gunnery.

Camlock. A breechlock, hinged at its forward end to permit opening the breech, and carrying a firing pin; formerly used on some types of breech-loading rifle, e.g., the early Springfield.

Camouflage. The disguising of an object or area by painting, covering with a net, or the like, for the purpose of deceiving an observer as to its existence, nature, or location. Any artificial means employed to deceive the enemy's visual or photographic observation from the ground or from the air. Work done for the purpose of deceiving the enemy as to the existence, nature, or location of materiel, troops, or military works. The materials used for such purpose. To make use of camouflage. To conceal by means of camouflage.

Camouflage Battalion. An organization which manufactures or prepares and installs camouflage materials.

Camouflage Discipline. Enforcement and observance of all rules of conduct designed to defeat the purposes of hostile observation. Efficient maintenance of camouflage installations and measures.

Camouflet. A land mine whose explosion does not rupture the surface of the ground.

Camoufleurs. Men organized and trained for camouflage purposes.

Camp. The condition of troops when sheltered by tents, other than shelter tents, or quartered in huts or other temporary structures specially constructed for military purposes. The ground on which troops are temporarily sheltered. A temporary location or station for troops. To put into camp. To establish a camp.

Camp and Garrison Equipage. The equipment required by troops for their personal welfare, rather than for battle, such as cots, tents, mess equipment, etc.

Camp Followers. Sutlers and dealers in small wares who attach themselves to and follow bodies of troops.

Camp Guard. A guard consisting of the sentinels placed around a camp and relieved at intervals.

Camp Hospital. A hospital maintained for the care of the sick belonging to the organizations in a camp.

Camp Kit. A kit containing cooking and mess utensils.

Camp Kitchen. A troop kitchen in a camp. An arrangement of mess tins or pans, usually 8 in number, placed with the opening in the direction of the wind, which requires no trench for cooking.

Camp of Instruction. An encampment of troops for the purpose of training and habituating them in the duties and fatigues of war.

Camp Stove. A light, portable stove used for heating huts or tents and for cooking purposes.

Campaign. A connected series of military operations forming in themselves a distinct stage in war. The time during which an army keeps the field. To serve in or go on a campaign.

Campilan. A two-handed, straight-bladed sword used by the Malay tribes.

Candle Bombs. Pasteboard shells filled with pyrotechnic compositions, which make a brilliant light on explosion.

Canevas De Tir. Fire control maps, usually on a scale of 1:20,000. (Fr.)

Canister. A projectile consisting of a metallic cylinder filled with lead or iron balls or slugs. Case shot. A can containing chemicals for filtering out poison gases; used with a gas mask.

Cannelure. A circumferential groove at the base of a bullet or the front of a cartridge case. It serves to contain a lubricant (for lead bullets), as a recess for metal stripped from the bullet in passing through the bore, to prevent the bullet being pushed too far into the case, or for crimping the case to the bullet.

Cannon. A piece of ordnance mounted upon a fixed or movable mount. All firearms too heavy or bulky to be carried by hand. Artillery firearms. A general name for large guns as distinguished from small arms. See Gun.

Cannon Balls. Solid cast-iron spherical projectiles used in smooth bore guns. Rifled guns use an elongated projectile in the shape of an ordinary bullet.

Cannonade. A continued discharge of cannon for the purpose of neutralizing or destroying troops, materiel, establishments, fortifications, etc. A bombardment. To fire cannon repeatedly or continuously.

Cannoneer. A soldier who serves a cannon. A member of a gun crew.

Cannon-shot. The extreme effective range of a cannon.

Cant. The angle between the axis of the trunnions of a cannon and the horizontal, or between the sight leaf of a rifle and the vertical, when the piece is tilted. To incline or tilt laterally.

Canteen. A small drinking flask for water or coffee, which forms part of the personal equipment of a soldier. A chest containing cooking equipment. A place, in or near a camp or station, where refreshments, especially liquids, are dispensed. A post exchange.

Cantle. The rear bow or protuberance of a saddle.

Cantonment. A camp of temporary buildings especially erected for the shelter of troops.

Cap. A blunt nose of soft steel fitted on the head of an armor-piercing projectile, as an aid in penetration. A type of covering for the head. A covering for the vent of a cannon. A copper capsule containing fulminate or other high explosive, which serves to ignite a cartridge, primer, fuze, mine, or demolition charge. The horizontal member forming the top of a trestle and which supports the stringers. See also Blasting and Electric Cap.

Capablities of the Enemy. The strength, morale, leadership, organization, equipment, national characteristics, and lines of action open to the enemy, as deduced by critical examination of all available information.

Caparison. The complete military equipment of a military mount; includes the bridle, saddle and housings. Trappings.

Capital. The principal horizontal axis of a fortification, or any salient thereof, which points in the direction of its principal front and divides it into two symmetrical parts. The bisector of a salient angle.

Capital Ships. Ships which possess great fire power and are generally heavily armored. This class includes battleships and battle cruisers, and requires the employment of primary armament to attack it.

Capitulate. To surrender to an enemy on conditions agreed upon.

Capitulation. An agreement entered into between commanders of belligerent forces for the surrender of a body of troops, a fortress, or other defended locality, or of a district of the theater of operations. The act of surrender.

Caponier, Caponiere or **Tambour.** A casement in or projecting from the scarp of a fortification at a salient or reentrant angle, or in the middle of a face or curtain, to provide for enfilade fire along the face, and the defense of the ditch; characteristic of the polygonal system of fortification.

Cap-square. The upper, removable portion of a trunnion bearing, which secures the trunnion and which may be removed to permit mounting or dismounting of a gun or cradle.

Captain. An officer whose proper command is a company, troop, or battery, or an officer of the same relative rank. An officer immediately above the grade of lieutenant. The commander or senior officer of a ship, regardless of actual rank.

Captain-general. An old term signifying the commander-in-chief of an army or group of armies.

Captive Balloons. Balloons, fastened to the ground, which constitute elevated observation posts and serve as a means of extending the field of view under continuous observation. Any balloons anchored to the ground. See also Balloon Barrage.

Captor. One who takes a person or thing by force or stratagem in time of war.

Capture. A prize taken in time of war. To take or seize by force, surprise, or stratagem. To gain by force or stratagem a position or place held by the enemy.

Car. A structure in, or suspended from, the hull or envelope of an airship for carrying crew, engineers, passengers, etc.

Car, Subcloud. An observation car which may be lowered from an airship to a position below the clouds.

Carabao. See Military Order of the Carabao.

Carbine. A rifle with a short barrel, especially suitable for the use of mounted troops. See United States Carbine, M1.

Carbine Scabbard. A scabbard which is suspended from the saddle, under the trooper's leg, for carrying the carbine.

Carbineer, Carabineer. A mounted man or trooper armed with a carbine.

Cargo Carriers. See Cross-country Cargo Carriers.

Carriage. A vehicle or support for guns. A mount for a gun. The undercarriage of an aircraft.

Carriage-bridge. A roller bridge capable of being rolled up a glacis to form a bridge across the ditch of a fortification.

Carrier. The hinged support of a breechblock, especially in a rapid fire gun. A person who spreads the germs of a disease, especially typhoid fever, without being himself ill of the disease. A small musketoon mounted on a swivel. A term broadly used to designate carrier wave, carrier current, or carrier voltage. One of a detail carrying supplies by hand. A truck used to carry a supporting weapon.

Carrier-block. Part of the breech mechanism of most machine guns.

Carronades. Short iron guns, named after the Carron Iron Works in Scotland, without trunnions, lighter than ordinary guns and valuable at close quarters; they discharged a ball weighing from 12 to 68 pounds.

Carry. To secure possession of by force. A position in the manual of the rifle, saber, or pistol.

Carry Arms. A former position in the manual of arms.

Carry On. A British expression meaning to continue as before or to resume a situation or occupation. To persevere unfalteringly in the line of duty.

Carrying Party. A detail carrying supplies of any kind by hand.

Carte. A map or chart. (Fr.)

Carte, or Quarte. A movement of the sword or foil in fencing.

Carte-blanche. A full and absolute power delegated to an officer to act according to his best judgment.

Cartel. An agreement for an exchange of prisoners, or the regulation of intercourse between belligerents. A written challenge.

Cartel Ship. A ship used in the exchange of prisoners, or for carrying a proposal to the enemy.

Cartography. The art of making maps or charts.

Cartouch, Cartouche. A roll or case of paper, etc., containing a charge for a firearm. A cartridge. A cartridge box. A wooden case filled with cannon balls. A gunner's ammunition bag. A pass for a soldier on furlough.

Cartridge. A complete charge for a firearm, contained in or held together by a case of some material, usually brass. Fixed ammunition.

Cartridge Bag. A silk or woolen bag used to contain the charge, or part of the charge, for a gun or mortar.

Cartridge Belt. A belt for carrying small arms ammunition.

Cartridge Box. A small leather case designed to contain cartridges and be carried on the belt.

Cartridge Case. A metallic container for the propelling charge, which conforms in shape and size to the powder chamber of the piece for which it is designed; in fixed ammunition, the case holds both the propelling charge and the projectile.

Cartridge Ejector. The mechanism of a firearm which ejects cartridge cases.

Cartridge Extractor. That part of a breechloading firearm which catches the flange of a cartridge case and draws it to the rear from the bore of the gun.

Cascabel. The projection in rear of the breech of a muzzle-loading cannon. It served as a point of attachment in mounting and dismounting, and was often provided with an eye through which the recoil rope passed.

Case. A tube to inclose explosives or a pyrotechnic composition. The charge holder of a submarine mine. A wooden frame used as a lining for a shaft or gallery in military mining. See also Cartridge Case.

Case I Pointing. Pointing in which direction and elevation are given to a gun by a sight. Used for all rapid fire guns.

Case I½ Pointing. Pointing in which direction is given to a gun by a sight, and elevation by a combination of a sight and an elevation scale or graduated drum; used in antiaircraft firing.

Case II Pointing. Pointing in which direction is given to a gun by a sight, and elevation by a quadrant or other device.

Case III Pointing. Pointing or laying in which direction is given to a gun by an azimuth circle or a panoramic sight pointed at an aiming point (not at the target), and elevation by a quadrant or other device. Indirect laying. Used for all mortars.

Case Shot. An assemblage of small balls or bullets inclosed in a cylindrical case or canister. Canister.

Cased Colors. The colors when furled and covered by a case.

Casemate. A shelter which protects a gun and its crew, the gun firing through a loophole. A loop-holed gallery in the scarp or counterscarp, for the defense of the ditch. See also Mine Casemate.

Casemate Battery. A battery emplaced in a bomb-proof chamber or casemate, with openings for the guns to fire through.

Casemate Truck. A low, 3-wheeled truck, used in moving cannon in a casemate or other restricted space.

Casemated. Furnished with, protected by, or built like a casemate.

Casemated Retrenchments. Works erected at the gorges of the bastions, provided with two tiers of loopholes, the upper for riflemen, the lower for small mortars.

Casern, or Caserne. A barrack in garrison towns, usually located near the fortifications. Buildings for the shelter and accommodation of a garrison. (Fr.)

Cashier. To dismiss with ignominy from the military service.

Casing. The outer metal case of converted guns. The lining of a subterranean shaft or gallery.

Casque, or Cask. An ancient head-piece of armor. The French word for helmet.

Castle. A fortified place or stronghold to defend a town. The insignia of the Corps of Engineers in the U. S. Army.

Castrametation. The art of laying out camps and of arranging troops therein. The act or art of encamping.

Casuals. Officers and soldiers awaiting assignment, or temporarily in a post.

Casualties. Losses in numerical strength by death, wounds, discharge, capture or desertion. The dead, wounded and sick in campaign and battle.

Casualty Agent. A chemical agent designed to cause casualties among personnel. Any chemical agent whose physiological action is sufficient to render a soldier unfit for duty.

Casus Belli. Any event or complication between sovereign powers which leads to a declaration of war.

Cat Walk. A narrow footway along the keel of a rigid airship.

Catapult. A device by which an airplane can be launched at flying speed when space is limited, as from the deck of a vessel. A ballistic machine of the crossbow type used in ancient and medieval warfare. It threw stones and derived its projectile force from the torsion of tightly twisted cordage. To launch or project, as from a catapult. See also Balista; Ballistic Machines; Trebuchet.

Caterpillar Club. A mythical organization among flying men to which only airmen who have jumped for life from aircraft during flight are eligible.

Caterpillar Tractor. A tractor which travels upon two endless belts, one on each side of the machine, actuated by sprocket driving wheels, which carry it over rough or swampy ground or up steep grades. A track-laying vehicle.

Cavalier. An armed horseman, especially a knight or gentleman. A caballero. A retrenchment or interior work in a fortification, sufficiently elevated to command the exterior ground.

Cavalry. A combatant arm of the service that marches and fights on horseback, or both on horseback and on foot. The history of warfare, until comparatively recent times, has been largely the story of the concomitant development of infantry and cavalry as combatant arms, their opposition and cooperation. The great importance of mobility in campaign and battle has long been recognized, and for thousands of years the horse provided the mobility which nature denied to man. Cavalry was naturally the dominant arm in those countries especially suited to the breeding of horses. These included the great Asiatic nations, Scythia, Assyria, Persia, Tartary and China, whose armies often consisted almost exclusively of cavalry. On the other hand, the early European nations, especially Greece and Rome, relied largely on infantry. The two arms early came into conflict; each in turn exercised a decisive effect on the conduct of war, enjoyed periods of dominance and suffered periods of decadence. It was recognized that both cavalry and infantry were essential components of a well balanced army, and that any nation that relied exclusively on either was at a disadvantage when opposed to one that made skilful use of both. Accordingly, all the great commanders, such as Alexander, Hannibal, Caesar, Gustavus Adolphus, Marlborough, Frederick the Great and Napoleon, gave careful attention to the organization, training, tactical and strategical employment of both infantry and cavalry.

    Cavalry has made use of nearly every known hand arm, and has been employed in every type of combat, including fire action, shock and hand-to-hand fighting, both mounted and dismounted. Perhaps its most important role is that of independent action, including tactical and strategical reconnaissance and counterreconnaissance, security, delaying action, pursuit, raiding, etc. In the past cavalry has too frequently been trained and utilized for a single type of combat or service to the neglect of others. It was the American Civil War that compelled recognition of the fact that to realize its greatest usefulness true cavalry must be organized, equipped and trained for every type of combat or service.

The development of firearms threatened to drive cavalry from the battlefield. An effort was made to meet this menace by increasing the armor of the already over-burdened knight. The error of this was soon manifest. The cavalry discarded its armor, regained its mobility and armed itself with the new weapons. It was able again to play a decisive part in battle by perceiving and taking prompt advantage of fleeting opportunities for effective mounted shock action or dismounted fire action. Continued improvement in weapons and the increasing power and importance of the infantry and artillery resulted in fewer and fewer opportunities for the effective employment of cavalry against infantry in battle. These conditions, with the increasing size of armies, brought about a progressive reduction in the proportion of cavalry in modern armies. Cavalry was recognized as an arm of opportunity in battle and reverted more and more to its important role of independent action.

The railroad and motor vehicle have now provided armies with a mobility which formerly the horse alone could supply; and these, with the airplane, telegraph, telephone, etc., now perform many of the functions which formerly were the natural role of the cavalryman and his horse. The modern battlefield, swept by a hail of fire, and intersected by trenches and obstacles, is not favorable to the battle use of cavalry. The great cavalry armies of the past will never be seen again. However, cavalry remains a part of all modern armies, and is in process of being in part motorized and mechanized, with a view to increasing its mobility, fighting power, and general combat usefulness.

Cavalry Charge. An impetuous mounted attack made at a rapid gait by cavalry.

Cavalry, Classifications of. According to the size of the horses and men, and the weight of their arms and armor, cavalry has been classified as light, medium, and heavy. The following are some of the descriptive terms that have been used: cuirassiers, dragoons, hussars, uhlans, chasseurs, mounted rifles, carbineers, lancers, etc. These terms formerly had a special significance which has now been largely lost.

Cavalry Corps. A corps consisting of two or more cavalry divisions and auxiliary troops.

Cavalry Division. A division in which the principal combat element is cavalry instead of infantry, and all auxiliary troops are sufficiently mobile (mounted, animal-drawn and motorized) to accompany the cavalry.

Cavalry Screen. A deployed cavalry force which covers the front of an army in campaign, to prevent the enemy from gaining information of its movements, etc.

Cavalry Tactics. The art or science of maneuvering cavalry and employing it either alone or in conjunction with other arms and services. Cavalry tactics include reconnaissance and screening, fire and shock action, mounted or dismounted, and combined mounted and dismounted action.

Cavalryman. A cavalry soldier. A trooper.

Cave Shelters. Shelters constructed entirely below the surface of the ground by mining methods and having a cover of undisturbed or virgin earth.

Cavity. The hollow interior of a shell, which receives the bursting charge.

Cease Firing. A command or signal signifying that firing is to cease immediately.

Ceiling. The height of the lower level of a bank of clouds above the ground. The height to which a particular aircraft can climb.

Ceiling, Absolute. The maximum height above sea level at which a particular airplane would be able to maintain horizontal flight under standard air conditions.

Ceiling Projector. A projector that produces an illuminated region on the under side of a cloud for the purpose of determining the height of that part of the cloud above the indicator.

Ceiling, Service. The height above sea level, under standard air conditions, at which a given airplane is unable to climb faster than a small specified rate (100 feet per minute in the United States and England). This rate may differ in different countries.

Ceiling, Static. The altitude in standard atmosphere, at which an aerostat is in static equilibrium after removal of all dischargeable weight.

Ceiling-height Indicator. A device that measures the height from the horizontal to the illuminated spot produced by a ceiling projector as seen from a fixed position.

Cellule, or Cell. In an airplane, the entire structure of the wings and wing trussing of the whole airplane on one side of the fuselage, or between fuselages and nacelles, where there are more than one.

Censorship. Measures taken within the theater of operations to prevent the leakage of military information; they are applied to private communications, photography, press dispatches, publications, and all communications.

Center. The middle point or element of a command. That portion or element of an army between the two wings. On a rifle target, the annular division between the bullseye and the next larger circle, in which the value of a hit is four. To bring to a center or to place on center.

Center Fire Cartridges. Cartridges in which the priming composition is contained in a cap in the center of the base of the case.

Center of Burst. The point about which the points of burst of several projectiles are evenly distributed. The mean position in space of a series of bursts.

Center of Gravity of a Shot Group. A point on any plane surface (horizontal, vertical, or inclined), whose rectangular coordinates, measured from any axes in the plane, are the algebraic means of the coordinates of all the separate shots. The center of impact.

Center of Impact. The mean position of several points of impact of projectiles fired with the same data. The center of gravity of a group of shots. The center of the rectangle of dispersion. See Apparent, True Center of Impact.

Center of Resistance. An old term meaning the defensive area of a rifle battalion.

Centered. A projectile is said to be centered when the axis of the projectile is in line with the axis of the bore when the piece is fired.

Centering Slope. The conical part of the bore of a cannon between the powder chamber and the forcing slope, whose purpose is to bring the axis of the projectile in line with the axis of the bore.

Centralization. A system of organization or administration in which all matters are directed and supervised by one headquarters.

Centralized Control. A system based on extreme systematization, prearranged, closely regulated, and harmonized action, with control centralized in the high command, and a minimum of initiative left to subordinates.

Centralized Direction. A system based on the assignment of definite missions, with the greatest degree of latitude possible left to subordinates in initiative in execution.

Centrical Rifling. A type of rifling having the grooves rounded to prevent violent shock when the bearing surface of the projectile strikes the rifling.

Ceremonies. Formal military formations and evolutions such as parades, reviews, inspections, escort of the color, guard mounting, etc.

Certificate of Capacity. An instrument in writing which certifies that an officer of the Organized Reserves, named therein, has met the professional qualifications prescribed for the grade and section specified in the certificate.

Certificate of Disability. A certificate stating that an officer or soldier is physically unfit for military service.

Certificate of Discharge. A certificate given to a soldier on termination of a term of service or enlistment, setting forth the fact of discharge, character of service rendered, etc. See Discharge.

Certificate of Merit. A certificate given by the President to enlisted men for acts of great personal bravery in the presence of the enemy.

Chain Ball. A light body attached to the rear of a projectile by a cord or chain to prevent tumbling.

Chain Mail. A kind of armor formed of hammered iron links, much used in the 12th and 13th centuries. Flexible armor made of interlinked metal rings.

Chain of Command. The successive commanders through whom responsibility is fixed and control is exercised.

Chain Shot. Two cannon balls connected by a short chain.

Challenge. The act of a sentinel in halting and demanding the countersign or other identification from those who appear on or near his post. A formal objection by the judge advocate or an accused before a court-martial to any member of the court; either peremptory (without stated reason), or for stated reason.

Chamber. That part of the breech of a cannon which holds the propelling charge. In the bolt-type and other repeating or automatic firearms, the space into which a cartridge is fed from the magazine and from which it is forced into the breech by the action of the bolt. A short cannon not provided with a carriage. In mine warfare, a place prepared for the reception of a large charge of explosive. An enlarged space in a subterranean shelter, used as a bunk room, or for other purposes. A recess or pocket in a wall, bridge or other structure in which a charge of explosive may be placed to destroy the structure.

Change Lever. A lever in an automatic rifle whereby it may be set for automatic or semi-automatic fire, or no fire ("safe").

Change Step. A drill movement to gain or lose a step in marching, as to gain cadence or get in step with another person or unit.

Channel. A narrow, navigable waterway, or the navigable portion thereof. A structural steel shape of channel cross section. The route of official intercourse. A narrow band of frequencies of sufficient width for a single radio transmission.

Channel of Communication. A channel for the transmission of official communications which passes through intermediate commanders. See also Military Channels.

Chape. The catch or piece by which an article of equipment is attached. The hook of a scabbard. The metallic part on the end of a scabbard.

Chapeau. A plumed hat worn as a part of an official costume or uniform. A cocked hat worn by general officers.

Chaplain. A clergyman with a military commission, who is responsible for the spiritual welfare of soldiers.

Characteristics. The distinctive and peculiar properties inherent in any person, class, or thing, as an arm of the service or weapon. Powers and limitations.

Charge. A sudden and rapid advance for the purpose of closing in hand-to-hand combat with the enemy. The assault or culmination of an attack. A mounted attack by cavalry. The amount of explosive for a single discharge, burst or blast. The amount of explosive mixture drawn into the cylinder of an internal combustion engine on any intake stroke. The quantity of electricity which a storage battery, condenser, etc., holds or is designed to hold. A formal statement of the crime or misdemeanor for which a person is brought before a court for trial. To make a charge or assault. To load with explosive. To bring a weapon, as a bayonet, to position for attack. To charge with electricity. To lay court charges against. See Assault; Cavalry Charge; Bursting Charge; Propelling Charge, etc.

Charge Bayonet. An offensive movement in bayonet exercise. The position of readiness for bayonet combat.

Charge of Quarters. Supervision of the care and police of barracks or camp by a noncommissioned officer, usually detailed by roster.

Charge Sheet. A form on which are entered the charges and specifications against an accused, together with certain data, reference to trial, certificate of service on the accused, and other pertinent items.

Charger. A military mount, particularly an officer's mount.

Charges and Specifications. The form of indictments tried by courts-martial.

Charlottesville Rifle. A highly esteemed rifle made at the shop established at Charlottesville in 1740.

Chart. A graphical representation or record of data, especially a graphical representation of a variable function or factor. A map, especially a map of or including water areas, for the use of navigators.

Chase or Chace. That part of a cannon in front of the trunnion band.

Chassepot Rifle. The rifle adopted by the French Army shortly after the Austro-Prussian War of 1866, which had a breech closing with a sliding bolt.

Chasseurs. A body of light troops, cavalry or infantry, trained for rapid movement.

Chassis. A movable base on which the top carriage of a gun moves backward and forward. The lower part of a gun carriage. An entire vehicle or automobile, exclusive of the body. See Landing Gear.

Chatellerault Automatic Rifle, Model 1924. The automatic rifle used in the French Army. It has a caliber of .295 inches, a magazine capacity of 25 cartridges, and weighs 19.84 pounds.

Chauchat Gun. A French light automatic rifle used in the World War.

Check Concentrations. Registration of fire on easily identified points throughout the zone of fire, from which transfers can be made to targets of opportunity.

Check Inspection or Roll Call. An inspection made to ascertain whether all men, not absent by proper authority, are present.

Check Point. A visible point on which fire is adjusted and from which a transfer of fire may be made. The check map range is the map range to the check point. The check adjusted range is the range corresponding to the adjusted elevation of the check point, less site.

Cheeks. The sides of a gun carriage, in which the trunnions of the gun rest. The sides of an embrasure.

Chemical Agent. A substance used for military purposes which produces by direct chemical action either: (1) A powerful physiological effect upon the human body; (2) An irritant or harassing effect upon the human body; (3) An obscuring smoke; or (4) An incendiary action.

Chemical Agent, Nonlethal or Nontoxic. A toxic material which exerts a profound irritating action at a concentration far below that required to produce death or seriously endanger health. A chemical agent whose maximum possible concentration in the field is not enough to produce death.

Chemical Agent, Non-persistent. A toxic chemical agent which dissipates or settles rapidly, so that its effects continue for a very short time, usually a few minutes only, after release from its container.

Chemical Agent, Persistent. The persistent group of harassing agents are liquids of high boiling points or solids which slowly vaporize when exposed to air. The rate of vaporization determines the relative persistency, which may vary from a few hours to two weeks or more.

Chemical Agent, Toxic. Any substance which, by its direct chemical action, either internally or externally, on the human or animal organism, is capable of destroying life or seriously impairing normal body functions.

Chemical Bomb. An airplane bomb filled with a chemical agent.

Chemical Cylinder. A metal cylinder from which a chemical agent may be discharged under pressure.

Chemical Grenade. A small container filled with a chemical agent, capable of being thrown by hand or fired from a rifle, and ignited or detonated by a time fuze during flight or after impact.

Chemical Land Mine. A form of mine consisting of a container of simple construction, filled with chemicals and equipped with a bursting or ejecting charge.

Chemical Mortar. A mortar for the discharge of chemical shells.

Chemical Security. All measures of protection against chemical agents, including individual, collective, and tactical gas defense.

Chemical Shell. An artillery or chemical mortar shell which contains a bursting charge and a chemical agent.

Chemical Spray. A discharge of a chemical agent into the air from airplane tanks.

Chemical Troops. Special troops organized, trained, and equipped with weapons designed especially for the projection of chemical agents.

Chemical Warfare. The tactics and technique of the use of chemical agents in warfare. Warfare conducted by the utilization of chemical agents, including gas, smoke and incendiaries. Protection or defense against similar measures. Combustible and poisonous substances to diminish the resistance of an opponent are probably as old as organized warfare. Thucydides, who lived in the 5th century B.C., describes two instances of the use of sulphur and pitch, and throughout classical times and the Middle Ages such methods were frequently employed, Greek fire being a mixture of this nature. As war became more mobile and the range of weapons increased the opportunities for using such means disappeared. The introduction of chemical agents in modern warfare dates from April 2, 1915, when the Germans discharged chlorine from cylinders. Later on in this same year lacrimatory shells were employed. From this time onward chemical agents were used in increasing quantities during the World War.

Chemical Warfare Service. That branch of the service which is charged with the development of offensive and defensive means and methods of chemical warfare.

Chess. The deck plank or floor boards of floating bridge equipage.

Cheval-de-frise. A portable obstacle in the form of a saw horse, having two or more sets of legs or cross pieces, which will stand of itself, sometimes covered with a network of barbed wire.

Chevalier. A knight. A cavalier. A horseman. A knight of the lowest grade of the Legion of Honor.

Chevron. A distinguishing mark in the shape of a caret, to represent rank or service, worn on the sleeve.

Chief. The commander or leader of any body of troops. The administrative head of any branch in the War Department, as Chief of Infantry, Chief of Ordnance, etc. The senior officer of any arm or service in a division or higher unit, who performs both command and staff duties, as chief of artillery, chief surgeon.

Chief of Artillery. The artillery commander and technical staff officer of a division or higher unit. See Chief.

Chief of Staff. The senior general staff officer, detailed as such, on duty with the staff of a division or higher unit.

Chief of Staff of the Army. The chief of the general staff corps of the military forces of a nation.

Chilled Projectiles. Steel projectiles hardened by chilling their points, used to penetrate armor plate.

Chinese Anchor. An improvised anchor made of heavy stones inclosed in a wooden crib.

Chinese Attack. An attack simulated or apparently threatened by the use of dummies.

Chlorination. The process of sterilizing water for drinking purposes by treatment with calcium hypochlorite.

Chock-a-block. The condition of a tackle rigging when two blocks are hard against each other so that further movement is impossible.

Choker. An instrument for bringing the ends of a fascine to its nearly intended girth when it is bound.

Chord. A straight line joining the ends or any two points of an arc or curve. The base of a trajectory, either horizontal or inclined. The straight line between the ends of a road or railroad curve (long chord), or one of the successive 100 foot chords. The upper or lower longitudinal member of a truss. A datum line from which ordinates and angles of an airfoil are measured.

Chronograph. An instrument for measuring and graphically recording time intervals, as in determining the velocity of projectiles.

Cincinnati, Order of the. A hereditary order, founded 1783, of male descendants of officers of the American Revolution.

Cipher. A system of changing the arrangement of letters (very occasionally words) of a message, or the substitution of other letters or figures for the original letters and figures. A prearranged alphabet, system of characters, or other mode of writing, contrived for the secret transmission of information. See also Code.

Cipher Device. An instrument used in enciphering and deciphering messages.

Cipher Messages. Messages sent in cipher in order to insure secrecy.

Circulars. Official publications which usually contain matter that is directive in nature, general in application, but temporary in duration.

Circulation. Free and continuous passage or transmission from point to point or from person to person. Regulated movement within a prescribed area.

Circulation Maps. Administrative maps issued to show the authorized direction of traffic on all roads that they cover.

Circulation Plans. Plans which prescribe the direction of traffic on different roads and such other restrictions on the use of roads as circumstances require.

Circumvallation. An inclosing line of fortifications. The line of investment surrounding a besieged place. Any encircling wall, rampart, trench, etc.

Citadel. A fortress or castle in or near a town in which the garrison was generally housed, and which served the double purpose of keeping the inhabitants in subjection and as a final refuge and place of defense. An interior central defense. A keep.

Citation. Specific mention in official orders for bravery or exceptional performance of duty. A written notification to appear at a time and place named.

Citizens Military Training Camps. Camps to provide citizenship instruction and military training for young men.

Civil Authority. Pertaining to the administration of civic life and affairs in distinction from military.

Civil War. A war between parties or factions of the same state; specifically, the American Civil War.

Class A Range. A range, more or less limited in extent, equipped for known distance firing.

Class B. Officers found unsuited to continue to hold commissions and recommended to be discharged or retired.

Class B Range. A range on more or less extended and diversified terrain used for combat firing.

Class A Property. New property.

Class B Property. Reclaimed property, or property which has been used but is still serviceable.

Class C Property. Unserviceable property the condition of which justifies reclamation.

Class D Property. Unserviceable property not fit for reclamation, including "waste."

Class I Supplies. All articles consumed at an approximately uniform daily rate irrespective of the nature of operations or terrain, and which do not require special adaptation to meet individual requirements, such as rations, forage, fuel, etc. All articles which can be handled in the field on the basis of a daily automatic supply.

Class II Supplies. Authorized articles for which allowances are established by Tables of Basic Allowances; such as clothing, gas masks, arms, trucks, radio sets, tools and instruments.

Class III Supplies. Engine fuels and lubricants for all vehicles and aircraft.

Class IV Supplies. Supplies not covered in Tables of Basic Allowances, the demands for which are directly related to the operations contemplated or in progress (except articles in Classes III and V); such as fortification and construction materials, and machinery.

Class V Supplies. Ammunition, pyrotechnics, antitank mines, chemicals, etc.

Class of Defense. The method of conduct of defensive combat, as determined or influenced by the mission of the defending force and by the enemy's situation. The usually recognized classes are active defense, passive defense, and delaying action.

Classification. The arrangement of the individuals of an organization in groups or classes according to the degree of skill displayed. The determination of the duty or duties an individual is best qualified to perform with a view to such assignment as will best meet the needs of the military service.

Classification Board. A board of officers convened to pass upon the qualifications of officers whose records indicate that they should be removed from the service, or transferred to another branch of the service.

Classification of Casualties. Casualties in campaign or battle are classified as dead, wounded, gassed and sick. Wounded are further classified as slight, medium and severe; and, for transportation purposes, as walking, transportable (sitting or recumbent) and nontransportable.

Classification Yard. A railroad yard where freight is classified and trains are made up.

Classify. To determine the classification or qualifications of a person, persons, or material.

Claymore. A large, generally two-handed, double-edged sword, about 43 inches long, formerly used by the Scottish Highlanders.

Cleaning Rod. A rod for use in cleaning the bore of a firearm.

Cleaning Up. Searching the ground which has been passed over by attacking troops for the purpose of capturing or destroying any hidden enemy. Mopping up.

Clear. A column of marching troops or transport is said to clear a certain point or locality when the tail of the last element of the column has passed the point. To depart; to leave a port.

Clearance. The elevation of a projectile in flight over any point or object, as a mask or friendly troops. The space by which two moving objects, or a fixed and a moving object, clear in passing. Quittance or release.

Clearance Angle. The angle of elevation necessary to clear a mask or assure safety of friendly troops in overhead fire.

Clearing Station. An installation where sick and wounded are assembled from the collecting stations, sorted, treated as necessary, and evacuated. See also Triage.

Clearing the Field of Fire. Removal of brush, trees and other obstacles that interfere with the view and fire from a defensive position; an important part of the organization of the ground for defense.

Climb Indicator. An instrument which indicates the rate of vertical ascent or descent of an aircraft.

Clinometer. A small hand instrument used for measuring slopes, elevations, or inclinations. An instrument for measuring the angle between the axis of the bore of a gun and the horizontal; similar to a quadrant.

Clinometer Rest. A support for a clinometer inserted in the muzzle of a gun; also called a bore rest.

Clip. A metallic holder for cartridges of repeating small arms to facilitate loading. A device for lifting heavy shells. A wire rope fitting which clamps two ropes together.

Clip Slots. Slots in the receiver of a repeating rifle which receive the clip.

Clock-face Method or **System.** A method used to designate horizontal or vertical directions by reference to the figures on a clock dial; the gun, observer or target being considered as at the center of the dial and the 6-12 line being along the line of fire, or vertical or horizontal axis of the target. For example, a 3 o'clock wind is one blowing from the right, squarely across a target range.

Close. To converge a battery sheaf of fire.

Close Arrest. A state of arrest involving actual confinement.

Close Billets. The disposition of the units with part in billets and the remainder in bivouac.

Close Column. A mass formation in infantry close order drill in which the companies are arranged in column of platoons at reduced distances, each platoon being in line.

Close Line. A mass formation in infantry close order drill in which the companies are arranged in line of platoons at diminished intervals, each platoon being in column of squads.

Close Order. A drill formation in which the elements are arranged in line or column with normal intervals and distances. Any formation for march, maneuver, or assembly in which normal intervals and distances are prescribed in such a manner that the individuals and other elements are closely formed in compact bodies for purposes of control, economic occupation of space, and discipline.

Close Reconnaissance. Reconnaissance conducted by commanders of all units when within striking distance of the enemy. Airplane reconnaissance which is minute in character and extends about 30 miles into the enemy's territory.

Closed in Mass. A formation in close order in which the units are arranged with intervals and distances less than normal.

Closed Traverse. A traverse, in sketching or surveying, which ends at its point of beginning, or at a point whose location is known or can be otherwise determined.

Closed Work. A fortification that is entirely inclosed by a continuous parapet and prepared to defend itself against attacks from any direction. A field emplacement for a weapon having a protective roof or overhead cover; e.g., a redoubt, a pillbox.

Clothing Account. A record of the clothing due and issued to a soldier or organization.

Clothing Allowance. An allowance based on the value of the clothing actually required to clothe the average soldier during an enlistment period.

Clothing Allowance, Initial. The clothing issued to a soldier upon enlistment; or the commuted money value thereof.

Clothing Allowance, Maintenance. A periodical allowance of clothing to a soldier during his term of enlistment, to replace worn-out articles; or the commuted money value thereof.

Clothing Roll. A combination clothing and bedding roll for use in the field.

Clothing Slip. A slip on which a soldier requests articles of clothing needed.

Coach. A special instructor charged with the duty of correcting errors and giving advice and information.

Coach and Pupil Method. A method of individual instruction, especially in marksmanship, in which the men are paired, and take turns in acting as coach and as pupil.

Coarse Sight. An aim in which a considerable portion of the front sight is seen through the notch in the rear sight.

Coast Artillery. Artillery mounted for the defense of important points along the coast. Includes harbor-defense, railway, and antiaircraft artillery, and tractor-drawn artillery especially assigned for coast-defense purposes; also, such sound-ranging units as are needed in the performance of its missions. The Coast Artillery Corps.

Coast Artillery Corps. That arm of the service which is charged with the care and use of the fixed and movable elements of the seaward and landward defense of the coast fortifications, including guns, mortars, submarine mines, and torpedoes, and the operation of mobile and antiaircraft artillery with an army in the field.

Coast Artillery District. A tactical command which includes all harbor defenses located within a specified area and such mobile coast artillery units as may be assigned thereto.

Coast Batteries. Batteries of large caliber, installed along a coast to protect harbors and ports from attack by sea.

Coast Defense. The dispositions and operations planned for the defense of any portion of the seacoast or coastwise sea lanes. It may be either by land forces acting alone or in conjunction with naval or other military forces.

Coast Defense Command. A fortified area constituting a strong point on the coast line and embracing one or more coast artillery forts, together with all personnel and means provided for their tactical employment, administration, and control. A group of fort commands provided for the defense of a harbor or point of the coast.

Coastal Frontier. A geographic division of the coastal area established (for organization and command purposes) to insure effective coordination between army and navy forces.

Coastal Zone. All navigable waters adjacent to the coast, extending seaward to include the coastwise sea lanes.

Coat of Arms. An emblazonment of armorial bearings on an escutcheon, used as the distinguishing mark of a state, organization or family.

Coat of Mail. A suit of armor made of metal scales, or rings linked one within the other, worn during the Middle Ages.

Cock. The hammer of a firearm. The notch of an arrow. To draw back the hammer or firing pin preparatory to firing.

Cocking Piece. The rear end of a long firing pin or striker in certain small arms, whereby the piece is cocked.

Cockpit. An open space in an airplane for the accommodation of pilots or passengers. When completely inclosed, such a space is usually called a cabin.

Code. A system of signals for communication by means of the telegraph, telephone, flags, or other signalling devices. A list of letters, syllables, words, and phrases, each of which is represented by an arbitrary group of letters or figures.

Code Message. A message sent in code to insure secrecy.

Coehorn Mortar. A small, muzzle-loading, smooth-bore mortar, cast in a single piece, and mounted on a block or platform.

Coincidence Adjustment. See Halving and Coincidence Adjustment.

Coincidence Type Range Finder. A self-contained horizontal base range finder which shows the range when the two images of the target are made to coincide.

Cold-worked Gun. See Autofrettage.

Collate. To examine and compare critically.

Collation. A bringing together for purposes of verification, coordination, or similar purpose.

Collecting Battalion. One of the battalions of the medical regiment, including a headquarters and 3 collecting companies. The functions of the battalion are to establish collecting stations and collect wounded from aid stations and from the field of battle.

Collecting Point. A point designated for the collection of sick or wounded, prisoners of war or stragglers.

Collecting Station. An advanced station of the medical department where casualties are collected and prepared for evacuation to the rear.

Collective Protection. Devices used and measures taken for the protection of the command as a whole, as distinguished from individual protection. As applied to protection against chemicals, the term includes gas-proofing of shelters, filtering of air, neutralizing agents, gas sentries and alarms, etc.

Collimate. To adjust the line of sight of an optical instrument so that it will coincide with the axis of the barrel or tube.

Collimator. An optical device used on gun sights whereby a sight is aligned on a target or aiming point, without the use of a telescope.

Collimator Sight. A sight provided with a collimator in lieu of a telescope.

Colonel. The highest authorized officer of a regiment. The grade immediately below that of brigadier general and above that of lieutenant colonel.

Colonel-general. Formerly, an officer in supreme command of an army; a field marshal.

Color. A flag carried by dismounted troops; the term includes the national color and the regimental color when both are carried. The flag of a high commander. See also Standard.

Color Bearer. A man who carries a color at ceremonies, etc.

Color Company. The center or right center company of the center or right center battalion of a regiment in line. The company with which the colors are posted in formations.

Color Guard. The guard of honor which escorts and carries the colors of an organization.

Color Line. The line on which the pieces are stacked, the colors, furled, sometimes being laid on top of the stacks.

Color, Map in. See Map in Colors.

Color Salute. A salute made by dipping the colors; the national colors are never dipped in salute.

Color Sentinel. A sentinel posted on or near the color line to guard the colors and stacks.

Color Sergeant. A sergeant detailed to carry the colors.

Colors. Certain types of flags carried by troops, such as standards, guidons, regimental flags, etc.; a term of general application.

Colors and Standards. The silken national and regimental colors or standards carried in battle, campaign, and on all occasions of ceremony.

Colt Automatic Pistol. The pistol used in the United States Army; invented by John M. Browning.

Colt Machine Gun. A hand operated and air cooled light machine gun, which weighs 35 pounds and fires about 500 rounds per minute.

Colt Repeating Rifle. The first repeating rifle, which appeared in the United States in 1840.

Columbiad. An American bronze, smooth-bore cannon cast about 1810. It fired a 50-pound ball to a distance of 600 yards.

Column. A formation of troops in which the elements are placed one behind another.

Column Half Right (Left). A movement by which a 45° change of direction is executed by a marching column.

Column of Close Columns. A dismounted formation with each battalion in close column of companies in close column of platoons in line.

Column of Close Lines. A dismounted formation in close column with rifle companies in close line of platoons in column of squads.

Column of Files or Twos. A column having a front of one or two men, as the case may be.

Column of Fours. Cavalry in column, four abreast. A column of squads.

Column of March. The formation of troops for a march.

Column of Platoons. A dismounted formation in which platoons in line are formed in column with normal distances.

Column of Route. A body of troops arranged in column in the prescribed order for a march.

Column of Squads. A formation in which the squads, four men abreast, are placed one behind the other.

Column of Troopers. Cavalry in single file.

Column Right (Left). A movement by which a change of direction to a flank is executed by the head of a column of troops, followed in turn by each succeeding element.

Comb. A part of the stock of a rifle. To search thoroughly by reconnaissance or fire.

Combat. A contest between armed forces, irrespective of their size. A fight. A battle. To fight with. To oppose by force.

Combat Arms. See Combatant Arms.

Combat Battalion. A battalion deployed with all or part of its elements in the front line.

Combat Car. An armed and armored track-laying or convertible track and wheel motor vehicle designed primarily for combat. A tank.

Combat Company. A company deployed with all or part of its elements in the front line.

Combat Echelon. The principal or most advanced element of offensive or defensive power.

Combat Elements. Those troops of a command which actually engage in combat or perform combat missions, as distinguished from supply elements. One of the three essential elements of military organization, the other two being command and supply.

Combat Emplacement. A prepared or selected position for a weapon, from which it can execute a combat mission.

Combat Exercise. An exercise to demonstrate the application of the principles of musketry or tactics.

Combat Formation. A formation assumed and suitable for combat or for entering the fire fight.

Combat Group or Post. A term formerly used to designate the smallest unit areas of a defensive position, whose garrisons varied from a squad to a platoon.

Combat Intelligence. The military intelligence obtained in the field after the outbreak of hostilities; generally, confined to the location, strength, composition, armament, equipment, supply, tactics, training, discipline, morale, movements, intentions, condition, and situation of the enemy forces opposing a combat unit, and the terrain over which a unit is operating, or is to operate; used as a basis of plans for future operations.

Combat Liaison. The means and methods employed to maintain contact and communication between the various units during combat. Liaison is the contact by which cooperation is insured, as distinct from signal communications, which is one of the means by which liaison is achieved.

Combat Orders. Oral, dictated, or written orders issued by a superior to a subordinate unit, covering any phase of operations in the field. Combat orders include field orders, letters of instruction, warning orders, movement orders, and administrative orders.

Combat Outpost. An outpost employed when the opposing forces are in close contact.

Combat Pack. A colloquial term applied to that portion of the pack carried into action; usually the full pack less the roll.

Combat Patrol. A patrol employed for combat purposes, for example, a detachment having the mission of protecting the flank of an organization, or guarding an interval between organizations in combat.

Combat Planes. Airplanes of high speed and great climbing power whose mission it is to seek out or drive back enemy planes.

Combat Practice. Training in the use of weapons in assumed tactical situations, the enemy being represented by suitable targets. The prescribed firing at targets which simulate the appearance of an enemy under conditions approaching those encountered in war. The application of fire to tactical exercises.

Combat Practice Firing. A form of training wherein a tactical unit solves a problem involving a tactical situation in which service ammunition is fired at an enemy represented by suitable targets.

Combat Principles. The tactical principles, or principles of war, practically applied to the actual conduct of any unit in tactical situations.

Combat Team. A nonorganic grouping of two or more units of different arms for combat purposes, as a regiment of infantry and a battalion of field artillery.

Combat Trains. Unit trains carrying, in general, the ammunition, rations, materiel, etc., required or that may be required for combat. See **Train**.

Combat Troops. See Combatant Arms.

Combat Unit. The smallest force that possesses the requisite fighting power, strength and capacity for subdivision and maneuver, that can sustain action independently, that solves minor problems of combat, and that remains an efficient combat unit after serious losses, generally considered to be the platoon.

Combat Vehicles. See Tactical Vehicles.

Combat Wagon. A wagon forming part of the combat train and transporting combat materiel for a certain organization.

Combat Zone. That part of a theater of operations in which the active operations of the combat units are conducted. Specifically, the area occupied by the field armies, between the front line and the forward boundary of the communications zone.

Combat Zone Depot. A depot located in the combat zone.

Combatant. A soldier. A fighting man. A member of the armed forces, as distinguished from a civilian. A member of a combatant arm, as distinguished from a member of one of the non-combatant services.

Combatant Arms. The arms which actually engage in combat with the enemy. Those branches of the service legally designated as such: viz., infantry, field artillery, cavalry, engineers, coast artillery, signal corps, chemical warfare service, armored force, and air forces. The combatant arms are distinguished from the services, which do not engage in combat.

Combination Fuze. A fuze which has both percussion and time elements, the percussion element functioning automatically in case of failure of the time element.

Combined Action, Cavalry. The use of mounted and dismounted troops in combination, for combat.

Combined Arms. The combatant arms which cooperate in battle.

Combined Observation. Observation of fire by several, usually two, observers, by axial, lateral and flank observation, or a combination of any two of these methods.

Combined Operations. The tactics of the combined or associated arms, as the infantry, cavalry, field artillery, air service, or any two or more of them. Joint operations, as by two or more allies, the army and navy, etc. Grand tactics.

Combined Sights. A method of engaging a deep target with two or more guns using the same aiming point with different elevations or ranges.

Combined Sketching. A method of area sketching in which an area is divided into sections to be sketched by individuals, the results being subsequently combined.

Combined Tactics. The science or art of assigning proper tactical missions to the different branches comprising a command in order to secure the best results. Grand tactics.

Combined Training. The training together of the separate combatant and administrative branches that usually operate together in the field in order to familiarize them with their proper functions and relations in combination.

Combustion Chamber. That part of the cylinder of a gas engine above the piston in which combustion is initiated. The powder chamber of a cannon.

Come-Alongs. A slang term applied to loops of barbed wire prepared to be thrown over the heads of prisoners to force them to come along. Devices used to take up the slack in a line of wire, as a fence or telephone line.

Command. The authority which an individual exercises over his subordinates by virtue of rank and assignment. The direction of a commander expressed orally and in the prescribed phraseology. A body of troops under the command of one officer. One of the essential elements of military organization, the other two being combat elements and supply. The vertical height of the fire crest of the parapet above the original natural surface of the ground. The vertical height of any ground over other ground in its vicinity. To order or exercise command.

Command Airplanes. Airplanes which observe the general progress of the combat and report all that occurs on the side of the enemy to the headquarters of the higher commands to which assigned.

Command and General Staff School. A school, located at Fort Leavenworth, Kansas, to prepare officers for command and general staff duty.

Command Car. A motor vehicle, usually armed and armored, equipped with radio and other facilities to assist in the execution of command functions therefrom.

Command Channel. A channel of communication between superior and subordinate commanders.

Command Elements. The commander, together with such assistants or staff officers as are necessary to relieve him of the burden of details and enable him to exercise adequate control over his command.

Command Group. That portion of a headquarters which is concerned with command and communications. In combat it forms the forward echelon of the headquarters. See also Administrative and Supply Group.

Command of Execution. The latter part of a drill command, which effects its execution.

Command Post. The place occupied by the commander of a unit and the forward echelon of his headquarters, for the purpose of exercising command and facilitating operations.

Command Post Exercise or Maneuver. A field exercise for the training of unit commanders, in which only the various commanders, their staffs and headquarters personnel participate or are represented.

Command Reconnaissance. Reconnaissance executed for the especial information of high commanders.

Command Tanks. Tanks occupied by tank company or higher commanders.

Commandant. The commanding officer of a fort, garrison, regiment, company, etc. A commander. The head of the military department of a school.

Commandeer. A commander. To compel the performance of military service. To seize stores or property for military purposes.

Commander. The ranking officer of a unit, organization, or station.

Commander in Chief. The officer who has supreme command of all the forces of a state or nation. An officer who exercises control over a theater of war, which may consist of two or more mutually dependent theaters of operations within easy communication with each other.

Commander in Chief of the Army of the United States. The President of the United States, who is the head of the military as well as the civil branch of the government.

Commanding Ground. A rising ground which overlooks a post or position.

Commanding Officer. The senior officer present for duty with a command or at a post or station.

Commando. In South Africa, a military body or command. A raid or expedition.

Commence Firing. A call, command, or signal, used to notify troops to begin firing at once.

Commissariat. The supply service of an army, taken as a whole.

Commissary. A place or establishment where subsistence stores are issued or sold. An officer charged with the sale or issue of subsistence stores.

Commissary General. Formerly, the chief of the subsistence branch of the army.

Commission. A certificate of appointment or rank given to a commissioned officer. A writing, in the form of a warrant or letter patent, authorizing one or more persons to exercise authority.

Commissioned Officers. Those persons in the military service who have received a commission of rank from proper authority.

Common Mine. A mine whose explosion produces a crater whose radius is equal to the line of least resistance.

Common Tent. A small tent having no side walls. An A-tent.

Communicable Disease. A disease caused by germs, which can be communicated from one person to another. According to the means of transmission (not always fully known) such diseases are classified as: (1) Respiratory; (2) Intestinal (water-borne); (3) Insect-borne; and (4) Direct contact. See also Prophylaxis.

Communicating Files. Men designated to march between the elements of a column in march in order to maintain communication, or for the rapid transmission of orders or information.

Communication. Interchange of information by message ·or otherwise. Means of communicating. A verbal or written message. Intercourse by means of words, letters, messages, or signals.

Communication Band. In radio the band of frequencies due to modulation (including keying) necessary for a given type of transmission. See also Service Band.

Communication Trenches. Trenches designed primarily to provide cover for personnel moving from one part of an intrenched position to another.

Communications. Prepared routes by which troops move from one locality or one part of a position to another. Lines of supply and reinforcement. All means by which a command maintains contact with adjacent units, higher echelons, its own elements or bases from which supplied. Letters, memoranda, reports, indorsements, telegrams, cablegrams and radiograms, used in official correspondence. See also Art of War.

Communications Net. The various means, mechanical and otherwise, coordinated and employed to transmit information, orders, and reports within an organization. The fabric formed by the command posts of an organization and all lines of signal communication connecting them.

Communications Platoon. A part of the headquarters or headquarters company of a unit, charged with the establishment of the message center and the maintenance of signal communications.

Communications Zone. That part of the theater of operations containing the establishments of supply and evacuation, lines of communication, and other agencies required for the immediate support and maintenance of the entire forces in the theater of operations. It includes all territory, between the zone of the interior and the rear boundary of the combat zone.

Communique. An official announcement of information.

Commutation. A conditional pardon vested in the President. A reduction in the severity of a sentence of a court-martial. The conversion of allowances into their money value. Exchange.

Commutation of Quarters. A money allowance, at a rate prescribed by law or regulations, made to persons in the military service for rental at places where no public quarters are available.

Commutation of Rations. The allowance of a prescribed amount of money per diem in lieu of rations in kind.

Commute. To give compensation or substitution. To substitute for. To reduce the severity of.

Company. A body of men ordinarily consisting of a headquarters and two or more platoons, which is the authorized command for a captain; in this connection, the terms battery and troop have the same significance. The lowest administrative unit of any branch.

Company (Battery, Troop) Aid Men. Men of the medical department assigned to a combat company (battery, troop) to render first aid to the wounded.

Company Clerk. A soldier who assists in performing the clerical work of a company.

Company Council. A council of administration consisting of all the officers on duty with a company.

Company Defensive Area. See Defensive Areas.

Company Discipline. Corrective measures, including light punishments, administered in the discretion of a company commander, without resort to a court-martial, under the 104th Article of War; the principal reliance for the maintenance of discipline.

Company Fund. A fund created from ration savings and receipts from all other sources, which is expended for the benefit of the members of a company or similar organization.

Company, Infantry, Types of. Infantry organization in the United States army includes companies of the following types: rifle; machine gun; heavy weapons; antitank; headquarters, and service.

Company, Troop, Battery Officers. Officers belonging to a company, troop, or battery, including lieutenants and captains.

Company, Troop, Battery Parade. An area, in front of or near to its quarters in camp or garrison, where a company, troop, or battery assembles for roll call, or in preparation for drill or ceremonies.

Company Proficiency. The proficiency of each individual in the company in his duties as a soldier, and the proficiency of each component unit in its duties and functions as a separate entity, and as an element of the next higher unit.

Company Punishment. See Company Discipline.

Company Street. The open space in front of or between the company tents.

Company Train. The transportation, together with its operating and accompanying personnel, allotted by Tables of Organization to machine gun, howitzer and headquarters companies of regiments.

Comparator. An instrument for indicating the relation between the data determined by the sound locator and the data indicated by the pointing of the searchlight. The searchlight is pointed in the same direction as the sound locator by matching the pointers on the instrument. It is used in connection with a distant controller in antiaircraft fire.

Compartment of Terrain. An area of generally open terrain inclosed on at least two opposite sides by terrain features (e.g. ridges) that limit observation and observed fire into the area. A relatively long and narrow compartment, or one bounded on two opposite sides only, is designated as a corridor or cross compartment according as its axis or long dimension is approximately parallel or approximately perpendicular to the front or direction of advance.

Compass. An instrument for determining direction by means of a magnetized bar or needle which turns freely on a pivot and points to the magnetic north.

Compass Azimuth. An azimuth as indicated by a compass. Due to the variations of the compass it is seldom the same as the magnetic azimuth.

Compass Bearing. A bearing (direction) as indicated by a compass.

Compass Declination. The pointing error of a compass as compared to the direction of true north. The angle between the compass needle and the true meridian.

Compass North. The north direction indicated by a properly adjusted compass. Due to variations of the compass it is not exactly the same as magnetic north.

Compensation. Any salary, pay, wage, allowance or other emolument for services rendered.

Complement. The full establishment of a regiment or other organization. To supplement.

Complete Type Trench. A standard type of trench for deliberate organization, which provides a fire step for riflemen, with a wide and deep portion in rear so that men can pass in an upright position, without interfering with the men firing. It is usually revetted and provided with trench boards.

Compliment. A prescribed mark of respect paid to a body of troops or to a person entitled to it, or to the national color or anthem.

Components. The parts which make up the whole; as the subordinate units of an organization, the food items of a ration, etc. Manufactured articles ready for assembly into items of issue.

Composite. Made up of distinct elements or parts, as of different arms, services or organizations. Combined.

Composite Battalion. A battalion formed of companies belonging to different battalions, regiments, arms, or services. A provisional battalion.

Composite Photograph (Vertical). A picture formed by joining together the vertical and transformed oblique photographs made by a multi-lens camera.

Composite Unit. A temporary or provisional unit of any size, formed of elements of various units of the same or different arms.

Compound. A place for the temporary confinement of prisoners.

Compound Guns. Built-up or wire-wrapped guns.

Compression System of Rifling. The system which comprises all projectiles which are forced by the action of powder gases through the bore of the gun, the diameter of which is less than the greatest diameter of the projectile.

Compulsory Service. Forced or involuntary service, especially military service, which a government may require of its citizens. Conscriptive or selective service.

Concealment. The state or condition of being hidden from the enemy's view. Any object affording protection from the view of the enemy. See also Cover.

Concentration. An assembly of troops in a particular locality, on mobilization, for training, attack or defense. Fire placed upon a definite limited area for a limited time. The amount of toxic chemical vapor in a given volume of air at any particular time and place.

Concentration Area. A locality, normally in the theater of operations, in which units are assembled preparatory to active operations.

Concentration Camp. A camp in which troops are assembled for immediate service against an enemy or for transportation to the theater of war. A camp or cantonment where prisoners, refugees, or unruly inhabitants are concentrated.

Concentration March. A march made by a body of troops to effect a concentration with other troops.

Concertina. A collapsible and extensible portable wire entanglement, cylindrical in form. Same as Brun Spiral.

Concussion Fuze. A fuze which detonates on impact. A percussion fuze.

Condemned Property. Property which has been officially inspected and pronounced unfit for further military use.

Conduct of War. Everything connected with the raising, equipping, training, and handling of troops in war. The making of war in all its aspects.

Conduct to the Prejudice of Good Order and Military Discipline. Disorders and neglects, not specifically mentioned in the Articles of War, which bring discredit upon the military service, and are prejudicial to discipline; punishable under the 96th Article of War.

Conduct Unbecoming an Officer and a Gentleman. Behavior in an official or personal capacity which stamps an officer as morally unfitted to hold a commission; punishable under the 95th Article of War.

Cone of Dispersion. The figure formed in space by the trajectories, considered together, of a series or number of shots fired with the same sight setting. The cone made by fixed fire by a machine gun, or by the dispersion of shrapnel bullets when the shrapnel bursts in the air.

Cone of Fire. The cone of dispersion. The group or groups of trajectories which form the cone of dispersion.

Cone of Silence. That space immediately above a station transmitting radio directional beams, in the form of an inverted cone, in which radio transmission cannot be heard.

Confederacy, Confederation. A league or alliance of states or individuals for mutual security or common action.

Confidential. A document will be marked confidential when it is of less importance and of a less secret nature than one requiring the mark secret, but which must, nevertheless, be guarded from hostile or indiscreet persons.

Confinement. Imprisonment. Actual physical restraint in an inclosed place of detention.

Conical Tent. A cone-shaped tent.

Connecting File. A man or element posted or marching between two bodies or elements of troops for the purpose of assisting in maintaining contact between them. A single man, or men in pairs, specially detailed to assist a detached body in keeping touch with its main body.

Connecting Group. A detachment posted or advancing on the flank of a unit for the purposes of protecting the flank and maintaining contact with a neighboring force. Any group used to maintain contact between separated forces or elements.

Connecting Lights. Lights emplaced to cover the areas between searchlight groups.

Connecting Pipe. A pipe connecting the throttling and equalizing pipes in the recoil system of a gun carriage, forming a by-pass between the ends of the recoil cylinders, and including a throttling valve by which the hydraulic pressure is regulated.

Connecting Trench. A narrow or shallow trench, approximately parallel to the front, connecting adjacent fire trenches, either for communication or as a dummy.

Conscientious Objector. One who, for conscience' sake, objects to warfare or to personal military service.

Conscript. A person compulsorily enrolled in the military service in accordance with the laws of the nation requiring its citizens to render service. To enroll for compulsory service.

Conscription. A compulsory draft or enrollment for military service.

Consolidate a Position. To prepare a captured position for defense against an enemy counter attack, or as a line of departure for a continuation of the attack.

Consolidated Report, Return. A report or return made by combining the reports or returns received from several organizations or stations.

Constructive Condonation. The release of an accused without trial by the authority competent to order his trial; it may be pleaded in bar of subsequent trial.

Contact. Proximity to friendly or hostile troops that permits constant communication or observation. To bring into contact.

Contact Firing. The firing of a controlled submarine mine when struck by a vessel.

Contact Mine. See Mine, Contact.

Contact Plane. An airplane which maintains contact with advancing troops in combat and reports observed enemy movements.

Contact Squadron. A cavalry squadron pushed well to the front to gain contact with the enemy.

Contact Troops. Troops pushed forward from the main body of the independent cavalry to support and furnish reliefs for reconnoitering cavalry.

Contain. To hold in place. To check. To restrain.

Containing Action. An attack designed to hold the enemy to his position or to prevent him from withdrawing any part or all of his forces for use elsewhere.

Containing Force. A body of troops, inferior in number, with the mission of holding a superior enemy force in check or position.

Continental Army. The army maintained by the Continental Congress during the Revolutionary War.

Contingent. The quota of troops furnished to a common army by each member of an alliance or confederation. The quota of armed men or pecuniary subsidy which one state gives to another. Possible but not certain. Dependent upon events that may occur, as a contingent barrage or contingent zone of fire.

Contingent Barrage. A barrage which may be employed in certain contingencies, and for which data are prepared.

Contingent Zone. An area within the field of fire, other than the normal zone, within which a unit may be called upon to fire under certain contingencies.

Continuity of Defense. The integrity of a defensive line or position.

Continuous Maneuver. A series of tactical exercises extending over a period of days, and designed to cover an entire operation or short campaign.

Continuous Pull Firing Mechanism. A mechanism in which a continuous pull on the lanyard first cocks and then releases the firing pin.

Contour. A line on a map, representing an imaginary line on the ground, all points of which are at the same elevation. A level or horizontal line. A convention by which the forms of nature, such as hills, valleys, ridges, etc., are depicted on a map.

Contour Interval. The difference in elevation of two adjacent contours. Vertical interval.

Contraband of War. All articles of commerce which cannot be supplied to a belligerent except at the risk of seizure by an aggrieved belligerent. In a broad sense, any supplies which may be used in the conduct of war.

Contract Surgeon. A civilian physician temporarily serving in the medical department.

Contradiction. The result obtained when two shots, fired with the same data, give impacts of opposite sense in either distance or direction.

Contravallation. See Line of Contravallation.

Contribution. An imposition or tax levied on the people of a conquered town or territory.

Contributory Plans. Comprises all the plans prepared to insure that the demands of a basic plan will be met.

Control. In surveying and sketching, the procedure of locating accurately a number of points of the terrain, some in horizontal position and some in elevation, by means of which other points and details may be subsequently located. The points so located are known as control points, triangulation points, benchmarks, etc.

Control of a Sketch. The control and critical points and elevations of a sketch, by reference to which the details of the terrain are fixed.

Control Panel. A panel mounting all the apparatus for controlling and firing submarine mines.

Control Point. A point, the coordinates of which have been accurately determined, that may be used in topographical operations. A point established to facilitate and regulate supply or traffic along a route. See Regulating Station.

Control Station. An arrangement which permits a searchlight to be pointed in elevation and azimuth from a distant point. See Distant Electric Control.

Control Stick. The vertical lever by means of which the longitudinal and lateral control surfaces of an airplane are operated. The elevator is operated by a fore-and-aft movement of the stick, the ailerons by a side-to-side movement.

Control Surface. See Surface, Control.

Controlled Maneuver. A continuous maneuver which is conducted according to a previously prepared scheme covering the entire operation.

Controlled Submarine Mine. See Mine, Controlled, Submarine.

Controller. See Distant Electric Control.

Controls. A general term applied to the means provided to enable the pilot to control the speed, direction of flight, altitude, power, etc., of an aircraft.

Convalescent. A soldier who, though discharged from the hospital, is not sufficiently recovered to return to duty. Recovering from sickness or injury.

Convalescent Camp. A camp for the care of convalescents.

Convalescent Hospitals. Hospitals located well to the rear of and central to the army area, which receive from evacuation hospitals, slight, convalescent and other cases offering prospect of early restoration to combat fitness.

Convenience of the Government. Signifies that an action has been taken in the best interests or for the advantage of the government.

Convention. An agreement entered into by belligerents, either for the evacuation of some place, the suspension of hostilities, an exchange of prisoners, or the like. Anything established by custom or general consent.

Conventional Signs. Signs, of general acceptance, used on maps and charts to indicate features of importance, such as rivers, railroads, forests, lakes, etc.

Converged Sheaf. A battery sheaf of fire converged on a point or small area.

Convergence Difference. An angular measurement, constant in any case, which may be progressively applied to all the pieces of a battery except the base piece in order to cause their lines of fire to converge toward that of the latter, either to intersect at a point or to distribute fire on a target of less width than the battery front.

Converging Attack. An attack delivered from different directions upon a single point or locality.

Converging Order of Battle. See Order of Battle, Converging.

Conversion. A movement in which the front of a line or column is changed to a new direction. A change of front, as by a body of troops attacked on a flank. Changing the unit in which any quantity is expressed, as from metric to English units.

Conversion Chart. See Angular Unit Method.

Converted Guns. Cast iron guns relined with steel tubes.

Convertible Vehicle. A vehicle which can be converted from the wheel type for movement on roads, to the track-laying (caterpillar) type for movement on the battlefield, thus combining strategical and tactical mobility.

Convoy. A column of vehicles, including the escort, transportation, and materiel or personnel being transported. Prisoners on the march with their attendant guards. To escort. To accompany for the purpose of protecting.

Convoy Camps. Camps in which the vehicles and loads of pack animals are utilized as a means of defense.

Cooperation. The act of cooperating or working together for the accomplishment of a common end.

Coordinate. To regulate and combine into harmonious action. To adjust. See Coordinates.

Coordinated Attack. A carefully planned and executed attack in which the efforts of the various elements are combined in such a manner as to utilize their powers to the greatest advantage.

Coordinates. Measurements from an origin by which the position of a point is fixed.

Coordinates, Polar. The direction and distance of any point from any other point as an origin.

Coordinates, Rectangular. The distances of a point from an origin, measured along two perpendicular axes known as X line and Y line; the two measurements determine the location of the point with reference to the origin.

Coordination. The act of supervising, regulating, and combining, so as to gain the best results.

Co-pilot. An assistant pilot.

Cordeau or **Detonating Cord.** Lead tubing filled with pulverized TNT, used for detonating blasting or demolition charges.

Cordite. A high explosive, used in shells and for blasting.

Cordon. A chain of posts, detachments or sentries, covering or inclosing any body of troops or locality. The coping of a scarp wall.

Cordon System of Outposts. A system in which the groups on the line of observation of an outpost are placed so close together that no considerable body can pass between any two without being detected.

Cored Shot. An elongated projectile having a cavity in its body for the purpose of throwing the center of gravity forward, thus insuring greater steadiness in flight.

Corporal. The lowest grade of noncommissioned officer in the United States Army, who commands a squad.

Corporal of the Guard. A noncommissioned officer of the guard who has the duty of instructing, posting and relieving the sentinels.

Corporal Punishment. Physical punishment, usually by stripes and lashes.

Corporal's Guard. A detachment of several men under arms.

Corps. A tactical unit comprising two or more divisions and corps troops, capable of conducting a major operation. A body of men organized under a common direction. A designation for certain branches of the service, as the Quartermaster Corps, Corps of Cadets, etc.

Corps Area. One of the nine military geographical districts of the United States, including several states, organized under a single commander for purposes of administration and training. On a basis of population each corps area is supposed to furnish one regular, two national guard, and three reserve divisions, with its quota of corps and army troops, in case of a general mobilization. The area occupied by a corps in the combat zone.

Corps Area Service Command. A group of agencies, collectively considered, available to a corps area commander for the primary purpose of executing his mobilization mission. It includes corps area headquarters and various installations subordinate thereto.

Corps Artillery. Artillery which is an organic part of a corps as distinguished from division or army artillery.

Corps Cavalry. Cavalry belonging to a corps, or attached for operations.

Corps Front. The extent of front covered by a corps in battle, or expectation of battle.

Corps of Cadets. The cadets of the United States Military Academy or other military school.

Corps of Engineers. A corps of officers and enlisted men trained in military engineering work, one of the combatant arms and supply services.

Corps Park. A place for the temporary storage of the supplies carried by the corps trains during periods when it is necessary to use the vehicles for other purposes.

Corps Sector. The area occupied by a corps in defense.

Corps Service Area. That portion of the army combat area in rear of the front line division areas.

Corps Troops. Troops and trains which are an organic part of the corps organization, but not a part of the component divisions.

Corrected Azimuth. The azimuth from the directing point to the target corrected for all known variations.

Corrected Map Shift. The map shift from a check point to a target, corrected for the difference in drift for the ranges of the check point and target.

Corrected Range. The map range or the actual range corrected for all known variations in conditions from those assumed as standard in the construction of firing tables. The computed range at which the sight should be set.

Correction for Tilt. Reducing a tilted aerial photograph to a true vertical projection.

Corrections. Changes applied to basic data to correct for conditions not standard, the effects of which are divided into two groups: position and materiel, and weather. Compensations for any errors, equal to the errors but of opposite sign or sense.

Corrector. A device on a time fuze setter by means of which minor changes in the time of burning, due to non-standard conditions, may be made without changing the setting of the principal scale. See also Acoustic Corrector.

Correspondence Book. A book, kept in units smaller than a regiment, in which is recorded in brief the subject matter and disposition of all correspondence matter sent and received.

Correspondence School. A school for conducting military or other instruction by means of correspondence or extension courses.

Corridor of Terrain. A compartment of terrain whose longest dimension lies generally in the direction of movement of a force, or leads toward an objective.

Cortege. The official staff, civil and military. A procession, as at a funeral.

Cossack Post. An outguard consisting of four men posted as an observation group, with a single sentinel in observation, the remaining men resting nearby and furnishing the reliefs for the sentinel.

Cossacks. Warlike tribes of southern Russia, noted for their skill as horsemen. Irregular cavalry of the Russian Army.

Cotta. A stone fort, used by natives of the tropics.

Council of War. An assembly of officers of high rank, convened to consult and discuss in regard to matters of military importance.

Councils of Administration. Councils which assemble at stated times, or on order, to ascertain and examine the sources from which the funds have accrued, and to recommend expenditures therefrom.

Count Off. A command for soldiers to count twos, fours, etc., preparatory to drill or other exercises.

Counter Approach. A trench, sap or work pushed forward to meet and check the approaches of besiegers.

Counter Barrage. A barrage laid down in opposition to an enemy's barrage.

Counter Offensive. A counter attack on a large scale for the purpose of diverting the enemy's attention or activities; often intended also to reverse the general situation by passing from the defensive to the offensive.

Counterattack. An offensive operation directed by the defender against the attacker, usually for the purpose of expelling him from any portion of the defended area which he may have penetrated. Local counterattacks are made by supports and local reserves. General counterattacks on a large scale are made by general reserves, and may exhibit all the characteristics of an attack. The act of making a counterattack. To deliver a counterattack. See Counter Offensive.

Counterbattery. Artillery fire directed against the hostile artillery, either in attack or in defense. It is the particular function of the corps artillery.

Counterclockwise. Circular motion in a direction opposite to that in which the hands of a clock move. Left hand rotation or angular measurement.

Counterespionage. Measures taken to prevent espionage by the enemy.

Counterfort. A supporting wall built in rear of and at right angles to another and larger wall. See also Buttress.

Counterguard. A low outwork built in front of a bastion or ravelin, consisting of two lines of ramparts parallel to the faces of the bastion.

Counterinformation, or Counterintelligence. Measures taken to prevent the enemy obtaining information of our troops, movements and intentions; it comprises counterespionage, censorship, counterreconnaissance, night movements, the use of covered approaches, camouflage, smoke screens, the enforcement of secrecy discipline, and the observance of secrecy measures in the preparation of plans of operation.

Countermand. To revoke a command or order already given.

Countermarch. An evolution by which a file of men reverses its direction of march, the individuals remaining in the same order from head to tail; used especially by military bands. To execute a countermarch.

Countermine. To push forward subterranean galleries and place charges of explosive to destroy like works of the enemy, and check his subterranean attack. To check or intercept by countermining.

Countermining. The defense or counterattack in subterranean warfare. Destroying submarine mines by means of explosives.

Counteroffensive. See Counter Offensive.

Counterpoise or Counterbalance. A spring cylinder which assists in opening and closing the breech of a large gun. A weight which acts in opposition. A system of wires or other conductors, elevated above and insulated from the ground, forming the lower system of conductors of an antenna. See also Counterweight.

Counterpreparation. Fire delivered by the defender prior to the launching of a hostile attack for the purpose of breaking up and checking or hampering and demoralizing the attack. It is delivered according to a schedule, and may be directed at any or all elements of the attack.

Counterrecoil. The movement of a cannon from its recoiled position after firing to its firing position.

Counterrecoil Buffer. That part of the counterrecoil system which controls the last movement of a cannon in counterrecoil and prevents shock as the cannon returns to its firing position.

Counterrecoil Mechanism. The mechanism which returns a cannon from its recoiled position to its firing position. A recuperator.

Counterreconnaissance. Means and measures taken to interfere with or prevent reconnaissance by the enemy. The effort made to defeat hostile reconnaissance; it is executed by the air service, the cavalry, and security detachments.

Counterscarp. The side of the ditch of a fortification opposite the parapet.

Counterscarp Gallery. A casemate or gallery, placed within the counterscarp at a salient, for the defense of the ditch and enceinte.

Counterscarp Wall. A masonry retaining wall for a counterscarp, or a detached wall at the foot of a counterscarp.

Countersign. A word given daily from the principal headquarters of a command to aid guards and sentinels in identifying persons who may be authorized to pass at night.

Counterslope. The first slope in rear of a slope which descends toward the enemy, and that is wholly or partially hidden from him by a covering ridge.

Countervallation. See Line of Countervallation.

Counterweight. A weight attached to the gun levers of a disappearing gun carriage, which takes up part of the recoil and serves to return the gun to its firing position. Any weight used to balance or partially balance another weight.

Counterworks. Any works constructed to oppose the operations of the enemy or to destroy or neutralize his works.

Courier. A messenger. One who carries a message in person. Couriers include runners; mounted, bicycle and motorcycle messengers; and those who travel by other means of transportation, as a motor vehicle, railway train, ship, or aircraft.

Court of Inquiry. A court, consisting of three or more officers, to examine into any transaction of or accusation or imputation against any officer or soldier.

Court-martial. A military court, in general restricted to the trial of persons in the military service for military offenses. A general, special or summary court-martial, or, formerly, a regimental court-martial.

Cover. Natural or artificial shelter or protection from fire or observation, or any object affording such protection. The vertical relief of a trench measured from the bottom, or from the trench board, to the top of the parapet. A distinction should be made between cover and concealment, the latter providing shelter from view, but no actual protection from fire. To screen, hide, or insure the security of another force, or a locality. To stand or march directly in front or in rear of another person or unit, as to cover in file or trace. To establish a correct alignment of files or units from front to rear. See also Concealment.

Cover in File or **in Trace.** To march or be directly in rear of the corresponding individual or unit in front.

Cover Position. A position immediately in rear of the fire position which affords protection to riflemen or a weapon from hostile flat-trajectory fire.

Cover Trench. A relatively deep and narrow trench placed closely in rear of a fire trench and connected thereto at short intervals, intended for the protection of troops held to man the firing line promptly in case of attack. Any trench designed primarily for protection rather than fire. A shell, slit or assembly trench. A trench provided near an emplacement for the protection of a gun crew.

Covered Approach March. An approach march protected by forces sufficiently strong to provide security against hostile ground attack. See Approach March.

Covered Communications. Any routes of movement, especially trenches, which afford any degree of concealment or protection for men passing along them.

Covered Defenses. Intrenchments and arrangements intended to protect troops and materiel from high angle fire.

Covered Sap. A sap provided with overhead cover. A blind or Russian sap.

Covered Way. A passage or road along the top of the counterscarp, covered by an embankment whose front slope forms the glacis, used in defense and as a line of departure for sorties.

Covering. The act of one or more men in placing themselves directly in rear of men in their front. Correctly aligned from front to rear. See also Covering Force.

Covering Detachment. A body of troops which screens others from observation or attack. A detachment used to cover an approach, organization of a position, river crossing, withdrawal, etc.

Covering Force. Any body or detachment of troops which provides security for a larger force by observation, reconnaissance, attack or defense, or any combination of these methods. Covering forces include advance, flank and rear guards, outposts, cavalry screens, combat patrols, forces posted to cover a withdrawal or for delaying action, etc. A mobile force which protects a besieging force against attacks from the outside. A covering detachment.

Covering Mask. An obstruction of any nature, such as a ridge or wood, which affords concealment from hostile observation and a degree of protection from fire.

Covering Mass. Any natural or artificial feature, such as hills, substantial buildings, etc., which affords shelter from view and also a certain amount of protection from fire.

Covering Position. A position occupied to protect and facilitate the movement of other troops. A designated position well to the rear and to the flank of the lines of withdrawal from action, close enough to permit it to cover the withdrawal of local reserves.

Cradle. A part of the carriage of certain guns which supports the gun, guides it in recoil, and houses the recoil mechanism. A device used in transporting heavy guns over short distances. A support used in making fascines. A support in which something rests.

Crash. To bring an airplane down in such a manner that in landing the craft is damaged. To disable an enemy aircraft to such an extent that it falls out of control and is wrecked on hitting the ground. The act or an instance of crashing. An aircraft accident in which the machine is completely or partially demolished.

Crater. A pit formed by the explosion of a mine or shell.

Crawl. To move while lying flat on the stomach, without raising the body from the ground. A slang term meaning to berate or abuse.

Credit. An allocation for a definite quantity of supplies which is placed at the disposal of the commander of an organization for a prescribed period of time.

Creep. To move on the hands and knees. To drag the bottom of a channel with grappling hooks to locate the cables connecting submarine mines with the shore. A movement of the trigger of a small arm, after taking up the slack, and before discharge.

Creeping Barrage. A barrage which advances slowly in front of advancing infantry. A rolling or jumping barrage.

Cremaillere. An irregular or broken trace, as of a parapet, designed to prevent enfilade of long sections of trenches. As used in the older forms of fortification the cremaillere usually consisted of relatively long faces, offset at intervals by short flanks nearly perpendicular to the faces.

Crest. The summit or highest line of a ridge embankment or parapet. The actual or topographical crest. A plume of feathers or other decoration worn on the helmet, indicating the rank of the wearer. A helmet.

Crest, Military. See Military Crest.

Crew, Landing. A detail of men necessary for the landing and handling of an airship on the ground. A ground crew.

Crib. A support, wall, pier, etc., made of layers of logs or timbers laid alternately lengthwise and crosswise.

Crisis of a Battle. The moment which comes in practically every battle when the issue of victory is in doubt and when the decisive blow should be struck.

Critical Point. A point whose possession is of tactical importance. A point where there is a change of direction, or change of slope, in a stream or ridge line; a hilltop, stream junction, etc.; used by a sketcher or topographer in mapping the contours of the terrain. A selected terrain feature along a route of march with respect to which instructions are given to serials for the purpose of controlling a movement.

Critical Raw Materials. See Materials, Critical Raw.

Critical Zone. The zone immediately beyond the bomb-release line of a defended area equal in width to the distance travelled by an airplane in one minute. It is the zone in which the defense must be capable of delivering an adequate volume of effective fire, against attacking aircraft.

Critique. A discussion, generally conducted by the officer in charge, which follows a combat exercise, field problem, drill or demonstration. A criticism. A review.

Croix de Guerre. A French military decoration.

Cross Belts. Belts worn over the shoulders and crossing over the breast.

Cross Compartment of Terrain. See Compartment of Terrain.

Cross Examination. The examination of a witness by the opposing side, following his direct examination by the side which called him.

Cross Wind. A wind which blows at right angles to the plane of fire, or nearly so.

Cross-bar Shot. A shot which folds into a sphere for loading, but on leaving the muzzle expands into a cross with sections of the shot at the extremities of the arms.

Crossbow. An ancient individual missile weapon consisting of a bow fixed transversely upon a stock that contained a groove to guide the missile, a notch to hold the taut cord, and a trigger to release it. Evidence tends to prove that the crossbow was introduced about the 10th century, A.D., and early in the 12th century its use became general throughout Europe in spite of the fact that it was condemned by the Lateran Council of 1139. It was in the Crusades that the crossbow made its reputation, and until about

1460 it was the favorite weapon of continental Europe. The perfected crossbow of this period was of steel and weighed 14-16 pounds. It had a range of around 400 yards and a point-blank range of about 70 yards. While it was lighter, cost much less, and could be fired 5 to 6 times as rapidly as the harquebus, the effectiveness of the latter caused the retirement of the crossbow. The last use of the crossbow was by the Chinese at Taku in 1860.

Cross-cannon, rifles, sabers. The distinctive insignia of the artillery, infantry, cavalry, displayed on coat collars and elsewhere.

Cross-country Cargo Carrier. A self-propelled vehicle, either armored or unarmored, for transporting ammunition and other supplies across country.

Cross-country Flight. See Flight, Cross-country.

Crossed Sheaf. A battery sheaf of fire converged at a point short of the target.

Crosshairs, Crosswires. Fine wires, spider web, or lines etched on a glass plate (reticle) placed at right angles in the focal plane of an observing instrument. They serve to define the line of collimation, the vertical and horizontal, and (if there be several parallel wires) to measure intervals or intercepts.

Crosshead. A transom or crosspiece which connects the lower ends of the gun levers of a disappearing carriage, and supports the counterweights.

Crosswire Illumination. Artificial illumination of the reticle of a telescope, whereby the crosswires and images are more clearly seen.

Crowning. The upper course of a revetment, placed for security, also in some cases to avoid dangerous splinters resulting from shellfire.

Crown-work. A work consisting of two or more fronts of fortification joined by two long branches to the ditch of another work, a river, village, etc.

Crowsfeet, or Caltrops. An obstacle consisting of multi-pointed steel burrs which lie on the ground with one point up.

Crow's Nest. A recess on a parapet or traverse, used as an observation station.

Cruise. A flight of considerable extent not directed to any particular landing place but rather for the purpose of test or experimentation.

Cruising Radius. The distance to which an aircraft can cruise, and return to its base, without replenishment of fuel.

Cryptanalysis. The deciphering of cryptograms.

Cryptanalytics. The science which embraces the principles, methods and means employed in the analysis of cryptograms.

Cryptograms. Messages translated into code or cipher.

Cryptography. The science which embraces the methods and devices used to convert a written message into code or cipher.

Cuartel. Barracks or quarters for soldiers.

Cuff Leggings. A common name for short canvas leggings.

Cuirass, or Curiet. Defensive armor for the body between the neck and the girdle, originally made in the form of a leather jerkin, but later consisting of a breastplate and a backplate hooked together. Generally, any of the ancient kinds of close-fitting body armor.

Cuirassiers. Heavy cavalry, which formerly wore body armor.

Culminating Point. The summit of a trajectory.

Cultural Features. Artificial features, or works of man, existing on the terrain or portrayed on a map or photograph.

Culverin. An ancient gun which was at first the lightest and shortest, but later the longest and heaviest of early cannon. The whole culverin fired a 40-60 pound shot. The culverin proper had a caliber of 5.2 inches and fired 18-35 pound shot. The demi-culverin had a caliber of 4.0 inches and fired 8-18 pound shot. The culverin bastard had a caliber of 4.5 inches and fired 11-20 pound shot.

Cupola. An armored turret, sometimes arranged to revolve.

Cupro-nickel. An alloy used for the jackets of small arms bullets.

Current. Now in force. Pertaining to the current year.

Current Series. Orders, bulletins, etc., published during the current year.

Current Supplies. Supplies held in storage to meet anticipated current requirements.

Curtain. That part of the rampart or parapet located between two bastions or salients or two gates.

Curtain Angle. The angle between the curtain and flank in the bastion trace.

Curve of Accidental Errors. A curve showing the frequency of occurrence or probability of errors of any magnitude. The probability curve.

Curve of Security. The curve tangent to all trajectories which are possible when firing a gun with fixed muzzle velocity and direction.

Curved Range. A measurement of range that takes into consideration the curvature of the earth.

Customs of the Service. The common law or practice of the service. The unwritten rules that govern intercourse between military men.

Customs of War. Customs which, through long usage in the military service, have become recognized and are accepted by civilized nations as rules for the government of the army in like cases.

Cut-and-cover Shelter. A protected shelter consisting of an open excavation in which the framework for the shelter is placed, after which the soil is filled back around and over the framework to the level of or somewhat above the original surface.

Cutlass. A heavy curved sword about 3 feet in length.

Cut-off. A device on modern rifles which changes them from a repeater to a single-shot weapon, and the reverse.

Cyclic Rate. The rate at which an automatic firearm delivers its fire while firing without interruption, expressed in shots per minute. The maximum rate of fire for an automatic weapon.

Cyclogiro. A type of rotor plane whose support in the air is normally derived from airfoils mechanically rotated about an axis perpendicular to the plane of symmetry of the aircraft, the angle of attack of the airfoils being always less than the angle at which the airfoils stall.

D

Dagger. A stabbing weapon resembling a sword, but much smaller. To stab.

Daily Telegram. The daily call made by a division or higher unit for its daily requirement of Class I supplies.

Daily Train. A railway train which arrives daily at a railhead with supplies for the troops which the railhead serves. A train which conveys Class I supplies daily from the regulating station to a railhead.

Damping. The process of successively reducing the amplitude or period of a number of oscillations, or tending to do so. Checking the motion of, as the vibrations of a compass needle.

Danger Range. The maximum range for any weapon which is all danger space.

Danger Space. That portion of the range within which a target of given height on the line of fire will be struck by a projectile; e.g., for rifle fire the danger space for a man standing is that portion of the trajectory in which the bullet is not higher than a man's head.

Danger Zone. The danger space plus the beaten zone.

Dart. A missile spear or javelin, much used in ancient times, and still found among uncivilized people. A short lance. The missile of a blow gun. An arrow.

Data Computer. A predicting instrument, of several types, for determining the firing data for the future position of an aerial target. The instrument functions when the target is followed or tracked in elevation and azimuth by sights mounted on the case and operated by handwheels, the speed of rotation of the handwheels being proportional to the components of the speed of the target.

Data Transmission System. The system, taken as a whole, by which firing data is transmitted from observers to the plotting room or data computer, and thence to the guns. See Electrical Data Transmission System.

Datum Plane or Level. An assumed level surface from which the vertical heights of contours, or any elevations are measured; usually mean sea level.

Datum Point. A fixed point whose azimuth and range from one or more observing stations have been accurately determined.

Day of Fire. The estimated average number of rounds that will be fired by a piece of artillery during one day of actual combat, as applied to larger commands. A more or less arbitrary and variable unit of measure on which the supply of ammunition in campaign is based.

Day of Supply. The estimated average expenditure of a particular item of supply by a particular organization or command during a day of campaign. A more or less arbitrary unit of measure, which varies with the article and the conditions of service. The aggregate of all items constituting a day of supply.

Day's March. The estimated or actual distance covered by a body of troops on the march in a single day, which varies with the class of troops, the size of the command, the state of training, and other conditions.

D Day. The day on which an attack or phase of operations is to be initiated.

Dead Angle. See Angle, Dead.

Dead Area or Ground. Any area within range of particular weapons or of a defensive position, which cannot be reached by their fire.

Dead Line. A line within a military prison which cannot be crossed by prisoners except under penalty of being instantly shot. A limiting date for the accomplishment of some act or purpose.

Dead March. Solemn music played for the march of a funeral procession. A funeral march.

Dead Space. Ground which cannot be covered by fire from a position, because of intervening obstacles. Air space surrounding an aircraft which cannot be reached by its fire.

Dead Space Chart. A chart (or overlay) showing dead spaces.

Dead Time. The time interval during which the target travels from its position at observation to its position when the gun is fired. It is the time necessary to compute and apply the firing data and fire the gun.

Deadman. A buried log, or timber, serving as an anchorage for a rope or cable.

Debouch. An outlet in works for the debouching of troops. To march from a defile, wood or other close country into open country. To cause to debouch.

Debussing Point. The point at which troops leave, or are to leave, the busses.

Decalage. The difference between the angular settings of the wings of a biplane or multiplane.

Decentralized Control. A method of administration in which higher authority outlines in very general terms the object to be attained, leaving details to subordinate commanders.

Decentralization. A system of organization and administration by means of which great independence of action is left to subordinate and local commanders, higher authority only outlining the objects to be attained.

Decherd Rifle. An early American rifle having a 48 inch barrel and weighing 12 pounds, first manufactured in Philadelphia in 1732.

Decimation. A kind of punishment in which every tenth man is punished. Heavy losses in battle.

Decision. The general plan of a commander, based on an estimate of the situation and expressed definitely and briefly. The ruling of a court-martial on any point. A decisive outcome of a battle, one side being decisively defeated.

Decisive Attack. The foreseen, determined and sudden effort, employing surprise, against a weak point in the enemy's line made with a view to deciding the issue of the battle. The main attack which is intended to decide the issue.

Decisive Counterattacks. Counterattacks delivered with a view to gaining the initiative.

Decisive Element. The dominating factor in tactical or strategical success. It may be leadership, direction of attack or concentration, preponderance of power, maneuverability, or other factors.

Decisive Elements. The elements or classes of enemy information which are most important to a commander. The essential elements of information.

Declaration of War. The formal announcement by a state or nation of its intention to wage war against another.

Declinate. To determine the declination constant for an instrument in any locality.

Declinating Point or **Station.** A point, free from local attraction, used for the declination of instruments. The declination constant is an average value determined from observation on several points.

Declination. The angle between any two north directions, true, grid, magnetic, compass. The process of determining the pointing error of a compass instrument.

Declination Constant. The angle between Y- or grid north and the north pointing of a particular compass in a particular locality.

Declination of the Needle. The angle between the compass needle and the true meridian.

Declination, Magnetic. See Magnetic Declination.

Declinator. A magnetic needle attached to a sketching board for the purpose of orientation.

Decode. To translate a code message into ordinary language.

Decorations. Medals, badges, ribbons, etc., bestowed for meritorious service in war.

Decorations, United States. Consist of the Medal of Honor, Distinguished Service Medal, Distinguished Service Cross, Silver Star, Purple Heart, The Soldiers Medal, Distinguished Flying Cross, and Oak Leaf Cluster. The Service Medals are: Civil War, Indian Campaign, Spanish Campaign, Spanish War Service, Army of Cuban Occupation, Army of Porto Rican Occupation, Philippine Campaign, Philippine Congressional, China Campaign, Army of Cuban Occupation, Mexican Service, Mexican Border Service, and Victory Medal (World War).

Decoy Batteries. Batteries (often dummy) installed to draw hostile attention for the purpose of keeping the enemy from firing at batteries whose position it is desired to keep concealed.

Defense. The means adopted for resisting attack. The act of defending, or state of being defended. That which defends or protects. A combatant acting a defensive role. A defender. The method of procedure adopted by the accused in answer to the charges in a court-martial case.

Defense Area. See Defensive Area.

Defense Command. One of the four main subdivisions of the continental United States for defensive purposes. The four areas are the Northeast Defense Command, the Southern Defense Command, the Central Defense Command, and the Western Defense Command.

Defense in Depth. The modern conception of defensive tactics, in which all the elements are distributed from front to rear, so that the attacker is not successful when he has ruptured a single line or even a single defensive position, but must fight his way through a deep zone or belt of resistance.

Defense Unit. The tactical unit best suited to form a unit in defense, generally considered to be the battalion.

Defenses. All the works of any nature, that defend a place against the enemy.

Defensive. The side on the defense. The condition of resisting attack. Resistance to attack. In a state of defense.

Defensive Area. An area occupied and organized for defense. Specifically, one of the unit areas of a battle position; i.e., combat group, strong point or center of resistance, or more properly, platoon, company, or battalion defensive area.

Defensive Coastal Areas. Those parts of the coastal zone, and adjacent coast, which, by reason of the presence of important harbors or industrial centers, require joint defense by the army and navy.

Defensive Position. Any area occupied and more or less organized for defense. A battle position. A system of mutually supporting defensive areas or tactical localities of varying size, each with a definite assignment of troops and a mission. A defensive position may be fortified, and otherwise organized to the extent that the situation requires or time allows. See Battle Position.

Defensive Sector. See Sector.

Defensive System. A series of organized areas arranged in depth for coordinated use in the defense of a certain position or area.

Defensive War. A war in which a country attempts only to repel invasion or the attacks of the enemy.

Defensive Weapons. See Weapons.

Defensive Zone. A belt of terrain, generally parallel to the front, which includes two or more organized or partially organized battle positions, and an outpost position or area in front.

Deferred Classification. Under a selective service law, a classification given a registrant for any of various specified reasons, which defers his call into the service.

Defilade. Concealment from enemy observation or protection from his fire by intervening obstacles, such as hills or ridges. The measure of defilade is the vertical distance from the defiladed position to the line of sighting or trajectory which just clears the covering mask. To protect by defilade. See Sight Defilade, Flash Defilade, Smoke Defilade.

Defiladed. Concealed from view or protected from fire by a mask.

Defiladed Route. A route concealed from view or protected from fire.

Defiladed Space. An area protected from fire or observation by either a natural or artificial mask.

Defile. Any narrow space or place which can be passed only when troops are undeployed, such as a ford, bridge, road through a village, a mountain pass, etc. To march off file by file.

Deflagration. The rapid burning of an explosive, specifically a low explosive, such as a propelling charge. See also Detonation.

Deflection. The deflection setting for a sight. The setting on the deflection scale of the rear sight such that when the line of sight is on the aiming point the piece is pointed for direction. The horizontal angle between the line of sighting to the target or aiming point or line of collimation of the sight, and the axis of the bore when the piece is pointed for direction. In antiaircraft firing, the angular amount the gun must lead the target, both laterally and vertically, to allow for its travel, for drift and wind, jump, and other conditions. The sag, dip or deflection in the cables of a suspension bridge.

Deflection Board. A device for determining the resultant of deflection corrections for wind, drift, and travel of the target during prediction interval and time of flight.

Deflection Computer. An instrument used in determining the lateral and vertical deflections to be applied to antiaircraft guns.

Deflection Constant. For some sights, a constant which must be applied to the deflection setting to bring the line of sighting parallel to the plane of fire.

Deflection Difference. The common converging or diverging difference applied to guns, other than the directing gun, necessary to bring their fire on the proper portion of the target. In a battery, the difference in deflection setting between adjacent pieces, when using a common aiming point.

Deflection Fan. See Range Deflection Fan.

Deflection Scale. A scale on a gun sight, in mils or degrees, for pointing a piece in direction or applying corrections in deflection.

Degassing. Chemical treatment of gassed areas in the field or storage area, which destroys or neutralizes chemical agents.

Degtyarev Light Machine Gun. A light machine gun used in the Russian Army. It has a caliber of .30 inches, a belt or clip capacity of 49 cartridges, and weighs 15.65 pounds.

Delay-action Fuze. A fuze which delays the explosion of the charge until some time after impact, to allow penetration. Such fuzes are classed as short-delay and long-delay, from .05 to .15 seconds. In some types the delay may be varied (selective delay).

Delaying Action. A form of defensive action employed to slow up the enemy's advance and gain time without becoming decisively engaged, characteristic of the tactics of a rear guard in a retreat.

Delaying Force. Troops interposed between the covering force and the battle force, in defensive combat, with the mission of forcing the enemy to develop his forces and intentions, and delaying his advance and attack. A force which fights a delaying action.

Delaying Position. A position taken up for the purpose of slowing up or interfering with the advance of the enemy.

Deliberate Intrenchments. Formal, complete and elaborate intrenchments constructed according to prescribed standards, when there is reason and opportunity to do so. See also Hasty Intrenchments.

Deliberate Organization. The thorough organization of the ground when the enemy is at a distance, and time and facilities are available.

Delivery Table. The table at the upper end of an ammunition hoist from which the projectiles are delivered to the trucks.

Delouse. To remove lice from, in a specially prepared plant.

Demilitarize. To do away with all military organization and preparation. To move all troops out of an area. To dismantle fortifications.

Demi-lune. A work of fortification consisting usually of two long faces and two short flanks, having the general shape of a crescent or spearhead. It was commonly employed as an outwork covering the curtain between two bastions.

Demobilization. A change from a war footing to a peace footing. The act of demobilizing.

Demobilize. To disband or discharge from the service. To change from war to peace footing.

Demolition. The destruction of works or materiel by any means. The creation of obstacles by the use of explosives, such as the destruction or blocking of roads, bridges, railroads, etc., justifiable as a military measure.

Demolition Bomb. A bomb containing a large charge of explosive, designed for the destruction of material objects.

Demonstrate. To make a demonstration. To teach by demonstration.

Demonstration. An attack delivered on a front where a decision is not sought and made with the object of deceiving the enemy. A show of force. A specially arranged exercise illustrating the tactical employment of troops, or the use of certain weapons or equipment.

Demountable. Capable of being dismounted, disassembled, or taken apart.

Density. Closeness of deployment. Often expressed as the number of infantrymen or of rifles per yard, mile, etc., of front. The weight of a volume of gas (or air), liquid or solid in terms of some standard specific gravity.

Density of Loading. The weight of a powder charge as compared to the weight of water which would fill the powder chamber.

Dental Corps. The dental branch of the medical department.

Department. A military geographical subdivision of the country. A subdivision of official duty or organization.

Departure, Line of. See Line of Departure.

Deploy. To extend the front, as in changing from column to line, or from close to extended order. To extend a unit both laterally and in depth, or to increase intervals or distances, or both.

Deployed Defense. A form of defense in which the troops are deployed in position, but the ground is not organized for defense, or is organized with hasty works for which less than 6 hours time is available.

Deploying Interval. The lateral space required between elements on the same line to permit proper deployment.

Deployment. The act of deploying.

Deployment in Depth. The extension of a unit or force from front to rear in a succession of lines, usually for or during combat. Formation in several successive lines. Distribution in depth. The elastic defense.

Deposition. A written declaration, under oath, containing replies to interrogatories and cross interrogatories, presented as evidence before a court.

Depot Brigade. A brigade charged with training replacements for a combat unit.

Depot Stocks. Supplies assembled in a depot and available for distribution to the troops.

Depots. Supply establishments maintained primarily for the purpose of receiving, storing, and distributing supplies; they may be charged with other functions, including procurement, as directed by regulations and orders.

Depots, Recruit. See Recruit Depots.

Depression. The pointing of a gun below the horizonal. A hollow or basin in the terrain.

Depression Position Finder. A position finder which measures the range to a floating target by means of the angle of depression from the upper end of a vertical base whose lower end is at sea level. Used only in seacoast defenses.

Depth. The space from front to rear of any formation, including the leading and rear elements; the depth of a man is assumed to be 12 inches. The vertical height from the bottom of a trench to the original natural surface.

Depth Bomb. A bomb designed to explode at any predetermined depth below the surface of the water, for use against submarines and other submerged objects.

Derringer. A pistol of large caliber having a very short barrel.

Descending Branch of the Trajectory. That part of the trajectory between the summit of the trajectory and the point of impact or fall.

Descriptive List. A paper containing a description and a brief history of a soldier and a statement of his accounts.

Desert. To leave the service without authority and with the intention of remaining absent. To abandon without leave.

Deserter. One who deserts.

Desertion. The act or condition of deserting.

Detached. Separated and serving away from the organization to which it (they) belongs, as a detached battalion or company, detached officers and enlisted men.

Detached Duty. Duty performed when detached from one's permanent organization or station.

Detached Enlisted Men's List. A list of enlisted men serving away from the units to which they belong, or who are not assigned to any unit.

Detached Officers' List. A list comprising the names of all officers not performing duty with the branch to which commissioned or to which detailed or assigned.

Detached Piece. A piece of artillery detached from its unit for the execution of a special mission; including roving guns and antitank guns. Upon the completion of its special mission a detached piece rejoins its battery.

Detached Post. A post established outside the limits of the outpost proper for a special mission, as to observe or guard some locality of special importance.

Detached Service. Service away from one's unit, organization or station.

Detached Unit. A unit temporarily detached from the command and control of the higher unit of which it is a part.

Detached Works. Works of fortification entirely separated from and not within supporting distance of the main work.

Detachment. A part of a unit separated from the main organization for a special purpose.

Detachment Warfare. Warfare conducted by more or less independent and relatively small commands, widely dispersed, as in a thinly settled country.

Detail. One or more men designated for a particular service, as range detail, guard detail, etc. A temporary assignment to another branch or duty. To select for such a duty or service.

Detailed Orders. Orders covering fully the details of the subject to which they pertain.

Determinate Error. An error of such nature that it may be evaluated and its effect corrected for, such as known variations from normal in muzzle velocity, known winds, or other atmospheric variations from normal or standard.

Detonating Fuze. A fuze which detonates an auxiliary charge called a booster charge, which in turn detonates a bursting charge.

Detonation. An explosion of high order. In a detonation, the chemical reactions are not confined to the surfaces of the substances exploded but appear to progress rapidly in all directions throughout the mass of the charge from the initial point in a very short time. The very rapid, almost instantaneous conversion of a high explosive into heated gases. See also Deflagration.

Detonator. A metal tube containing a high explosive detonating compound. A high explosive used to detonate a charge. A gun fired by a percussion cap.

Develop. The act of developing. To break up into smaller units. To extend a column from mass or route column. To work out or learn in detail, as to develop the enemy's position and strength.

Developed Probable Armament Error. The probable armament error as computed from a finite series of shots; it is the average armament error of a particular series of shots multiplied by an empirical factor.

Developed Trajectory. The actual trajectory, resulting from non-standard conditions, as distinguished from the firing-table trajectory.

Developing Attack. An attack made for the purpose of obtaining information as to the enemy's strength and dispositions, and causing him to reveal his intentions.

Development. The breaking up of a command into fractional columns and the marching of them on a designated terrain objective, or on the enemy, to facilitate the approach march and deployment; the initial phase of deployment for battle.

Development of Offensive Combat. The progressive advance from each step of an offensive to the next.

Deviation The absolute distance between a point or center of impact or burst, and the target; or any component thereof, as longitudinal, lateral or vertical (range, direction or height). The horizontal angle between a point of impact and the target, measured at the observation post.

Deviation of Projectiles. The deflection, right or left, from the line of fire, due to wind, drift, etc. The difference between the ranges of like projectiles fired under similar conditions.

Diamond Hitch. A rope lashing used to secure the load of a pack animal.

Diaphragm Gas Mask. A special type of gas mask which permits the use of the voice.

Diaphragm Shell. A shell having a partition or diaphragm, which separates the bursting charge from the bullets.

Diary. See Military Diary.

Dictated Order. An order delivered orally, of which each recipient records in writing such parts as pertain to his own command, a complete record being made by a staff officer.

Difference Chart. A chart from which the range or azimuth of a target from a gun or station may be determined when the range or azimuth from some other gun or station is known.

Dig In. To excavate a trench or system of trenches for immediate defense or occupancy, as during a temporary halt in an attack.

Dihedral Angle. The acute angle between a line normal to the plane of symmetry and the projection of the wing axis on a plane normal to the longitudinal axis of an airplane. The divergence of the wings of an airplane from the lateral axis, which is helpful in identification.

Dip. Deflection of a compass or magnetic needle in a vertical plane, which increases as the magnetic pole is approached, but may also be caused by local attraction. A rapid drop or dive of an airplane, followed by a climb. The inclination of the sole of an embrasure.

Dip the Flag. To lower the flag and quickly restore it to its former position as a mark of respect. The national emblem is never lowered in salute.

Direct Laying or Pointing. Pointing a piece for direction, or in both range and direction, by means of a sight directed at the target. See Pointing.

Direct Laying Position. An artillery position from which direct laying on the target area is possible.

Direct Pursuit. Pursuit conducted against the rear of the retreating columns, and including envelopment thereof.

Direct Support. Fire, especially artillery fire, delivered in support of a subordinate unit of the division, corps or army of which the fire unit or artillery is an organic part, or to which it has been assigned. Artillery in direct support has a dual role of general support of the entire organization of which it is a part, and direct support of a subordinate unit thereof.

Directing Gun. The gun from whose line of direction and fire that of other guns is regulated. The gun which is used as a directing point. The gun for which the initial data are computed. Ordinarily there is one directing gun for each battery. The base piece.

Directing Point. The point in or near the battery from which all ranges and deflections are initially computed. If a gun of the battery is the directing point, it is called the base piece or directing gun.

Direction. The line or path along which a body is moving. Azimuth. The horizontal pointing of a firearm.

Direction Finder. An apparatus which, by the application of radiotelegraphy, permits the location by the operator of an aircraft during flight, of any ground station, in fog, darkness, etc. A radio receiving device permitting determination of the line of travel of radio waves as received. See also Radio Beam.

Direction of March. The direction in which the base of the command in question, whether actually in march or at a halt, is facing at any instant.

Direction Probable Error. The probable lateral error, measured perpendicular to the line of position.

Directional Gyro. A gyroscopic instrument which indicates direction of flight or angular deviation from a prescribed course. A gyro compass.

Directional Radio. A method of determining position by means of radio bearings. See also Radio Beacon.

Directive. A statement, oral or written, prescribing a general line of action or conduct. A letter of instructions.

Director. A data computer for artillery fire, especially antiaircraft fire. An officer who conducts a tactical exercise or war game.

Directrix. The principal axis of a front of fortifications. The bisector of a salient angle. The center line of the field or sector of fire of a gun. A capital.

Dirigible. Capable of being directed or steered. A balloon or airship which can be propelled and steered.

Disability. Condition of being disabled. Physical incapacity for active service.

Disappearing Gun Carriage. A seacoast mount on which the gun is concealed below the parapet when not in the firing position.

Disappearing Target. A target which can be temporarily exposed to view and then withdrawn. A target that falls when hit. See also Revolving Target.

Disarm. To deprive of weapons or armor. To deprive of the means of attack and defense. To reduce the military establishment materially.

Disarmament. The act of disarming.

Disassemble. To take apart, as a weapon or machine. To strip.

Disband. To discharge from the military service. To break up an organization.

Disbursing Officer. An officer who accounts for and disburses funds. An accountable disbursing officer is any commissioned officer of the Army who receives and disburses public money in his own name.

Discharge. The act of discharging. Release. A document certifying the termination of a period of service. To release from service or confinement. To fire a charge from a weapon.

Discharge by Purchase. A discharge obtained prior to the normal expiration of a term of enlistment by payment to the government of a prescribed amount; authorized in time of peace after one year of service in any enlistment.

Discharge, Dishonorable. A discharge from service not honestly and faithfully rendered, involving loss of certain rights of citizenship.

Discharge, Honorable. A discharge certifying that the soldier's service has been honest and faithful.

Discharge Without Honor. An unfavorable discharge from the service without loss of any rights of citizenship.

Disciplinarian. One who pays particular regard to and maintains strict discipline within his command.

Disciplinary Exercises. Exercises designed to teach precise and soldierly movements and to inculcate prompt and unhesitating obedience to orders from proper authority.

Discipline. The spirit inculcated in well trained soldiers which is manifested by willing and cheerful obedience to orders, scrupulous conformity to standardized procedure, and unremitting effort in the appropriate sphere of activity. The system, mental, moral, and physical, used for the training of military personnel. Subjection to control. To teach subordination to. To train to act together under orders. To inculcate the habit of obedience. To subject to discipline or punishment.

Diseases of Animals. The more serious of the diseases of the horse and mule are as follows: Communicable, glanders-farcy or glanders, lymphangitis, stomatitis, inflammation of the lungs, strangles, influenza, surra, anthrax, mange, lockjaw, and ringworm; Non-communicable, colic, gripes, diarrhea, staggers, thumps, azoturia, scratches, sand cracks, navicular disease, and founder.

Disembark. To land troops or supplies from a boat or ship.

Disengage. To extricate a command from a critical situation. To break suddenly from any particular formation. To quit the guarded side of an opponent's bayonet or blade.

Disinfection. Destruction of pathogenic germs by the use of chemicals. Disinfectants are applied to inanimate objects, but not to the human body. See Antiseptics.

Disinfestation. Treatment of clothing, bedding, etc., designed to kill pathogenic germs and vermin.

Dismiss. To discharge from the service with dishonor. To remove from office or employment. To release an organization from ranks at the conclusion of a drill or ceremony.

Dismissal. The separation of an officer or cadet from the service without honor as a result of conduct unbecoming an officer and gentleman, or more serious offenses.

Dismount. To alight from a horse or vehicle. To remove a gun from its carriage. To disassemble or strip.

Dismounted. Not mounted on horseback, in tank, on artillery carriage or other vehicle. Removed from a carriage or other mount.

Dismounted Combat. Combat on foot, specifically by dismounted cavalry, in which the rifle is the principal weapon.

Dismounted Defilade. Defilade from view or fire for a dismounted man standing. A position in which a dismounted man can just see over a mask

Dismounted Units. Units which habitually operate on foot. The following units are considered as dismounted and are entitled to carry a color (1) Chemical regiments; (2) Coast Artillery regiments (harbor defense) (3) Engineer regiments and separate engineer battalions in infantry divisions; (4) Infantry regiments and separate battalions; (5) Military polic battalions; (6) Service regiments and separate service battalions, quarter master corps, in infantry divisions.

Dispatch. A term which includes all written messages, orders, reports, and instructions in plain language or code, except those transmitted by mail or by direct personal agency. To send a message, etc.

Dispatch Rider. A motorcycle, bicycle, or horse messenger.

Dispensary. A place where medicines are dispensed and medical and dental treatment is furnished, without hospitalization; usually outside the combat zone.

Dispersion. A condition in which the elements of a command are more widely deployed or separated than the situation warrants. The scattering of shots intended to strike or burst in the same place. The more or less irregular distribution of the points of impact of projectiles fired with the same data about a certain point called the center of impact. The scattering of shots fired with the same data.

Dispersion Diagram. A diagram made up by superimposing the dispersion ladder for direction on the dispersion ladder for range and indicating in each resulting rectangle the percentage of shots expected to fall therein See Area of Dispersion.

Dispersion Ladder. A diagram made up of eight successive zones, each equal in width to one probable error, either lateral or longitudinal. The center of impact is on the line between the two central zones and in each zone is indicated the percentage of shots expected to fall therein.

Dispersion Scale. See Dispersion Ladder.

Displaced. A British term applied to officers removed from a particular regiment for misconduct, but who are at liberty to serve elsewhere.

Displacement. A tactical movement of artillery to the front or rear to new firing positions necessitated by the progress of the attack or defense, or by other changes in the situation. The horizontal distance of a gun from the directing point of a battery. The horizontal distance of any point from another point. The weight of water displaced by a vessel or other floating object, which equals the weight of the object. See also Parallax.

Distance. The horizontal space between an observer and any point or between any two points. Space between elements in the direction of depth; distance is measured from the back of the man in front to the breast of the man in rear; the distance between ranks in the United States Army is 40 inches in both line and column; facing distance is 14 inches; it is the distance between men who are faced to the flank, from a position in line with 4 inch intervals. Range.

Distance Rings. Bearing and separating rings which hold the traversing rollers of a gun carriage in their proper radial positions.

Distant Defense. A defense in which the enemy's movements are interrupted some distance from the defender's main position.

Distant Electric Control. A system by which the pointing of a searchlight may be controlled from a distance. There are two types of apparatus, the impulse type and the brush-shifting type.

Distant Range. See Ranges, Classification of.

Distant Reconnaissance. Reconnaissance directed against distant objectives to procure the information upon which strategical and operative plans and decisions of the high command are based; it is a particular mission of the cavalry and air service.

Distinguished Flying Cross. A decoration awarded to any person who, while serving in any capacity in the Air Corps of the Army of the United States, including the National Guard and the Organized Reserve, has distinguished himself by heroism or extraordinary achievement while participating in aerial flight.

Distinguished Marksman. A rifle shot who has won three of the authorized medals in rifle competitions or as a member actually firing on a prize winning team in the national team match.

Distinguished Pistol Shot. A pistol shot who has won any three of the authorized badges or medals in pistol competitions.

Distinguished Service. Exceptionally meritorious service of a military nature, usually in war, but sometimes in peace. It includes outstanding gallantry in battle and also exceptional service other than in actual conflict.

Distinguished Service Cross. A decoration awarded by the President to any person who has distinguished himself by extraordinary heroism in military operations against an armed enemy of the United States and under conditions which do not justify the award of the medal of honor.

Distinguished Service Medal. A decoration awarded by the President to any person who has distinguished himself by exceptionally meritorious service in a duty of great responsibility in connection with military operations against an armed enemy of the United States in time of war.

Distributing Point. A place or point at which the supplies are distributed to the field and combat trains of the troops. A corps area installation in the chain of supply that may be interposed between a supply point and posts or stations.

Distribution. The manner in which troops are disposed for any particular purpose, as battle, march or maneuver. Dispersion of projectiles. An intentional dispersion of fire for the purpose of covering a desired frontage or depth, accomplished in various ways. A delivery of supplies, specifically by the supply officer of a higher unit to subordinate units, or to individuals. See also Fire Distribution; Issue.

Distribution Box. A metal case resting on the bottom below a group of submarine mines, marked by a buoy, and connected by electric conductor cable to each mine and to the shore. It contains a distributor by which the detonating current may be directed to any mine of the group.

Distribution Difference. The angle through which the piece next to the base piece must turn, assuming that both are laid on the same aiming point, in order to be laid properly on its portion of a target having width; for the other pieces this angle is applied progressively in the ratio 2, 3, etc.

Distribution in Depth. See Deployment in Depth.

Distribution of Troops. The tactical components into which a command is divided and the troops assigned to each. The manner in which the elements of a command are disposed for battle, march, or maneuver. The order of march of the elements of a command as set forth in a march order.

Distribution, Storage and Issue. Embraces the receipt, warehousing, and furnishing of supplies and equipment in an orderly and efficient manner to troops or stations as needed.

Ditch. An excavation made around or in front of a fortification. An excavated drain at the side of a road, around a tent, etc. To dig a ditch.

Dive. A steep descent, with or without power, in which the air speed is greater than the maximum speed in horizontal flight. To descend or fall precipitously in an aircraft at an angle greater than the gliding angle so that the descent is with increasing momentum. To descend steeply with the nose of an aircraft down. To cause an aircraft to plunge head foremost.

Dive Bomber. A fast bombing plane designed to dive toward an objective and to release its bomb load or fire on ground troops or installations as it pulls out of the dive at a low elevation.

Diverge. To turn aside or deviate. To lead away from another. To tend to spread apart. To cause to diverge.

Divergence Difference. An angular measurement in direction which may be applied to all pieces of a battery except the base piece to cause their lines of fire to diverge from that of the base piece in order to place fire upon their proper portions of a target having greater width than the battery front.

Divergent Retreat. A retreat in which the retreating columns march on diverging lines.

Diversion. An attack or feint which is intended to divert the enemy's attention and draw his troops from the point where the principal attack is to be made. An attack made upon a weak spot in the enemy's line in order to draw off his forces from making an attack elsewhere.

Division. The largest permanent combat unit in most armies, varying in strength from about 8,000 to about 30,000 officers and men. The infantry division in the United States army at present is known as the triangular or streamlined division. It has a strength of about 15,000, and includes 3 infantry regiments (not brigaded) the divisional artillery, and necessary complements of engineer, quartermaster, medical troops, etc. The division includes no animals, and can be completely motorized by the addition of personnel carriers and trucks. Other types of divisions are the old square (4 regiment) division, now obsolete, the cavalry division, and the armored division. A military area, usually less than a corps area. A part of an organization, usually one having a definite function.

Division Area. The extreme forward part of the theater of operations in contact with the enemy. The front of the army combat area, occupied by divisions in the front line. The portion of the area occupied by a particular division is given its designation, as the 4th Division area.

Division Artillery. The artillery which is an organic part of a division.

Division Cavalry. The troops, squadron, regiment, etc., attached to or forming an organic part of an infantry division.

Division Dump. A dump in the division area, designated by the division commander, where reserve supplies normally carried in the division trains are stored when it becomes necessary to use the vehicles for other purposes.

Dock. A large shed used in housing airships. A wharf. A slang term meaning forfeiture of pay due to sentence by a court-martial. To haul an airship or vessel into its dock.

Doctrine. That which is taught or set forth for belief. The body of principles involved in a particular subject, as tactics, training, etc.

Doctrine of the Attack. To strike by surprise with maximum force, at a decisive point and in a decisive direction. Decisive results are achieved only by the offensive. To gain a decision it is necessary to drive the defender out of his trenches, into the open, where the infantry, supported by all the auxiliary arms, determines the issue.

Document File. A file in which the originals or copies of all communications sent and received by an organization or headquarters are kept; used in connection with a correspondence book.

Dog Leg. A slang term applied to the first stripe received on promotion. A sharp angle in a line of trench, as in the echelon trace. Two sharp turns in a mine gallery leading to a mine chamber to prevent the explosion from destroying the gallery.

Double Action. Said of a revolver that is both cocked and fired by pressing the trigger.

Double Envelopment. A simultaneous envelopment of both the hostile flanks.

Double Gang. A method of conducting intrenching or other work when men are plentiful and tools limited in number, by assigning two men to each tool. They change positions frequently, one resting while the other works.

Double Rank. Two ranks, one behind the other.

Double Sap. Two heads of sap pushed forward abreast, forming a single trench with a parapet as a protection at the head and on both sides.

Double Section Column or Line. Artillery formations in which each caisson is abreast of and close to its piece.

Double Sentry. Two men posted and acting together as observers or listeners, as in an outguard.

Double Shelter Tent. A tent formed by buttoning together the square ends of two single shelter tents.

Double Time. A step with a length of 36 inches and cadence of 180 steps per minute. To move at double time.

Double-apron Fence A wire entanglement consisting of a central four-wire fence about three feet high, with parallel rows of small pickets on either side placed opposite the intervals between the posts in the fence, to which diagonal wires supporting the aprons are strung; each of the two sets of diagonal inclined wires carries three horizontal wires.

Draft. A selecting of soldiers for any purpose. A selecting of men from the country at large for compulsory military service. The body of men so drafted. Power applied to draw a vehicle. Traction. Draft is classified as unit, tandem, multiple, and towing cable. Passage of heated air, gas and smoke up a chimney or smokestack.

Draft Dodger. One who attempts to evade military service under the draft or selective service law.

Draftee. One who has been drafted for military service.

Drag. A device for moving earth by dragging along the surface. A device for smoothing a road by breaking down the ridges and filling the ruts and holes. A scraper. A device for slackening the speed of vehicles. The component of the total air force on a body parallel to the relative wind. A slang term for influence.

Drag Line. A guide rope for aircraft. An operating line for an excavator.

Drag Rope. A rope used for pulling vehicles. A prolonge. A long rope which can be hung overboard from a balloon so as to act as a brake and a variable ballast in making a landing. Sometimes called trail rope or guide rope.

Dragon. An ancient cannon which fired a 40 pound shot. A small blunderbuss. A short hand gun of great bore to carry several balls or slugs. A soldier armed with a dragon.

Dragon Balloon. A kite balloon.

Dragoon. A soldier who fought either on foot or horseback. Heavy cavalry, capable of both mounted and dismounted combat; the cavalry of the present day. The term has been applied also to mounted infantry. To harass. To coerce.

Dragoon Guards. An early name for certain regiments of British heavy cavalry.

Drainage Lines. The channels of the terrain in which water flows, either continuously or intermittently. They form the skeleton of the terrain for the purposes of the map maker.

Draw. To pull from a sheath or scabbard. A small valley or drainage line.

Drawn Battle. A battle in which neither combatant secures or claims victory. An indecisive engagement.

Dress. The act of taking correct lateral alignment by men in rank. To cause to take alignment.

Dress Parade. A ceremony held at retreat at which soldiers appear in dress or full dress uniform, and in formation under arms.

Dress Uniform. A uniform that may be worn on occasions other than formations with troops. See also Full Dress; Special Evening Dress.

Dressing Stations. Stations established during combat for the reception of casualties and the treatment of minor wounds.

Drift. The lateral deviation of a projectile from the vertical plane in which it is fired, caused by the resistance of the air and the rotation of the projectile around its longer axis. The angle between the plane of fire and the vertical plane containing the muzzle and the point of fall. Leeway made by an aircraft in flight.

Drift Slide. A movable piece in the rear sight leaf of a rifle which carries the open and peep sights, and which when set at the desired range automatically corrects for drift by a slight lateral motion.

Drill. A form of instruction which has for its object the attainment of skill in the application of the mechanism of formations, methods, movements and performance of habitual duties, and the inculcation and maintenance of correct discipline. The exercises and evolutions taught on the drill ground and practiced for the purpose of instilling discipline, control, and flexibility.

Drive. An attack on a large scale which is pushed forward with great force and violence. To push forward with all possible force.

Drop. The vertical distance at any point of the trajectory from the line of departure. That part of a ditch which is sunk deeper than the remainder. A small metal plate on a telephone switchboard which drops as a signal that a call has come in over a particular line. To remove from a roll or return, as of a soldier in desertion, or property expended.

Drop-block Breech Mechanism. See Sliding Wedge Breech Mechanism.

Dropped Message. A message dropped from an aircraft.

Drum. A hollow cylinder of wood or metal with skin stretched tightly over the openings, used as a musical instrument and for drum calls or signals. One of many mechanical devices having the shape of a drum or spool, such as a circular micrometer, a reel for wire or cable, etc.

Drum Major. Originally, the chief drummer of a regiment. Now, the leader who directs the march of a band or drum corps. In the United States Army he generally acts as first sergeant of the organization.

Drumhead Court-martial. A summary court-martial called to try offenses on the march or battlefield, when sometimes a drumhead was used as a table.

Drumming Out. The ceremony of ignominiously discharging a soldier from the service, in which the culprit is marched out of garrison at the point of the bayonet, the drummers playing the Rogues March.

Dual Control. The provision of a duplicate control lever and rudder bar whereby either of two people may pilot an airplane.

Dual-status Officers and Warrant Officers. Officers and warrant officers of the National Guard who are also appointed in the National Guard of the United States in corresponding grades and sections under authority of the National Defense Act.

Duckboards. Pieces of sectional board walk used in the bottom of wet and muddy trenches or other muddy places. Trench boards.

Dugout. A subterranean excavation for protection from shell fire. A cave shelter.

Dumdum Bullet. A soft-point or mushroom small arms bullet, which expands on impact and inflicts a dangerous wound; originally produced at Dumdum, India. Steel-jacketed bullets are dumdummed by cutting the nose of the jacket. The use of these bullets in civilized war is prohibited.

Dummies. Devices used in bayonet and saber practice. In general, any device made to resemble something else, used in training or to deceive the enemy.

Dummy Cartridge. A cartridge, designed for use in drills and exercises, or on the target range, which cannot be exploded. Practice dummies are made of a different shape and color so that they are readily distinguished from live cartridges. Range dummies resemble live cartridges.

Dummy Gun. A device made to resemble a gun in order to deceive the enemy.

Dummy Obstacles. Devices intended to deceive the enemy in regard to the accessory methods of defense of a position.

Dummy Position or Emplacement. A position or emplacement prepared to simulate an occupied position.

Dummy Trenches. Trenches intended to deceive the enemy in regard to the location of occupied trenches, or as to the way in which a position is actually manned.

Dump. A place where supplies are heaped or stored for distribution in the field. A place for the temporary disposal of supplies, lacking the storage facilities of a depot.

Dutch Oven. An oven used for camp cooking.

Duty. The exercise or performance of those functions that pertain to a soldier. Any ordered or prescribed service; yet with this nice distinction, however, that duty is considered the mounting of the guard, or the like, while there is not an enemy directly to be engaged; while operations against an enemy are called field service.

Duty by Roster. The practice under which the man longest off any particular duty becomes next in line for such duty.

Duty Roster. A list of officers or enlisted men by name that is kept in an organization for the purpose of recording the duty performed by each person, from which record their status as to availability for duty may be determined.

Duty Sergeant. In general, any sergeant of a company other than the first sergeant.

Duty, Status of. The particular duty to which an individual is assigned. Present and available for duty.

Dynamite. A commercial high explosive, used chiefly in mining and quarrying. It will continue in military use because of its availability in large quantities, though inferior in many respects to TNT.

Dynamite Gun. An air or steam gun which throws dynamite or other high explosive.

E

Eagle. A design in which the figure of an eagle is prominent, as in the coat of arms and military standards of various nations, e.g., France and ancient Rome.

Earthworks. A general term for intrenchments or fortifications constructed chiefly of earth.

Eccentric Projectile. A projectile in which the center of gravity is not at the geometrical center or on the long axis.

Eccentric Screw Breechblock. A type of breechblock which is not removed from the breech in loading. It is provided with a loading hole which is centered with the bore for loading and extracting, and moves eccentrically to another position for firing; used only with fixed ammunition.

Echelon. A formation in which the subdivisions are placed one behind another extending beyond and unmasking one another wholly or in part. In battle formations, the different fractions of a command in the direction of depth, to each of which a principal combat mission is assigned; e.g., the assault echelon, support echelon, and reserve echelon; also the various subdivisions of a headquarters, as, forward echelon, rear echelon. To give a special deflection to each of the guns of a battery so that the increments of deflection increase by successive and equal amounts. To form or dispose in echelon.

Echelon Trace. The line of a trench used when it is necessary to gain ground to the front or rear, or for a sharp change of direction with a minimum exposure to enfilade; commonly employed on a flank. A dog-leg trace.

Echelonment. The act of forming echelon.

Echelonment in Depth. Distribution or deployment in depth.

Echelonment of Supplies. The storage of supplies in depots echeloned from the front to the base in increasing amounts and ready to be used in case of any unexpected interruption to insure the continuous flow of supplies to the troops at the front.

Economic Blockade. A kind of blockade by which a country is subjected to the severance of all trade relations with other countries. The application of economic sanctions.

Economy of Ammunition. The act of saving ammunition by firing at a slower rate than the most effective rate.

Economy of Force. A principle of war which requires the use of the minimum force absolutely necessary to the accomplishment of any subordinate purpose or mission in order to conserve strength for the main effort.

Effective. One who is capable of active service. Fit for service. Present for duty. Producing a desired or decisive result. Efficient.

Effective Area of Burst. The area within which the fragments or explosive force of a shell, bomb or grenade will incapacitate personnel or destroy materiel.

Effective Beaten Zone. That part of the beaten zone containing an arbitrary percentage of the total impacts, as the 75% zone.

Effective Forces. All the efficient elements of an army that may be brought into battle. All the efficient troops of a nation that are available for service in the field.

Effective Range. The range at which, for different guns, effective results may be expected. It is often well inside the extreme range. See Ranges.

Effectiveness of Fire. The results obtained by fire.

Effectives. The number of men actually available for full duty.

Effects. The property of officers or soldiers, specifically of those deceased.

Efficiency. The condition of being capable or qualified, or the degree thereof.

Efficiency Report. A report, rendered at stated times, which forms the basis for the classification of an individual, both as to his efficiency in his own branch and his qualifications for special duties.

Efficiency Tests. Tests conducted by commanders to determine whether the units under their command have attained the prescribed standards of efficiency.

Egg Grenade. A hand grenade in the shape of an egg.

Ejector. A device used on breechloading small arms to throw out the empty cartridge case after firing. See also Extractor.

Elastic Defense. Defense in depth, which see.

Elasticity Effect. A resistance to the movement of a projectile due to elasticity of the air, which varies with the temperature of the air and the velocity of the projectile.

Elbow Rest. A berm provided in a trench on which riflemen rest their elbows when firing.

Electric Fuze or **Primer.** A device which detonates a demolition, bursting, or propelling charge by means of heat developed by an electric current.

Electrical Data Transmission System. A system by which firing data is electrically transmitted from a director to dials at the guns; used with Case III pointing. See also Pointers, Electrical and Mechanical.

Electrified Wire. A wire entanglement in which the wire is charged with electric current.

Element. One of the component subdivisions of a command, such as a trooper, file, squad, section, platoon, battery, or any unit forming part of a larger unit. A first principle of any art or science.

Elementary Gun Drill. Drill in mounting, dismounting, loading, and unloading the machine gun.

Elementary Tactics. Minor Tactics.

Elementary Unit of Organization. The smallest unit in any arm or service provided with a leader by the Tables of Organization.

Elements of Data. The various items included in firing data, especially for antiaircraft guns.

Elements of the Trajectory. The various features of or pertaining to the trajectory, with their nomenclature.

Elephant Iron. Arches of corrugated steel, used as linings for shelters.

Elevate. To increase the angle of elevation. To promote to higher rank.

Elevating Arc, Rack or **Segment.** A toothed or geared circular rack attached to a gun or carriage and actuated by a pinion or worm gear, by means of which the gun is elevated or depressed.

Elevating Band. A band around a gun near the breech to which the elevating mechanism is attached.

Elevating Bar. An iron bar used in elevating and depressing certain guns.

Elevating Handwheel. A handwheel by which a gun is rotated in elevation about its trunnions.

Elevating Mechanism. Mechanism for elevating or lowering the breech of certain guns.

Elevating Screw. A screw beneath the breech of a gun to raise or lower the breech.

Elevation. The vertical height of any point above mean sea level or other assumed datum plane or position. An angle of site or position. The angle between the axis of a piece when pointed, and the horizontal plane or plane of site. The range in yards, degrees, or mils, as set off on a sight or quadrant. See Angle of Elevation; Angle of Site.

Elevation Setter. See Range Setter.

Elevation Table. A table of ranges with the corresponding elevations. A range table.

Elevator. A movable auxiliary airfoil, the function of which is to impress a pitching movement on an aircraft; it is usually hinged to the stabilizer.

Embargo. An order prohibiting the arrival or departure of ships, the sale of certain commodities, or any common or usual activities. A prohibition or check.

Embarkation. The act of going or putting aboard a vessel for military reasons; applied to personnel.

Embarkation Camps. Assembly camps located at convenient distances from points of embarkation.

Embarkation Center. An establishment in the vicinity of a port of embarkation where troops are sheltered and cared for while awaiting embarkation.

Embattle. To arrange in order of battle. To prepare or arm for battle.

Embattled. Having indentations like a battlement. Disposed or armed for battle.

Embrasure. An opening in a wall or parapet, especially one through which a cannon is fired. The sides are called the cheeks, the interior opening is the throat, and the bottom is the sole. An embrasure may or may not be open at the top. See also Loophole.

Emergency Barrage. A standing barrage for which data have been prepared by artillery or machine guns for the purpose of reinforcing the normal barrage of some other unit, or covering a portion of the front not included in the normal barrage. A contingent or eventual barrage.

Emergency Case. A case carried by men in the medical department containing material for the emergency treatment of sick or wounded soldiers.

Emergency Flotation Gear. A device attached to a landplane to provide buoyancy in case of an emergency landing on the water.

Emergency Landing. A landing made as the result of necessity by reason of accident, or derangement, or loss of motive power. To bring an aircraft to the ground when its static condition has departed from one of equilibrium or when the power units fail.

Emergency Medical Tag. A tag attached to a wounded man on which is recorded his identification, the nature of his injury, and the treatment given him at each station.

Emergency Ration. A ration intended for use when other rations are not available.

Emergency Sector of Fire. That part of the field of fire of a gun or unit, on the flanks of the normal sector of fire, within which emergency missions may be assigned. See also Emergency Barrage.

Empennage. See Tail Group.

Emplace. To put in place or position, as a weapon. To provide with a prepared emplacement.

Emplacement. A prepared position from which a unit or weapon executes its fire mission. A space in a fortification allotted to a gun. See Firing Position.

Emplacement Officer. An officer who is in charge of one or more emplacements of coast defense guns.

En, or In Barbette. Said of guns so mounted or elevated as to fire over a parapet instead of through an embrasure.

Encamp. To make camp or go into camp. To place in a camp.

Encampment. A living place for soldiers in the field. A camp.

Enceinte. The line of works inclosing the whole area of a fortified place. The area or place inclosed within the main wall.

Encipher. To translate ordinary language into cipher. To prepare a message in cipher.

Encircling Force. A pursuing force which endeavors to reach the heads of retreating enemy columns and to cut their routes of retreat or engage them and bring them to a halt.

Encircling Maneuver. A maneuver which outflanks the adversary and threatens his flank, or seeks to reach the head of a retreating enemy force.

Encode. To translate ordinary language into code. To prepare a message in code.

Encounter. A meeting with hostile purposes. A battle. To meet in combat.

Ending of a Field Order. That part which contains the authentication or signature of the order, a list of appended documents, and the distribution.

Enemy. One with whom we are at war. A military foe.

Enemy Capabilities. See Capabilities of the Enemy.

Enemy's Front Line. The line occupied by the enemy's most advanced elements, beyond which patrols generally cannot penetrate, and which may include important detached posts. It may constitute the enemy's main line of resistance, the line of resistance of his covering forces, or a line of observation.

Enemy Order of Battle Map. A map or overlay on which all known facts concerning the dispositions of hostile forces are graphically recorded.

Enemy Situation Map. A map or overlay on which information, of one or more classes, concerning the enemy is graphically recorded.

Enfilade. A firing in the direction of the length of a trench, line or column of troops. A raking fire. To subject to enfilade fire.

Enfilading Battery. A battery in position to deliver enfilade fire.

Engage. A movement in fencing and bayonet exercise. To enter into conflict. To join battle. To open fire upon a target.

Engagement. An encounter between two hostile forces. A general action or battle.

Engineer Aviation Units. Includes separate aviation companies, battalions and regiments. Their function is to furnish the engineer service required at airdromes and air depots.

Engineer Combat Battalion. Specifically, the engineer component of the triangular infantry division.

Engineer Combat Regiment. An organization trained and equipped for engineer service with the combat troops. In the United States service it includes a headquarters and service company and two battalions, each of 3 companies.

Engineer Combat Squadron. A mounted organization for combat engineer service with a cavalry division.

Engineer Department. A bureau of the War Department which has charge of the corps of engineers of the army, and of various civil works, especially river and harbor improvements, carried out by the corps of engineers.

Engineer General Service Regiment. An organization equipped for general engineer service, and usually attached to a corps or an army.

Engineer Troops, Classification of. Engineer troops are organized, trained and equipped for combat, general and special services. Amongst the special services in the United States Army are included: bridge, railroad, topographic, highway, forestry, water supply, depot, shop and camouflage units. General engineer troops include combat battalions, regiments and squadrons, general service regiments, and separate (labor) battalions.

Engineering. One of the five great branches of the art of war, which includes fortification, military construction of all kinds (roads, railroads, bridges, etc.), and the design and construction of all weapons, machines, etc., utilized in warfare. The application of the engineering art to the conduct of war.

Engineers. That branch of the military service which conducts or supervises engineering operations incident to the conduct of war. The early history of the military engineer is the history of primeval warfare, when every man was a soldier, and every soldier, to some extent, was an engineer. Early efforts were directed mainly to providing protection to the individual and the construction of machines for the destruction of the enemy. In addition there was road and bridge building to provide communication with newly conquered provinces.

With the introduction of gunpowder came an era of strongly fortified cities and the principal duties of the engineers were the construction of fortifications and the conduct of siege operations. Scientific discoveries and developments of the succeeding centuries have brought responsibilities in an ever-increasing degree, so that there is scarcely a problem of civil, mechanical or electrical engineering that the military engineer may not be called upon to solve.

Warfare gave birth to the art of engineering. The earliest engineers were military engineers, they were the forerunners of the civilian engineers. In ancient times there was no distinct organization of engineers. The necessary technicians and laborers were drawn from the army at large, and from civil life, the entire force engaging in important engineering work when necessary.

There was a corps of topographical engineers in Alexander's army. There was a headquarters of engineers in the English army prior to 1350, which later developed into the Office of Ordnance. In the 16th century combat engineers were separated from technical engineers, and a corps of pioneers was organized. In 1757 military rank as engineers was adopted.

In the French army the Corps du Genie was formed by Vauban about 1690. The officers were detailed from other branches or from civilians. The title was conferred on officers in 1766, and on organizations of sappers and miners in 1801.

In the United States three companies of sappers and miners were organized and served during the Revolutionary War, but were mustered out at its close. A Corps of Artillerists and Engineers, organized in 1794, had a brief existence, but was discontinued in 1802, and the Corps of Engineers was formally established. It was first stationed at West Point, N. Y., where it organized the United States Military Academy, which it conducted until 1866. A Corps of Topographical Engineers, organized in 1838, was merged with the Corps of Engineers in 1863. For many years the military engineers were the best trained engineers in the country and their services were utilized on many important public works. In recent years these have included all river and harbor improvements, the construction and operation of the Panama Canal, etc. In addition to the construction of permanent fortifications the present duties of engineers in the United States service include the construction and maintenance of roads, railroads, bridges and waterways; surveying, mapping and map reproduction; field fortification; the construction and maintenance of structures of every kind (except for signal communications); the procurement, storage, and issue of engineer supplies, etc.

Enlist. To engage in military service by subscribing to articles and enrolling one's name as a soldier.

Enlisted Men. Noncommissioned officers and privates.

Enlisted Reserve Corps. Enlisted men who belong to the reserve corps.

Enlisted Specialists. Enlisted men who perform technical or specialized work; rated for pay in several classes according to the nature of the work performed.

Enlistment. Voluntary enrollment to serve as a soldier or sailor. The act of enlisting, or state of being enlisted. The period for which one agrees to serve as a soldier or sailor.

Enlistment Allowance. An allowance paid to every honorably discharged enlisted man who reenlists within three months from the date of his discharge.

Ensign. A flag flown on airships, ships, tenders, launches, and small boats. A junior officer of the navy corresponding in rank to a second lieutenant. Formerly, a commissioned officer who carried the ensign or colors of a company or regiment. A cavalry troop of the 16th century.

Entanglement. An obstacle. See Wire Entanglements.

Entering Edge. See Leading Edge.

Entrain, Entruck. To go aboard or put aboard a train or truck.

Entraining Point. A point designated for entraining.

Envelop. To attack in front and on one or both flanks at the same time. An earthwork in the form of a single parapet or small rampart.

Envelope. The outer covering of an aerostat, usually of fabric. The bag containing the aerostatic gas of a free balloon, kite balloon, or nonrigid airship. A case or container.

Enveloping Attack or **Envelopment.** An attack made upon the enemy's front and one or both flanks at the same time.

Epaule. The shoulder of a bastion or other work; the point where face and flank meet and form a shoulder angle.

Epaulet, or **Epaulette.** A shoulder ornament, usually fringed, on which is indicated the rank of the wearer.

Epaulement. An embankment, usually revetted, which protects a gun and its crew from frontal, oblique and enfilade fire. Any protective bank of earth. A breastwork.

Equalization of the Bounty. A practice in the early wars of the United States of paying a bounty to soldiers who remained in the service, equal to that paid to new men entering the service.

Equalizing Pipe. A pipe connecting two recoil cylinders to equalize pressures therein.

Equilibrator. A device, as a pneumatic cylinder, used to balance unequal weight in front and in rear of the trunnions of a gun.

Equip. To furnish with the necessary and prescribed equipment. To furnish for service. To fit out.

Equipage. Includes everything needful for troops on the march or operating in the field. The necessaries of troops.

Equipment. The complete service outfit of an individual or organization, including arms, clothing, instruments, utensils, and supplies of every nature required for service in the field. The authorized allowances of necessities. The act or process of equipping.

Equipment "A". The equipment prescribed for use in campaign, in field exercises, or on the march; it is limited to the actual necessities for the particular service.

Equipment "B". The equipment which, in addition to Equipment "A" is prescribed for the use of troops in posts, in mobilization, concentration and maneuver camps, etc.

Equipment, Engineer. Equipment of engineer troops is commonly made up in compact sets for the following purposes: Demolition, carpenter, map reproduction, intrenching, sketching, surveying, road and railroad construction, quarrying, shop, water supply and purification, lumbering, sign painting, drafting and duplicating, tinsmith, cobbler, blacksmith, wheelwright, horseshoer, saddler, photographic, pipe fitting, footbridge, lithographic, telephone, illuminating, veterinary, etc.

Equipment Fund. A fund, accumulated by regular deductions from their pay, to be used in the purchase of uniforms and equipment by cadets graduating from the United States Military Academy.

Equipment, Individual. Items of issue necessary to enable the individual to function as a soldier. See Infantry Equipment.

Equipment, Organizational. Items of issue necessary for an organization to function as such.

Equitation. The art of riding and managing horses.

Erosion. The gradual enlargement of the bore of a gun due to the action of the powder gases. The wearing away of the earth's surface due to the action of natural forces, especially rain and running water.

Error. In general, a variation from the truth, as in a measurement, resulting from instrumental or mechanical defects or limitations, or other conditions, as distinguished from a mistake or blunder. Specifically, in gunnery, the divergence of a point of impact from the center of impact, or any component of this divergence, as lateral, longitudinal or vertical.

Error Along the Trajectory. In antiaircraft fire, the magnitude of any error, measured parallel to the mean trajectory of a series of shots.

Error of Closure. See Adjusting a Traverse.

Error of the Moment. In the preparation of firing data, the aggregate of errors due to meteorological conditions.

Error of the Rifle. Any error inherent in the rifle due to fault of manufacture or assembly.

Errors and Corrections. Errors and corrections are opposite in sense; i.e., if an error is positive the compensating correction is negative.

Ersatz. Spare; substitute; reserve; exempted. (Ger.)

Escalade. The act of scaling the walls of a fortification by means of scaling ladders. To scale. To mount by means of ladders.

Escarpment. A steep artificial slope, especially in a fortification. A scarp.

Escort. A body of troops which attends or guards a person or party. An armed guard which accompanies a convoy, trains, prisoners, etc. An armed unit or detachment designated to guard other troops who are not able to defend themselves at close quarters, e.g., work troops, artillery in an exposed position, etc. To accompany or convoy.

Escort of Honor. A body of troops detailed to receive and escort personages of high rank.

Escort of the Color. The ceremony of sending for and receiving the color.

Escort Wagon. A wagon used in the United States Army for general freighting purposes.

Escutcheon. A shield on which the heraldic arms of a family are emblazoned.

Esprit de Corps. A spirit of common devotion, pride, honor, and interest, which binds the men belonging to an organization together.

Essayons. "Let us try," the motto of the Corps of Engineers, U. S. Army, derived from its French instructors of Revolutionary War times.

Essential Elements of Information. Information of the enemy, of the terrain not under our control, or of meteorological conditions in enemy territory, needed by a commander in order to make a sound decision, formulate a plan, conduct a maneuver, or avoid surprise.

Essentially Military Schools. Schools maintaining units of the Junior Division of the Reserve Officers Training Corps, and which are organized on a military basis.

Establishment. The materiel and authorized quota of officers and men in an army or unit thereof. Any military installation, such as a depot, hospital, railhead, etc.

Estimate of the Situation. A logical process of reasoning by which a commander considers all available data affecting the military situation and arrives at a decision as to a course of action, including the expression of his decision. The steps in the estimate of the situation are as follows: (1) The mission; (2) The opposing forces; (3) The enemy situation and probable intentions; (4) Own situation and courses of action open; (5) The decision.

Estimating Distance or **Range.** Judging, without actual measurement, the distance to an object.

Etappen. Lines of communication. (Ger.)

Etat Major. A staff. (Fr.)

Etat Major General. General staff. (Fr.)

Evacuate. To withdraw from a position or fortress. To clear personnel, animals, or materiel from a position. To free or clear a place. To vacate. To remove sick, wounded, or prisoners to the rear.

Evacuation. The movement or transportation of casualties to hospitals in the rear. The act of evacuating a place or position, including removal of troops, civilians, materiel and supplies.

Evacuation Hospitals. Hospitals established on railways, whenever possible, at a distance of from 6 to 16 miles from the battle front; they provide facilities for the care of all classes of casualties and serve as centers from which evacuation from the combat zone to the rear is carried out.

Evacuation Points. Places at which the sick and wounded of a division are transferred to evacuation units.

Evaluation of Information. The critical and systematic analysis of enemy information for the purpose of determining its probable accuracy, significance and importance; information subjected to this process becomes military intelligence.

Evening Gun. A gun fired every night at permanent posts and stations, about sunset, at which time the flag is lowered.

Eventual Barrage. An emergency or contingent barrage.

Evolution. Any movement designed to effect, by passing from one formation to another, a new arrangement of troops. A movement by which a command changes its position or formation. A maneuver. A progressive development.

Examining Post. A small detachment stationed at some convenient point in an outpost to examine strangers and receive bearers of flags of truce, etc.

Executive. An officer charged with the responsibility of supervising the work of the staff, in a command not provided with a general staff, and of executing the orders of the commander; generally, the second-in-command.

Exemption Boards. Local and district boards, appointed under a selective service act to examine such drafted men as may be ordered to report to them and determine their availability or otherwise for service.

Exercises. Drills, ceremonies, maneuvers, problems, etc., conducted for instructional and disciplinary purposes. See also Tactical Exercises.

Expedient. A stratagem in war.

Expedition. A small army or body of troops sent on a special mission, the success of which generally depends on rapid or unexpected movements.

Expeditionary Force. A force raised or maintained for foreign service.

Expendable Property. Property which is normally used up or consumed in service, and which may be dropped on a certificate of expenditure.

Expert Gunner. The highest grade of classification for men firing guns other than small arms or machine guns.

Expert Machine Gunner. The highest grade of classification for men firing the machine gun.

Expert Pistol Shot. The highest grade of classification for men firing the pistol. See also Distinguished Pistol Shot.

Expert Rifleman. The highest grade of classification for men firing the rifle. See also Distinguished Marksman.

Expiration of Service. The termination of a term of enlistment.

Exploitation. The action of utilizing information or of taking prompt advantage of any favorable conditions. The analysis, for strategical or tactical purposes, of information obtained from any source.

Exploitation Objective. The most advanced position occupied as a result of or immediately following a successful attack.

Exploitation of a Success. The act of extending or making decisive a success in combat. Injury or damage inflicted upon the enemy following a tactical success.

Explosive. Any mixture or chemical compound which, under the influence of heat or mechanical action, undergoes a sudden chemical change with the liberation of energy in the form of heat and the development of high gas pressure. See High and Low Explosives.

Explosive Bullets. Bullets containing small charges of powder which explode on impact.

Explosives, Military. Important military explosives include nitrocellulose powders and gunpowder as propelling charges; and TNT, amatol, picric acid and picrates, nitroglycerin, cordite, melinite, dynamite, fulminate, tetryl, etc., as bursting and demolition charges, and in caps and fuzes.

Express Bullet. A light explosive bullet of large caliber. A form of dumdum bullet having a hollow point.

Extend. To increase the front or depth, or both. In drill, to change from a mass formation to one with normal intervals and distances. To deploy.

Extended Distance, Interval. Any distance or interval greater than normal or usual; e.g., in line of skirmishers, ponton bridge with extended intervals, etc.

Extended Order. A formation in which the individuals or units are separated by intervals or distances, or both, greater than those in close order.

Extended Order Drill. A series of exercises appropriate to the battlefield employed to train the soldier in the formations and movements used in combat.

Exterior Ballistics. That branch of ballistics which concerns the movement of a projectile from the time it leaves the muzzle of a piece until it comes to rest.

Exterior Corridor. A passage immediately in rear of a scarp wall.

Exterior Crest. The intersection of the superior and exterior slopes of a parapet.

Exterior Defenses. Detached works of considerable strength.

Exterior Guards. Guards used to prevent surprise, to delay or impede the enemy, and otherwise to provide for the security of the main force.

Exterior Lines. A term used to signify that the lines or routes available to or used by a combatant are such that he is unable to concentrate at any point of contact as quickly as his adversary. To illustrate, if the line of battle be curved the combatant on the outside or convex side cannot concentrate as rapidly at any point as the combatant on the inside; the former is then said to be operating on exterior lines. See also Interior Lines.

Exterior Slope. The outer or most advanced slope of a parapet, from the exterior crest to the front. See also Superior Slope.

Extra Duty. Continuous special duty formerly performed by an enlisted man.

Extra Duty Pay. Additional pay formerly allowed enlisted men performing certain particular duties, superseded by pay based on specialist ratings.

Extractor. A device for withdrawing a cartridge case from the chamber of a weapon. See also Ejector.

Extreme Range. The extreme limit of range for any weapon. The limit of effective range may be well inside the extreme range.

Eyes Right (Left). A command at which the head and eyes are turned to the right (left), as in passing in review.

F

Face. A term of varied application in fortifications. That portion of a work forming one side of a salient angle. The front between two neighboring bastions or other salient works. One of the sides of a formation. To confront. To be in front of. To turn on the heels, as right face, left face, etc.

Face Cover. An earthen mask constructed in front of a wall of a fortification to protect it from the artillery fire of the enemy.

Facing Distance. The difference between the front of a man in ranks, including his interval, and his depth; it is 14 inches in the United States Army.

Facings. The prescribed movements of soldiers executed by turning on their heels. The lapels, cuffs and collars of military coats or inserts of colored cloth.

Factor of Safety. In transfer of fire, an increase in the area fired upon, to allow for inaccuracies. In determining minimum range, an increase of elevation to allow for dispersion. As applied to a structure, such as a rope, bridge or gun, the ratio between the safe or allowable load or stress and the load or stress which would cause rupture or collapse.

Fall Back. To retire or withdraw from a position previously occupied.

Fall Back Upon. To retire or withdraw to a stronger position in rear.

Fall In. To form in ranks or other formation, in preparation for drill, ceremonies, or other duties.

Fall Out. To break ranks or column.

False Attack. An attack or demonstration intended to deceive the enemy in regard to the place or object of the main attack. A feint made to divert the attention of the enemy from the point of the main attack.

False Muster. An incorrect return of the number of effective men and animals.

False Return. A wilful report intended to deceive a higher commander as to the actual effective force of a body of troops.

Farrier. A horseshoer. One trained in the care of animals.

Fascine. A long cylindrical bundle of brush or small wood bound tightly together and used in constructing revetments. To cover, protect, or strengthen with fascines.

Fascine Choker. A device used to compress the brush forming a fascine so that the whole may be bound together tightly.

Fascine Cradle. One of the supports used in making fascines, consisting of two stout stakes driven obliquely in the ground so as to cross each other, and firmly bound together at the point of crossing to support the fascine in making.

Fascine Revetment. A revetment made of fascines.

Fatigue. The labor of soldiers in the upkeep of a post, etc., distinct from instruction or training.

Fatigue Call. The call for forming for fatigue duty.

Fatigue Clothes. Outer clothing made of denim or other similar material worn by soldiers on fatigue duty.

Fatigue Duty. Duty required from soldiers aside from the use of arms or training in the duties of a soldier.

Fatigue Party. A detachment of soldiers on fatigue duty.

Favorable Target. A target suited to the characteristics of a particular weapon.

Federal. Belonging or pertaining to the government of a federation of states. A soldier in the Union Armies during the Civil War.

Federal Recognition. The acceptance by the War Department of an organized militia unit, as complying with the requirements of the National Defense Act. Upon being federally recognized the unit becomes a part of the National Guard, subject to a measure of federal control and entitled to a measure of federal support, as provided by law.

Feed Belt. A belt, with loops for holding cartridges, by means of which they are fed into an automatic weapon.

Feed System. See Feeding.

Feed-drum. A magazine of circular shape by which cartridges are supplied to certain machine guns.

Feeding. The operation of supplying cartridges to a firearm, as from a magazine or belt. The process of supplying gas or lubricating oil to an engine or other machine. The continuous bringing together of the carbons of an arc light, to compensate for the shortening due to their combustion. Gravity, vacuum or pressure feed, force feed, splash feed, belt feed, etc.

Feint. An attack or demonstration intended to deceive the enemy. A pretense. A stratagem. To make a feint.

Fence. The art of fencing. Sword play. Fencing. To fence.

Feodorov Automatic Rifle. The automatic rifle used in the Russian Army. It has a caliber of .264 inches, a belt or clip capacity of 25 cartridges, and weighs 9.92 pounds.

Fermeture. The mechanism closing the breech of a firearm.

Ferrule. A metal ring connecting the spearhead and the pike or lance bearing a flag or standard. A metal thimble on the bottom of a staff, pike, etc.

Ferry. A means of crossing a stream in boats. The locality where such a crossing is made. See Rope, Trail, and Flying Ferry.

Fiat Machine Gun. A machine gun made at the Fiat works in Italy and used in the Italian Army. It is made in both light and heavy models, and has a caliber of .256 inches.

Fid. A block of wood used in mounting heavy guns, and other mechanical maneuvers.

Field. A battlefield. Any area in which military training or military operations are conducted. Outdoors. Outside of a permanent post or station.

Field Ambulance. A light ambulance designed for use in the field.

Field Army. The largest unit in the U. S. organization, commonly called an army. It is a temporary or changeable organization, which includes a headquarters and special troops, two or more army corps, and if necessary one or more cavalry divisions.

Field Artillery. See Artillery, Field.

Field Artillery School. A special service school, now located at Fort Sill, Oklahoma, for the higher education of officers in matters pertaining to the employment of field artillery.

Field Bakery. A bakery designed for use in the field.

Field Carriage. A carriage adapted to the transport of light guns.

Field Clerks. Civilian clerks of the War Department; now replaced by warrant officers.

Field Desk. A small, portable desk, for use in the field.

Field Duty. Service, under orders, with troops operating against an enemy, actual or potential.

Field Equipage. Equipage of all kinds for field service.

Field Exercise. A tactical exercise in which a military situation is laid out and worked out on the ground, the troops and armament of one side being actually present, either wholly or in part, the troops of the opposing side being imaginary or outlined only, the conditions of actual war being simulated as closely as possible.

Field Fortification. The art of increasing the natural strength of a defensive position by works of an engineering nature designed to permit the fullest possible scope to the fire and movement of the defender, and to restrict to the greatest possible extent the movement and the effects of the fire of the attacker. Defensive works of a temporary nature used in the field by both the attack and defense. See also Organization of the Ground.

Field Fortification, Items of. The following operations and structures are the principal items of field fortification, each to be given such weight as the situation calls for: clearing the field of fire; trenches; emplacements for weapons; obstacles; command and observation posts; facilities for movement; shelters; camouflage.

Field Glass. A binocular telescope in compact form, often provided with range and mil scales on its reticle.

Field Gun. A small caliber gun for use in the field.

Field Hospital. The materiel and personnel for the surgical treatment of the wounded in the field in a certain area or organization. A hospital station.

Field Intrenchment. See Field Fortification; Intrenchment.

Field Kit. The clothing and articles of equipment intended for the personal comfort of the soldier, carried on field service.

Field Maneuver. A tactical exercise in which a military situation is laid and worked out on the ground, the troops and armament on both sides being actually present, either wholly or in part, and the conditions of actual war simulated as closely as possible.

Field Manuals. A series of pocket-size volumes, containing in convenient and condensed form for ready reference in the field the approved principles, doctrines, and methods governing the training and employment of the arms and services.

Field Marshal. The highest rank in the French and other armies, above that of general.

Field Message. A message sent in the field by field methods of communication.

Field Music. The music of an organization other than the band. Buglers or trumpeters.

Field Notes. Notes made in the field for future reference. Technical instructions issued for the use of troops in the field.

Field of Fire. See Fire, Field of.

Field Officer. An officer above the rank of captain and below that of brigadier general, e.g., major, lieutenant colonel, and colonel.

Field Order. An order issued to regulate the operative and tactical actions of troops and such strategical dispositions as are not covered by letters of instruction.

Field Order, Paragraphs of. See Paragraphs of a Field Order.

Field Order, Subdivisions of. A written field order includes the following parts: (a) Heading; official designation of the command; place, date and hour of issue; serial number; map reference. (b) Distribution of troops (when appropriate). (c) Body; information and instructions to the command, set forth in 5 paragraphs (see Field Order, Paragraphs of). (d) Ending; signature, authentication; list of annexes; distribution of the order.

Field Oven. A part of the field bakery equipment. An oven used in the field.

Field Piece. A piece of field artillery. A gun mounted on wheels for use with troops in the field. A light cannon for use in the field.

Field Range. A small, light range for use on marches and in the field.

Field Ration. The ration prescribed for all persons entitled to a ration in time of war or field service and wherever the ration savings privilege is suspended.

Field Redoubts. Works entirely closed by parapets and capable of all around fire.

Field Remount Depot. A depot charged with the reception, care, conditioning or reconditioning and issue of animal replacements for the field forces. One such depot is ordinarily established for each corps and army, and in the communications zone as required.

Field Service. Service performed by troops in the field.

Field Service Regulations. A set of regulations for the information and government of the Army of the United States in the theater of operations and as the basis of instruction of the combined arms and services for war.

Field Targets. Targets used for field firing. See Silhouette.

Field Telegraph. A portable telegraph quickly set up in the field.

Field Telephone. A light, portable telephone for use in the field.

Field Training. Practical training in the military art conducted in the field, and including field exercises and maneuvers under simulated war conditions.

Field Trains. Unit trains which transport the authorized allowance of baggage, rations and forage, in contradistinction to the trains which carry ammunition and other combat equipment. See Trains.

Field Transport. All kinds of land transportation used by an army.

Field Wire. A twisted pair of weatherproof conducting wires, used for telegraph and telephone lines.

Field Wireless. Portable wireless equipment for use in the field.

Fieldwork. Any temporary fortification constructed by troops in the field.

Fifth Column. A term applied to spies, saboteurs, and others who furnish information to the country that employs them, and who strive in various ways to undermine the morale of the people of another country.

Fifty Per Cent Zone. The zone extending one probable error on each side of the center of impact (or burst) within which it is expected that fifty per cent of the projectiles fired will fall. On a fire distribution target, a horizontal zone within which fifty per cent of all shots will strike. See Beaten Zone; Dispersion Diagram; Zone of Dispersion.

Figure of Merit. The figure denoting the efficiency of a unit; arrived at by prescribed methods.

File. Two men, a front rank man and the corresponding rear rank man. A single man in ranks. A number of men one behind the other. A column of files. An orderly arrangement of letters or other documents for convenience of reference. To form in files. ¡To march in file. To place on file, as a document.

File Leader. The front rank man of a file or column of files.

Filipino Ration. The ration prescribed for use in time of peace by the Philippine Scouts.

Fill In. To fill in details after establishing the control of a sketch or map.

Fin. A fixed or adjustable airfoil, attached to an aircraft approximately parallel to the plane of symmetry, to afford directional stability; for example, tail fin, skid fin, etc. One of the thin pieces of metal on the outside of a gun barrel, or the cylinder of a motor, for air cooling.

Final Protective Line. For machine gun fire, a predetermined line, parallel or oblique to the front of a position, on which, in order to check an assault, it is intended to place a grazing fire, fixed as to direction and elevation, having a continuous danger space, and capable of delivery under all conditions of visibility. An extended front is thus defended by interlocking the final protective lines of a number of guns to form a more or less continuous belt of fire. See Fire, Band of; Belt of.

Final Statement. A full statement of a soldier's accounts furnished him on discharge from the service, and on which final payment is made.

Final Velocity. The velocity of a projectile at the instant of impact.

Finance Department. That branch of the service which is charged with the handling and disbursement of the funds appropriated for the support of the military establishment.

Fine. A pecuniary penalty imposed by a court-martial.

Fine Sight. An aim in which only the top of the front sight is seen through the notch or aperture in the rear sight.

Finger Width. The angle of 50 mils subtended by a finger held at a certain distance, about 15 inches, from the eye, used as a substitute for the sight leaf in target designation.

Fire. The discharge of a firearm or the collective discharges of a number of firearms. To discharge a firearm.

Fire, Accompanying. That portion of supporting fire which advances from time to time, keeping pace with the progress of the attacking troops. Accompanying fire may also be employed in withdrawal and retreat.

Fire Action. Artillery or small arms fire considered as an element of attack or defense.

Fire, Accuracy of. The relative effectiveness of fire. It is determined by dispersion and is measured by the closeness of the grouping of the points of impact about their center of impact.

Fire Adjustment Board. A plotting board used to determine quickly the stripped center of impact, the deviations of which can be applied to a percentage corrector.

Fire, Adjustment of. Bringing the center of impact of the fire on or within effective distance of the target. The process of applying the necessary corrections, based on observation or spotting, to bring the center of impact on or very near the target.

Fire, Aerial. Fire delivered from an aircraft. It is distinct from antiaircraft fire, which is directed from the ground against aircraft.

Fire, All-around. Fire which is, or may be, delivered in any direction from a single position.

Fire and Movement. The usual procedure in advancing the infantry attack in which the fire of its own and supporting weapons is used to gain fire superiority, and the movement of its own and supporting units is used to advance the assaulting and fire elements to successive positions from which fire can be successfully delivered and maintained on the hostile position. Fire superiority permits movement, and the advance culminates in an assault on the hostile position, unless the enemy sooner retreats.

Fire, Antiaircraft. Fire against aerial targets.

Fire, Application of. A very general term meaning the procedure involved in placing fire of any kind upon desired targets. It includes the selection of proper targets, range estimation, target designation, fire distribution, fire control and fire discipline. See also Fires, Tactical.

Fire, Assault. Fire delivered from the standing position, the firer advancing after each shot, which is aimed and fired from the halt; automatic riflemen fire from the hip. Marching fire.

Fire at Will. Fire in which, within the restrictions of the command for firing, the individuals deliver their fire independently of the commander and of each other, the usual method in battle.

Fire, Automatic. Fire delivered on the automatic principle, either continuously or in bursts. See Automatic Weapons.

Fire, Axial, Conduct of. Conduct of fire where the observer is on or near the line gun-target.

Fire, Axial-lateral, Conduct of. Combined conduct of fire where one observer is on or near the line gun-target and one is displaced more than 300 mils.

Fire, Band of. Fire, grazing through part or all of its trajectory (having a practically continuous danger space) usually by an automatic weapon that produces a cone of fire so dense as to make it probable that a man attempting to cross the line of fire at full speed will be hit. A final protective line.

Fire, Barrage. Fire having for its purpose the placing of a curtain or barrier of fire, executed on predetermined firing data, across the probable course of enemy troops or aircraft. A barrier or curtain of fire. The firing of any barrage. See Barrage.

Fire, Base of. The fire weapons which directly support the attack of an infantry unit and are under the control of the infantry commander.

Fire, Battery. Fire delivered by a battery as a unit. A fire in which the guns are fired successively at certain intervals.

Fire Bays. The salient portions of a fire trench, between traverses or re-entrants, from which the greater portion of the rifle fire is delivered.

Fire, Belt of. A barrier of fire produced by a combination of two or more interlocking bands of fire, having no gaps of sufficient width to endanger the tactical integrity of the defense. A continuous danger zone covering a certain front and created by various methods of interlocking beaten zones and bands of machine gun fire.

Fire, Bilateral, Conduct of. Combined conduct of artillery fire with two lateral observers, one on each side of the line gun-target.

Fire, Bracketing Method of Adjustment of. A method of inclosing a target in a bracket of any desired length, without determining the magnitude of deviations, by arbitrary increases or decreases of range, etc. The target is bracketed for range when one short and one over have been obtained.

Fire, Burst of. A series of shots fired by one pressure on the trigger mechanism of a machine gun or other automatic weapon. An intensive application of fire for a limited time.

Fire, Calibration. Preparatory fire having for its purpose the determination of the separate corrections to be applied to the individual guns of a battery in order to cause all the guns to hit a certain point if laid on that point.

Fire Call. A signal for men to fall in, without arms, in case of fire.

Fire, Classes of, Machine Gun. (1) With respect to the gun: (a) fixed, (b) traversing, (c) searching; (2) With respect to the target: (a) frontal, (b) oblique, (c) enfilade, (d) flanking, (e) reverse; (3) With respect to the ground: (a) plunging, (b) grazing.

Fire, Classification of. Fire has a great number of classifications. See all definitions under Fire.

Fire, Clip. The contents of one clip of cartridges fired at will.

Fire, Collective. The combined fire of a group of riflemen, automatic riflemen and machine gunners. The fire of a number of such weapons combined for a definite purpose.

Fire, Combined, Conduct of. Conduct of fire where there are two or more observers placed so that their observing lines intersect at an appreciable angle, as an axial and flank observer.

Fire, Combined Sweeping and Searching. Fire distributed both laterally and in depth, by successive changes in direction and elevation of the gun; as against oblique targets.

Fire Command. A tactical unit within a fort command in seacoast artillery, which consists of two or more battery commands, the additional fire control station and accessories, and the personnel. An order for opening fire, including the necessary firing data. A fire order.

Fire Commander. The senior officer for duty with a fire command.

Fire, Concentrated. Fire of a number of pieces directed upon a single point or a limited area for a limited time.

Fire Concentrations. See Concentrations; Fire, Concentrated.

Fire, Conduct of. The technical procedure involved in placing fire of the desired nature upon a designated target, including fire for adjustment and fire for effect.

Fire, Cone of. See Fire, Sheaf of.

Fire, Continuous. Fire conducted at the normal rate, without interruptions for the application of adjustment corrections or for other causes. A succession of salvos with the pieces fired consecutively at regular intervals. Fire at a particular aerial target in which the maximum rate is maintained by all guns until the target is destroyed or is out of range.

Fire, Continuously Pointed. Antiaircraft fire in which the fire-control devices are directed on the target and the data vary continuously with the position of the target.

Fire Control. The selection of targets and regulation of fire by fire unit commanders so as to bring the most effective fire, of the proper kinds and amounts, to bear upon the designated targets at the proper times. The exercise, by the commander of one or more fire units, of all technical and tactical functions necessary to apply fire effectively. Fire control includes conduct of fire and fire direction.

Fire Control Diagram. A diagram showing the fire missions and duties of all units belonging to any coast defense command.

Fire Control Installation. The material used in the fire control, fire direction, and position-finding service of any fire unit.

Fire Control Map. An accurate large scale map, showing the terrain in great detail, on which may be shown the enemy's works. A map designed for and used in control of fire.

Fire Control System. The appliances and materiel which are intended for use in controlling the fire of any coast defense command.

Fire Control Tables. Tables of ballistic data based on standard conditions.

Fire, Converging. Fire from different directions brought to bear upon a single point or area.

Fire, Coordinated. A planned fire in which every important target or area is covered by the fire of an appropriate weapon, which includes rifles and machine guns as well as artillery.

Fire, Counterbattery. Fire on enemy artillery. The mission of destruction or neutralization of hostile batteries, usually assigned to the artillery of the corps.

Fire, Covering. Any fire intended for the immediate protection of combat troops, or to facilitate their movements.

Fire, Creeping Method of Adjustment of. A method of fire adjustment used in cases where, because of difficulties of observation or presence of friendly troops, impacts must approach the target from one direction only. The magnitude of deviations must be estimated, and for each successive shot the distance from the target of the previous point of impact is halved.

Fire Crest. The inner upper edge of the parapet of a fire trench from which rifle fire is delivered.

Fire, Cross. Fire delivered from different positions in which the trajectories cross on or near the target.

Fire, Curved. Fire with curved trajectories, especially that of howitzers and mortars, which can search reverse slopes and reach areas defiladed from flat trajectory fires. Specifically, fire with elevations between 360 mils and the elevation for maximum range. See also Fire, Flat Trajectory; Fire, High Angle.

Fire, Deliberate. A slow, aimed fire. Fire which is conducted at a rate intentionally less than the normal rate of fire of the battery, in order that adjustment corrections may be applied between series, or for tactical reasons.

Fire, Deliberate Preparation of. The determination of basic data for artillery fire from maps or firing charts and surveys, and the application of accurate corrections for conditions not standard.

Fire, Demolition. Artillery fire, usually precision fire, delivered for the express purpose of destroying hostile works or establishments.

Fire, Density of. The number of shots falling on a particular target or area in a given time. The total volume of fire, depending on the number of guns and their rates of fire.

Fire, Destruction. Fire intended to cause the destruction of the target attacked. Fire concentration on a materiel object, which it is desired to damage physically to such an extent that it is neutralized or rendered useless.

Fire, Direct. Fire in which the location of the target is visible to the gunner. Fire conducted with direct pointing or laying.

Fire, Direct Overhead. Fire delivered over the heads of friendly troops when the target and the point to which troops can advance with safety are always visible to the gunner.

Fire Direction. The tactical control exercised by the commander of one or more fire units for the purpose of applying their fire to the desired targets at the appropriate times. The tactical measures involved in the application of fire, including the plan of artillery support, assignment of objectives, allocation of ammunition, etc.

Fire Direction Chart. A gridded sheet on which are plotted accurately the battalion base point, the directing gun of each battery, the battalion observation post, the base line of each battery, angular deflection scales for the battalion observation posts and for each battery, and all check and reference points; used in directing fire on targets of opportunity.

Fire, Direction of. According to the direction, relative to the target, from which it comes, fire is classed as frontal, oblique, flanking, enfilade, reverse, converging.

Fire Director. An instrument used in directing the fire of automatic anti-aircraft weapons.

Fire Discipline. That condition resulting from training and practice which insures an orderly and efficient conduct of the personnel in the delivery of fire.

Fire, Dispersion of. See Dispersion.

Fire, Distributed. Fire which is distributed along a linear target or over an area.

Fire Distribution. The application of fire to the whole of a linear target or area. The manner in which the fire is applied to such a target.

Fire, Drum. A continuous, heavy fire of artillery or machine guns, so-called because of the resemblance in sound to the roll of a drum.

Fire Effect. The effect of fire upon the enemy's personnel and materiel.

Fire, Effective. Fire which produces results of the desired nature.

Fire, Effectiveness of. The results obtained by fire.

Fire, Employment of. The tactical use of fire. See Fire Direction; Fire, Tactical.

Fire, Enfilade or Enfilading. Fire whose direction is parallel to the long dimension of a target, as a line of trenches or a line or column of troops. Raking fire.

Fire, Field of. The area in the direction of the enemy which can be effectively covered by the fire of a rifle unit, gun, or group of guns, from a given position. That portion of the terrain or water area covered by the fire of a gun, battery, or other unit. See Fire, Zone of.

Fire Fight. The struggle for fire superiority, a phase of battle.

Fire, Final Protective. A belt or curtain of fire, or a combination of interlocking belts and curtains placed along the final protective line to check an assault. See Final Protective Line; Fire, Belt of.

Fire, Fixed. Fire directed at a single point, without traversing or searching.

Fire, Flanking. Fire coming from a flank or directed in prolongation of a flank. Fire used to sweep along the front of a defensive line and thus enfilade the assailants as they approach the position.

Fire, Flat Trajectory. Artillery fire delivered at elevations not greater than 360 mils, usually by guns of relatively high muzzle velocity. See also Fire, Curved.

Fire for Adjustment. Fire delivered primarily for the purpose of correcting, by observation, inaccuracies in the firing data.

Fire for Destruction or Demolition. Accurately adjusted fire delivered with the object of destroying enemy works or materiel.

Fire for Effect. Fire delivered for the purpose of neutralizing or destroying a target, or the accomplishment of the tactical effect sought. Any fire against a hostile target, other than for registration.

Fire for Registration. See Fire, Registration of.

Fire, Frontal. Fire delivered approximately at right angles to the front of the enemy's line, or other linear target.

Fire, Grazing. Fire which is approximately parallel to the surface of the ground and does not rise higher above it than the height of a man standing. Fire with a long or continuous danger space.

Fire, Harassing. Fire delivered for the purpose of annoying or disturbing the rest of the enemy, causing casualties, curtailing movement, and lowering his morale; it is usually delivered upon camps, billets, shelters, front lines, assembly places, roads, working parties, etc.

Fire, High-Angle. Fire delivered at an elevation greater than the elevation for the maximum range. Fire with low velocity cannon at angles of elevation exceeding 45 degrees. Fire in which the ranges decrease with an increase in angles of elevation. See also Fire, Curved.

Fire, Horizontal. Fire delivered at low angles of elevation. Flat trajectory fire.

Fire, Improvement. Fire conducted any time after initial preparation of data, for the purpose of determining or verifying corrections necessary to place the center of impact on or very close to a target. All fire that can be observed is used to improve data, as far as practicable.

Fire, Indirect. Fire with indirect pointing or laying.

Fire, Individual. Fire opened without orders from a fire leader. Fire delivered at will by individuals.

Fire, Interdiction. Fire delivered for the purpose of preventing the use of certain routes or areas by the enemy.

Fire, Intermittent or **Interrupted.** Fire at a particular target in short series of bursts, the maximum rate in each series being maintained by all guns. Fire which is not continuously maintained.

Fire, Lateral Conduct of. Conduct of fire when the observer is displaced from the line of fire more than 100 mils.

Fire, Leading. Fire delivered to hit a moving target by aiming ahead of the target.

Fire, Lifting. The operation of shifting fire from one target to another at greater range. The successive jumps of a jumping barrage.

Fire, Limits of. Lines marking the limits of areas on which fire is to be delivered.

Fire, Line of. The trace or projection of the trajectory or line of site on the ground.

Fire, Low-Angle. Fire in which ranges increase with increases in angles of elevation. Fire delivered at angles of elevation at and below that required for maximum range.

Fire, Magazine. Fire in which the piece is reloaded from the magazine.

Fire Mission. The targets assigned to each fire unit, and the instructions as to the time and method of delivering fire on them in accordance with a fire plan.

Fire, Mixed. A salvo or volley including unequal numbers of shots sensed over and short.

Fire, Neutralization. Fire with the object of causing severe losses and destroying the combat efficiency of enemy personnel or weapons; established by a heavy burst of fire, and maintained by fire at intervals.

Fire, Normal Sector of. That part of the possible field of fire of a gun or unit within which it is expected, through tactical considerations, to employ its fire.

Fire, Oblique. Fire delivered at an oblique angle to the front of the enemy's line or other lineal target.

Fire, Oblique Reverse. Fire inclined to the front of the target and coming from the rear.

Fire, Oblique Traversing. Fire which is a combination of traversing and searching fire. Traversing fire along a line or target which is oblique to the direction of fire.

Fire, Observation of. Viewing the bursts or impacts of projectiles in order to determine their location with respect to the target.

Fire, Observed. Fire in which the firing data are computed for initial firing, after which the corrections are based on observation of fire.

Fire of Position. Fire, executed by infantry riflemen in position over the heads or to the flanks of attacking troops to assist the latter in advancing against the enemy.

Fire on Targets of Opportunity. Fire on targets located subsequent to the time the schedule fire is prepared, or appearing during the course of an engagement.

Fire, Open. To commence firing or a bombardment.

Fire Order. An order conveying the will of the leader to the gunner or unit. A fire command.

Fire, Order of. The manner in which the guns of a battery fire. The order in which individuals will fire on the range.

Fire, Organized. A prearranged fire which can be placed mechanically without observation if necessary; when practicable it should be observed.

Fire, Overhead. Fire that is delivered over the heads of friendly troops.

Fire, Percussion. The firing of high explosive or chemical shell equipped with percussion fuzes, or combination fuzes in which the time element is not employed, the projectiles bursting on impact.

Fire, Percussion Bracket. Bracket adjustment and zone fire for effect with ammunition fuzed to burst on impact.

Fire Plan. The prearranged plan of the unit artillery, as a whole. The arrangements made by a higher commander to secure cooperation in the fire of the various arms and units in defense or attack. The general scheme of fire support, indicating the general missions of all units.

Fire, Plane of. The vertical plane containing the axis of the bore when the piece is aimed.

Fire, Plunging. Fire in which the angle of impact is relatively large so that the danger zone is virtually limited to the beaten zone.

Fire, Point-blank. Fire delivered at very close ranges with no elevation of the sights.

Fire Positions, Primary, Alternate, Supplementary, Initial, Subsequent. These terms refer to the firing positions of supporting weapons. A primary position is one from which the weapon can execute its assigned mission. An alternate position is one to which the weapon may move from its primary position and perform its primary mission. A supplementary position is one from which the weapon can execute a mission not possible from the primary position. Initial and subsequent positions refer to the forward or retrograde movements of the supporting weapons in combat.

Fire Power. The volume, accuracy, distribution, and control of fire, considered together. The potential capacity of an organization for delivering fire.

Fire, Prearranged. Fire for which the need has been foreseen and for which data have been prepared in advance; it includes such fires as counterpreparation, concentrations, rolling barrages, and, in general, any scheduled fire.

Fire, Precision. A method of conduct of fire whose object is to place the center of impact at the target. Very accurate fire.

Fire, Preparation. See Artillery Preparation.

Fire, Preparation of. The technical determination of the data necessary for opening fire on a given objective, and to change from fire for registration to fire for effect. Preparation of fire may be either deliberate or rapid.

Fire, Preparatory. Fire delivered prior to an attack with the object of preventing enemy activity. Fire for the purpose of determining or verifying corrections to firing data, prior to conducting fire for effect.

Fire, Probability of. The probable results of firing, based on the laws of chance or theory of probability.

Fire Problems. Problems in combat practice firing under simulated battle conditions, whether by small arms or artillery.

Fire, Progressive. Any accompanying fire which precedes and conforms to the infantry advance.

Fire, Protective. Fire delivered in front of friendly troops with the object of denying ground to the enemy, or of neutralizing any enemy occupying such ground.

Fire, Purpose of. The object to be attained by fire; e.g., adjustment, counter-battery, destruction, effect, interdiction, neutralization, etc.

Fire, Quick. The class of fire employed in instruction and record fire when bobbing targets are specified.

Fire Raft. A raft laden with combustibles, used for setting fire to enemy ships or river and harbor defenses.

Fire, Raking. Enfilade fire.

Fire, Ranging. Fire used in determining or verifying the range.

Fire, Rapid. Fire in which a short time is set for completing a score. The fire used at the decisive moment of the battle. See also Rapid Fire Gun.

Fire, Rapidity of. The rapidity with which guns can be or are being fired. The number of shots fired, or to be fired per minute. Rate of fire.

Fire, Rate of. The average number of shots fired per minute during any particular firing.

Fire, Record of. An adjustment on an invisible target made of record by an adjustment on a check point.

Fire, Registration of. The adjustment of artillery fire on a number of selected registration or check points throughout the objective zone, in order that these points may later serve as transfer points for targets in their vicinities.

Fire, Retaliation. Fire delivered in retaliation for enemy harassing fire.

Fire, Reverse. Fire delivered so as to strike troops or lines of defense from the rear.

Fire, Ricochet. Any fire in which the projectiles bound along the ground; often delivered by the use of small charges and low angles of elevation.

Fire, Salvo. A single discharge of each gun of a fire unit, fired in order from either flank. A fire in which the pieces of a battery are fired at the command of their chiefs of sections in order from either flank at prescribed intervals.

Fire, Schedule or **Scheduled.** Fire delivered according to a plan, at designated times, or on call or signal from supported troops.

Fire, Scissors Sweeping. A method of sweeping fire by machine guns in which the front is covered by the superimposed fire of two pieces traversing in opposite directions, each covering the entire target.

Fire, Searching. A fire which is distributed in the direction of depth by successive changes in the elevation of the gun. Zone fire.

Fire, Searching and Traversing. Fire which is both searching and traversing at the same time.

Fire, Sector of. The area to be covered by the fire of a gun or unit.

Fire, Semi-automatic. Fire delivered on the automatic principle, one shot at each pressure of the trigger, each shot separately aimed.

Fire, Sheaf of. The cone formed by the trajectories of a number of projectiles fired as nearly as possible under the same conditions. The planes of fire of two or more pieces of a battery, considered as a group. Sheafs may be parallel, narrow, wide or open, converged, crossed, which see.

Fire, Shift of. Specifically, a change of fire from one objective to another by methods more rapid and less accurate than those used in a transfer of fire. In a more general sense, any change of fire from one objective to another.

Fire, Shifting. The attack of a target of wide front by successive open sheafs.

Fire, Slow. That class of fire in which no time limit is set for completing a score.

Fire Step. A berm on which the riflemen stand while firing from a trench.

Fire Superiority. A fire which is superior to that of the enemy in volume or accuracy or both, and whose effect is to render the enemy's fire less effective. Fire superiority is relative and is a moral phenomenon although largely dependent on physical effect. Fire superiority for the attacker is implied if he is able to advance against the defender without ruinous losses. Fire superiority for the defender is implied if he is able to hold his ground and check the attacker's advance.

Fire, Supporting. Artillery or machine gun fire delivered while the supported troops are engaged, either in attack or defense, and after the launching of the attack, as distinguished from preparation and counterpreparation fires. Barrages or concentrations for the immediate support or assistance of an infantry unit in attack or defense. See also Fire, Covering.

Fire, Surprise. Fire of opportunity. Fire delivered with great rapidity upon transient targets and which comes in the nature of a surprise.

Fire, Sweeping. Fire, especially of an automatic weapon, in which the weapon is rotated from side to side on its vertical axis, without change of range, so as to cover a considerable front with its fire. See Fire, Traversing.

Fire, Tactical. Fire executed for a definite tactical purpose. Tactical artillery fires are classified as follows: (1) As to effect, destruction and neutralization; (2) As to form, concentrations and barrages; (3) As to prearrangement, schedule fires and fire on targets of opportunity; (4) As to tactical purpose, supporting fire, preparation, counterpreparation, counterbattery, interdiction, harassing and retaliation.

Fire, Tactical Employment of. The application of the proper kind of fire in various situations.

Fire, Time. Fire with projectiles, either shell or shrapnel, equipped with time fuzes set to give bursts in the air before the projectile strikes. Fire in which the projectiles burst or are intended to burst before impact. Fire with ammunition giving air bursts.

Fire, Time Bracket. Bracket adjustment and fire for effect using ammunition equipped with a time fuze, intended to obtain effect with air bursts.

Fire, Transfer of. A shift of fire from one objective to another, or from a check or registration point to a target in the vicinity. The corrections for the check point or first objective are applied proportionally to the data for the new target. See K-transfer; VE-transfer; Map Transfer; Record Transfer; Air Photo Transfer; High-burst Transfer.

Fire, Traversing. Fire in which the piece is traversed a specific amount after each round. Fire used when a target has little depth and considerable width. Fire distributed laterally.

Fire Trenches. Trenches designed primarily to provide cover for personnel when delivering rifle fire.

Fire, Trial. Fire which has for its object the placing of the target within the zone or volume of dispersion. The first phase of adjustment fire at fixed targets, and the only phase of adjustment fire at moving targets.

Fire, Unaimed. Fire which strikes an object to the rear of the one it was aimed at. Fire delivered without the piece being aimed at a target.

Fire Unit. A unit whose fire is controlled by one commander; in the infantry and cavalry fire units are the rifle squad; in the artillery the fire unit is the battery.

Fire, Unit of. The average amount of ammunition, bombs, grenades, and pyrotechnics which a designated organization or weapon may be expected to expend in one day of combat. See also Day of Fire.

Fire, Unobserved. Fire which is not, or cannot be, observed and which, in general, covers a large area.

Fire, Verification. Preparatory fire having for its purpose the test of the mechanical adjustment of all guns and fire-control equipment of the battery and of the accuracy of the corrections determined as a result of calibration and trial fire.

Fire, Vertical. Fire at very great angles of elevation, usually applied to fire at aerial targets.

Fire, Volley. The delivery of fire by volleys. In artillery fire, each piece included in the command fires the specified number of rounds without regard to the other pieces and as rapidly as is consistent with accuracy.

Fire, Volume of. The amount of fire that can be or is being delivered.

Fire with Counted Cartridges. A method of controlling fire, formerly employed, by causing each rifleman to count out a certain number of cartridges, to fire at will and cease firing when the indicated number of rounds has been fired.

Fire, Zone. Fire delivered at successive ranges to cover a target or area having considerable depth. Fire for effect, searching a bracket or area. Fire with successively increased propelling charges between the minimum and the maximum elevation of a piece, giving a series of overlapping zones.

Fire, Zone of. The area within which the fire of an artillery unit is or may be delivered. In a broad sense, the objective zone is a circle whose center is the position of the unit, and whose radius is the extreme effective range of the pieces concerned. Actually, the zone of fire is usually limited laterally by the zone of action or sector of the division, corps or army of which the artillery unit is a part; and in cases of direct support it may be limited to the front of the subordinate unit thus supported. The zone of fire may be divided into a normal zone and, if necessary, one or more contingent zones, according to the missions assigned by the fire plan. See Fire, Zone of, Normal and Contingent; Zone, Objective.

Fire, Zone of, Contingent. Any area, within its objective zone and outside its normal zone, within which the fire of an artillery unit may be delivered in certain contingencies or emergencies. There may be one or more contingent zones, not necessarily contiguous with the normal zone or with each other. The contingent zone of any unit may include the whole or a portion of the normal zone of another unit. Contingent zones are not exclusive. The assignment of contingent zones makes possible the concentration of fire where needed.

Fire, Zone of, Effective. The area of ground beaten by the best 75% of the shots fired.

Fire, Zone of, Normal. That area, within its objective zone, in which the fire of an artillery unit is ordinarily delivered, in accordance with its mission. A unit is assigned but one normal zone, and may or may not have also one or more contingent zones. Normal zones are not exclusive. The assignment of normal zones insures the application of fire to all important objectives. See also Fire, Zone of, Contingent.

Firearms. Weapons used for the propulsion of projectiles by the force of gunpowder or other explosive; they include cannon and small arms. There is no trace of the use of firearms prior to 1300 A.D. Although Roger Bacon is given credit for the composition of gunpowder in 1248, it was Bernard Schwartz, a German monk, who, in 1313, first applied it to the propulsion of projectiles. However, the scarcity and great cost of iron, bronze, and the ingredients of gunpowder greatly retarded the development of firearms. The earliest guns, made in 1313, were pot-shaped vessels, from which cannon were evolved. The hand-gun was derived from the cannon late in the 14th century. The matchlock, invented about this same time, was the typical weapon of the individual soldier for two centuries. The invention of the wheellock in 1515 was a distinct step forward, although it was too uncertain in action and too expensive for general use. The flintlock, invented early in the 17th century, was open to the same objection, but by 1700 it was a satisfactory weapon and continued to be the principal weapon of the infantry soldier until about 1840.

Cannon, which had become very large by the end of the 14th century, were gradually reduced in size as the efficiency of individual weapons increased. The 17th century witnessed the invention of the elevating screw in 1650 by a Jesuit, and the elongated projectile in 1662 by the Bishop of Munster. Little further development in firearms occurred until 1800 when a rifled musket was adopted in the British Army. The invention of the percussion lock in 1807 by Forsythe, a Scottish minister, simplified firing mechanisms and by 1842 all the leading powers had adopted a percussion lock, rifled small arm. This was superseded by breechloading rifles during the period 1841-1872, which, in turn, were replaced by magazine rifles during the period 1884-1903. The pistol, revolver, and machine gun, in workable types, were developments of the 19th century. There was little change in cannon during the 17th, 18th and first half of the 19th centuries. In 1854 elongated projectiles and rifled cannon began to be adopted and from these the effective weapons of the present day were evolved. See Artillery; Flintlock; Grenade; Gun; Gunpowder; Harquebus; Machine Gun; Matchlock; Musket; Pistol; Revolver; Rifles, Military; Wheellock.

Fire-order Problems. Theoretical problems designed to teach the preparation and interpretation of fire orders, and the application of technical and tactical principles to the delivery of fire in typical combat situations.

Firelock. A musket discharged by any device producing sparks by friction or concussion, formerly used. See Flintlock.

Fireman's Carry. A method by which one man carries on his shoulder a wounded man who is unable to stand or walk.

Firing Angle. The horizontal angle between the plane of fire and the line of sighting. The horizontal angle between the target and the aiming point, whose vertex is at the piece.

Firing Azimuth. The azimuth at which the gun is laid in order to hit the target in its future position.

Firing Battery. That part of a gun battery so-designated by the Tables of Organization. It usually consists of the four gun sections and one ammunition section. That part of the battery normally at the position of the pieces when the battery is in position prepared for action.

Firing Chart. A fire control map, or, more specifically, a grid sheet on which are shown the elements used in the computation and preparation of the firing data. A substitute for a fire control map.

Firing Crest, Step. See Fire Crest; Fire Step.

Firing Data. All data necessary to give a gun correct elevation, direction, and fuze setting for a given target.

Firing, Field. Firing at field targets under simulated battle conditions.

Firing Jacks, Stability Jacks. Jacks used with a mobile mount to steady it during firing, prevent overturning and relieve strain on the carriage.

Firing Line. The line or echelon of rifle troops, in attack or defense, which delivers or is intended to deliver the principal volume of fire. The line of resistance of a battle position. In general, any line from which fire is or may be delivered.

Firing Lock. The assembly of mechanisms which fire a cannon, extract empty shells, lock the breech, etc.; a part of the breech mechanism. A type of firing mechanism.

Firing Mechanism. The machinery of any firearm which is used to ignite the propelling charge.

Firing Mechanism Block. A firing mechanism of a cannon, with its housing, which is detachable from the piece.

Firing or Lead Wire. A light, weatherproofed, electric wire used in firing charges by electricity.

Firing Order. The order in which the explosions occur in the several cylinders of an internal combustion engine, or in which the guns of a battery are fired.

Firing Party. A detail selected to fire over the grave of any person buried with full military orders. A detail selected to shoot to death a person sentenced thereto by a military tribunal. A firing squad.

Firing Pin. That part of the firing mechanism which strikes the primer and explodes the propelling charge in a firearm.

Firing Position. The position assumed by an individual for firing; the usual positions are standing, kneeling, sitting, and prone. A locality or emplacement from which a unit or weapon executes fire missions. See Fire Positions.

Firing Squad. See Firing Party.

Firing Tables. Ballistic tables giving the elements of firing data for standard conditions and corrections for conditions not standard. Range tables.

First Aid. Prompt emergency treatment given to wounded and injured.

First Aid Packet. A packet containing sterilized bandages and dressings for use in rendering first aid treatment to those needing it.

First Aid Pouch. A pouch for carrying the first aid packet, worn attached to the belt by all soldiers.

First Aid Station. A dressing station located close to the front lines, where emergency treatment is given.

First Call. The first or warning call or signal for a formation or assembly.

First Catch. A point in the descending branch of its trajectory where a bullet reaches the height of a man's head.

First Class Gunner. A grade of classification just below that of expert gunner.

First Class Pistol Shot. A grade of classification just below that of expert pistol shot.

First Classman. A grade of rifleman just below that of marksman. A cadet in the highest class at the U. S. Military Academy.

First Defense Guns. Machine guns which are sited to deliver fire on the enemy from the time he begins his attack until he penetrates the forward lines of the battle position.

First Firing Position. A position designated as such in an attack order from which fire upon the enemy will first be delivered by the assaulting infantry troops.

First Graze. The point where a projectile, if not interfered with, will first strike the ground.

First Line. The line on which the leading units of an organization are deployed.

First Parallel. The first trench, generally parallel to the line of investment, and as close as possible to the enemy's main works, opened by the besiegers as the first step in gaining ground to the front. See Parallels and Approaches.

First Sergeant. The ranking noncommissioned officer of a company or similar organization.

Fiscal Year. The fiscal year begins on July 1 and ends on June 30 of the following year. It is the period for which regular appropriations are made by the Congress of the United States.

Fishnet. A net of cordage, used as a support for camouflage materials.

Fishtail Wind. A wind, especially on a target range, whose direction constantly changes.

Fix. To immobilize. To pin in place.

Fixed Ammunition. Ammunition which is handled and loaded into the piece as a unit; the cartridge case contains the primer, igniter and propelling charge; the projectile is secured in the mouth of the cartridge case and is not removable therefrom by hand; the cartridge case serves also as a gas check.

Fixed Armament. Seacoast guns, howitzers and mortars mounted in permanent emplacements, incapable of being moved or readily transferred, and designed to fire over limited areas.

Fixed Mount. A gun carriage which cannot be readily moved from place to place.

Fixed Obstacles. See Fixed Wire Entanglements.

Fixed Pivot. The fixed point about which a line of troops turns in changing from a formation in line to one in column, or the reverse. The man on the pivot flank.

Fixed Searchlights. Searchlights mounted in fixed positions or capable of but limited movement for purposes of protection or concealment. They are ordinarily furnished power from fixed plants.

Fixed Wire Entanglements. Entanglements erected in place, the wire being attached to pickets driven or screwed into the ground.

Flag. A piece of cloth, either plain or bearing a device, used as a standard, symbol, or signal. The national emblem of any country. A general term for flags, ensigns, colors or standards.

Flag Kits. Kits containing flags for signalling.

Flag, Garrison, Post, Storm. See Garrison, Post, Storm Flag.

Flag of the United States. The national emblem includes six horizontal white stripes, alternating with seven red stripes, representing the original thirteen colonies, and in the upper corner, next the staff, a blue field on which is one five-pointed white star for each state in the Union at any time. See Color; Standard; Garrison, Post, Storm Flag.

Flag Signals. Signals made by means of a flag.

Flag-staff. A mast or pole on which a flag is displayed.

Flame Projectors. Devices for projecting flames. Flammenwerfer.

Flammenwerfer. Flame projectors. (Ger.)

Flanged Projectile. A projectile having projections on the cylindrical portion in the form of studs, ribs or flanges which are intended to be fitted into corresponding grooves in the gun; an early form of projectile for muzzle loading rifled cannon.

Flank. The right or left of a command in line, column, or battle formation, or the terrain in the immediate vicinity of its right or left. The element on the right or left. To pass around or turn a flank. To threaten a flank or on the flank. In speaking of the enemy, the term right or left flank is used to designate the flank that would be so designated by him.

Flank Attack. An attack upon the flank of an enemy's formation or position from a position oblique or parallel to the enemy's flank.

Flank Column. An artillery formation in which the pieces are in one column, and the caissons in a parallel column, with sufficient interval for wheeling by section into line.

Flank Company. A company on the flank of any formation in line.

Flank Defense. The means adopted to protect a flank or preserve all parts of a work from being unduly exposed to direct fire from the enemy.

Flank Element. The first element on the right (left) and the last element on the left (right) of an organization or formation in line.

Flank Files. The first men on the right and the last men on the left of an organization or formation.

Flank Guard. A security detachment which protects the flank of a moving force.

Flank March. A march across or nearly parallel to the front of the enemy.

Flank Observation or **Spotting.** Observation from a point on or nearly on the flank of a target or position.

Flank Patrols. Patrols which cover the flanks of a column on the march, to protect it against surprise, and observe the movements of the enemy.

Flank Position. A position on the flank of the enemy's line of advance occupied for the purpose of diverting or delaying him.

Flank Security. Measures taken for the protection of the flanks of a marching column, or of a force in line of battle.

Flankers. Men posted or marching so as to protect the flank of a column in march.

Flanking Group or **Party.** Any armed group placed on the flank of an organization for observation or security. A group designated to observe or harry a flank of the enemy.

Flankment. A band of grazing fire from an automatic weapon, approximately parallel to the front. (Old.)

Flare. A light used as an illumination and in signalling. A guide for a night landing of an airplane.

Flare, Parachute. A pyrotechnic device attached to a parachute and designed to illuminate a large area when released from an aircraft or shell at an altitude.

Flare Pistol. A large pistol used to fire flares.

Flare, Signal. A pyrotechnic signalling device of distinctive color and characteristics.

Flare, Wing-tip. A pyrotechnic device attached to an aircraft for illuminating the ground while landing.

Flare-back. A burst of flame from the breech of a gun when opened after firing, resulting from the ignition of the remaining powder gases when mixed with air and in contact with sparks remaining in the bore.

Flash. The flame which issues from the muzzle of a firearm on discharge.

Flash Defilade. A position that affords concealment for the flash of the guns.

Flash Hider. A device for attachment to the muzzle of a firearm to hide the flash of discharge.

Flash Message. A message giving warning of the approach or presence of hostile aircraft, and data concerning them.

Flash Ranging. Location of hostile artillery by observations of the flashes of the guns, from two or more accurately located stations. Adjustment or improvement of fire by observation of the flashes of bursting projectiles, including air bursts, from base-line stations equipped with observing instruments.

Flashless Powder. Powder which shows no flash on discharge.

Flasks. The side pieces of the trail of a gun carriage.

Flat Trajectory. A trajectory having little curvature, as of a high power gun or at short range.

Flatten Out. To bring the longitudinal axis of an aircraft parallel to the ground after a dive or climb. To assume such a position.

Fleeting Target. A target of opportunity which is moving or capable of moving rapidly.

Flexible Defense. A kind of defense which aims at maintaining its position by counterattacks delivered after the force of the enemy's attack has been spent in penetrating the first line. An active defense.

Flick. A momentary illumination of an aerial target in the beam of a searchlight.

Flight. A rapid and disorderly retreat of a body of troops from a victorious enemy. The passage of a projectile from the muzzle of a gun to the point of first graze. An elementary organization of the air service composed of a small number of airplanes of the same type. The support of a moving body in the air. The act of flying. Any change in position, in or through the air, of an aircraft involving time or distance in any degree.

Flight Commander. The commander of a flight of airplanes.

Flight, Cross-country. A flight of an aircraft which necessitates leaving the vicinity of a regular landing field. Any flight other than one made in a closed circuit within gliding distance of the airdrome from which ascent is made.

Flight Diagram. A map or chart showing the lines of flight of an airplane for topographical mapping, furnished for the guidance of the pilot and photographer.

Flight Log. A record kept by the pilot or navigator which shows the time of departure, speed and direction of travel, time of arrival and other incidents of an aerial flight.

Flight Pay. Increased pay for regular and frequent participation in aerial flights.

Flight Plan. A plan for an aerial flight, setting forth the probable time of departure, direction of flight as determined from a consideration of the direction and velocity of the wind and air speed of the aircraft, with estimated time of arrival.

Flinching. An involuntary movement of the muscles or body immediately before firing a small arm, which causes a derangement of the aim of the piece.

Flintlock or Firelock. A small arm, produced in Spain early in the 17th century, which was at first an adaptation of the wheellock, a piece of flint being substituted for the pyrite. The wheel was soon discarded and the inside of the pan made rough so as to cause the flint to emit sparks when striking against it and thus ignite the priming compound. It was not until 1670-1680 that it was sufficiently improved to become generally used. It was used as a military weapon until about 1840. A snaphance. See also Matchlock; Wheellock.

Float. A completely inclosed, watertight structure attached to an aircraft to give it buoyancy and stability when in contact with the water. A bearing block for a jack or outrigger. A support for a floating bridge. A ponton.

Floating Bridge. Any bridge with floating supports. A ponton bridge.

Floating Piston. A piston having no rod or connection, used in a hydropneumatic cylinder, with air or gas on one side and liquid on the other.

Flock. An irregular fan-shaped, deployed formation of the cavalry squad, in which the individual troopers are disposed in rear of the leader, but separated from him and from each other by distances and intervals of from 10 to 20 yards.

Flood Lighting. Artificial lighting in which the light is projected in such quantity and from such directions as to give uniformity in illumination over a designated area.

Floodlight System, Landing-area. A complete installation of floodlighting equipment designed to illuminate a landing area.

Flourish. An irregular or fanciful strain of music, by way of ornamentation or prelude, used in paying military compliments. The waving of a weapon. To execute a fanciful strain of music.

Fly. The length of an extended flag, measured perpendicularly to the staff to which it is attached. A canvas cover for a tent. To operate an aircraft in flight. To ride as a passenger in an aircraft. To display a flag.

Flying Blind. See Instrument Flying.

Flying Boat. A seaplane whose main body or hull provides flotation.

Flying Bridge. A temporary, hasty bridge. A flying ferry.

Flying Cadet. An enlisted man of the air corps undergoing flying instruction with a view to commission as a reserve pilot.

Flying Colors. Colors unfurled to fly in the breeze.

Flying Ferry. A ferry in which the force of the current is utilized to propel a raft or boat moored by a single line to a point well upstream.

Flying Force. The combined air force of a nation or a body of troops.

Flying Machine. Any heavier-than-air apparatus, without gas support, for aerial locomotion. An airplane.

Flying Officer. One who has been rated as a pilot of service types of aircraft, or as a pilot and observer.

Flying Party. A detachment of troops employed in hovering about an enemy and observing his movements and dispositions.

Flying Sap. Narrow trenches rapidly constructed by sapping under the protection of gabions or sandbags.

Flying Velocity. The speed required to enable an airplane to take off.

Fog of War. The uncertainty as to the strength, dispositions, and intentions of the enemy, which attends all the operations of war.

Foil. A blunt weapon used in fencing. To frustrate or defeat.

Follow. To guide upon or maintain a prescribed distance from a unit or individual marching in front.

Follow Me. A combat order of the infantry squad meaning that the squad is to follow its leader.

Follow Up. To pursue vigorously an advantage gained.

Follower. An adherent or subordinate. A plate, actuated by a spring, at the bottom of a magazine which forces the cartridges up as they are fed out at the top.

Following Edge. The trailing edge of an airplane wing.

Follow-the-Pointer Laying. A system of laying, used in antiaircraft fire, in which the guns and fuze-setter are kept constantly set to the data electrically transmitted from a director, as indicated by the electrical pointers. See Pointers, Electrical, Mechanical.

Foot. Foot soldiers. The infantry.

Foot Soldiers, Troops. Troops that march, maneuver, and fight on foot. Infantry.

For Official Use Only. A document will be so marked when it contains information which is not to be communicated to the public or to the press, but which may be communicated to any person known to be in the service of the United States whose duties it concerns, or to persons of undoubted loyalty and discretion who are cooperating in government work.

For the Duration. To extend throughout the course of the war; e.g., it is now provided by law that all enlistments in force during a war shall continue until six months after its close.

For the Good of the Service. A term used to signify that the action taken is considered to be for the best interests of the service.

For or **During the War.** A policy that all enlistments in force at the outbreak of a war or entered into during a war shall continue in force until 6 months after the end of the war.

Forage. Food for animals. To collect supplies for men and animals.

Forage Cap. A small low cap with a round, flat crown, and horizontal visor, formerly worn by soldiers when not in full dress.

Foragers. Mounted troopers abreast of each other with intervals greater than those prescribed for close order.

Forced Landing. A landing from ships made against enemy opposition. An emergency landing of an aircraft.

Forced March. Any march in which the distance covered is appreciably greater than is normal. A long and exhausting march made in a case of urgent necessity.

Forced Sensing. In lateral conduct of fire, a sensing of either range or deflection which is indicated by a positive sensing of the other element.

Forces. The organized troops of a nation.

Forces in the Field. The whole of the military forces mobilized in the theater of operations.

Forcing Slope. The part of the bore of a cannon in front of the centering slope, in which the diameter of the bore is slightly tapered so that the rotating bands are gradually forced into the grooves.

Forearm. To arm or prepare for hostilities before the necessity arises. A wooden casing covering part of the barrel of an automatic rifle and protecting the firer's hand.

Foreground. The terrain immediately in front of a fortification or a defensive position, or of a body of troops in attack or defense.

Foreign Legion. A voluntary corps of foreign sympathizers or mercenaries in the service of a nation at war, such as the Foreign Legion in the service of France during the World War; this legion had formerly been used in colonial service only.

Foreign Service. Service in any part of the world outside the continental limits of the United States. Military duties performed abroad.

Forfeiture. A pecuniary penalty resulting from conviction and sentence by a military tribunal. A stoppage of pay.

Fork. The difference in range or elevation, or in direction, required to change the center of impact in range by four probable errors.

Forlorn Hope. An extremely hazardous and doubtful operation entered upon as a last resort.

Form Lines. Wavy lines resembling contours, used in hasty sketching to indicate roughly the contour of the ground.

Formation. The arrangement of the elements of a command so that all fractions are in their proper order in line, in column, or for battle. An assembly for drills, exercises, or ceremonies. Troops disposed in any prescribed manner. In air tactics, a number of planes led and maneuvered as a unit, the dead space of each plane being covered by the fire of others.

Forming-up Place. An assembly place for smaller units.

Forms of Attack. The methods, or general plan, of executing an attack. The usual forms are frontal attack, enveloping attack, and turning movement.

Forms of Defense. The varying extent or degree of the organization of the ground for defense; as deployed defense, position defense, and zone defense.

Fort. A strong or fortified place. An inclosed work of the higher class of fieldwork. An area within a harbor defense wherein are located defense elements capable of offensive action against hostile warcraft, and which is organized to provide for such action and for its own protection and administration. A permanent post. See Bastion System; Polygonal System.

Fort Command. All the means of seaward and landward defense, including personnel and materiel, located at any coast artillery post. A fortified area within a coast defense command.

Forte. The strongest part of a sword blade. The half nearest the hilt.

Fortification. A military defensive work. The art of fortifying a place, position, or area in order to increase the defensive power of the troops sheltered thereby. There are various kinds of fortifications, such as defensive and offensive, field and permanent, etc. The act of fortifying. See Field Fortification.

Fortify. To strengthen a locality or defensive position by constructing military works of any nature.

Fortress. A large fort or group of forts. An area or city defended by forts.

Forward. To the front. Moving toward a position in front.

Forward Area. An area largely or wholly within range of hostile medium artillery, and in which combat troops are tactically disposed for battle.

Forward Area Sight. A sight for determining approximate lateral and vertical leads in firing machine guns at aerial targets.

Forward Echelon. The most advanced echelon which contains the elements necessary for combat purposes. The most advanced line or element of a unit or force in attack or defense. The advanced echelon of a staff or other organization.

Forward Observation Post. An observation post located well to the front; one which is in advance of the battery to which it pertains.

Forward Slope. Ground descending toward the enemy. The front slope of an occupied ridge or high ground. The exterior or most advanced slope of an embankment or rampart.

Fosse. The ditch or moat of a fortification, either with or without water. A hole or pit.

Fougasse. A land mine which projects a charge of broken stone.

Fouling. A deposit in the bore of a firearm resulting from the action of the primer, propelling charge, or metal stripped from the projectile; known respectively as primer, powder and metal fouling.

Fourragere. A braided cord decoration worn over the shoulder by men belonging to an organization which has been decorated with it.

Fox Holes. Small, individual shelter or rifle pits, usually dug during the course of a combat.

Fragmentary Order. An order issued in part to two or more subordinates at two or more different times or places.

Fragmentation. The breaking up and scattering of the fragments of a shell, bomb or grenade.

Fragmentation Bomb. A bomb intended primarily for use against personnel on the ground by fragmentation of the bomb case.

Frank. A printed statement "Official Business," used on official envelopes of the federal government in lieu of postage stamps.

Frapping. Drawing together several returns of rope, as in a lashing, by passing several turns around them.

Fraudulent Enlistment. An enlistment in the military service procured by means of wilful misrepresentation or concealment as to qualifications for enlistment.

Free Balloons. Balloons which float wherever the wind may carry them.

Free Lance. One who gives or sells his services without regard to country. A soldier of fortune.

Free Maneuver. A maneuver in which only the initial situation is stated, the action being allowed to develop according to the skill and initiative of the opposing commanders.

Frettage. The process of strengthening the breech of a gun by shrinking on hoops or frettes. The frettes used.

Frette. A hoop shrunk on a cast-iron gun to strengthen it.

Friction Bars. A device for absorbing the shock of recoil in some types of cannon, by sliding friction.

Friction Fuze or Primer. A fuze or primer that is ignited by the heat generated by friction.

Frog. A button or toggle which is passed through a loop on the other side of a military cloak or overcoat to fasten the two breasts together. The loop of a bayonet or sword scabbard. A slang name for a French soldier. The triangular protuberance in the sole of a horse's or mule's foot. A crossing of two rails of a railroad.

From Battery. The position of a cannon when withdrawn from its firing position.

Front. The lateral space occupied by any element or unit in line, column or battle formation (the front of a man is 22 inches). The direction in which a unit faces. The direction of the enemy. The battle-front or front line. The line of contact of two opposing forces, or the portion thereof occupied by an organization. The lateral dimension of a zone of action or sector in defense. Any geographical section of the battle line. One side or face of a fortification. All the works comprised between the capitals of two adjacent bastions or salients.

Front Line. In mobile warfare, the most advanced line held by a unit with an organized force. It is, in general, the line that would be protected, in case of attack, by the divisional or other artillery barrage. In stabilized warfare, it is the line which connects the most advanced strong points in the defensive organization. The line of battle.

Front Line Unit. A unit located on the line of the most advanced elements, exclusive of local security detachments; as a front line battalion or division.

Front of Operations. The geographical area comprised between the strategic fronts of the opposing armies.

Front Sight. A sight attached to a gun near the muzzle, commonly used on small arms.

Frontage. The extent of front occupied or covered by a unit in any drill or battle formation.

Frontages in Attack. The frontage assigned a unit in attack varies with its mission and the amount of resistance expected. Thus a platoon of three squads, in the front line, may be made responsible for a zone from 100 to 200 yards wide. The strength or density of attack of larger units is measured by the proportion of supports and reserves initially withheld.

Frontages in Defense. The usual frontages that may be defended by units in an active defense are about as follows: Platoon, 150 to 400 yards; Company, 300 to 600 yards; Battalion, 600 to 1200 yards. In passive defense or delaying action these frontages may be exceeded.

Frontal. Pertaining to the front. A metal face guard.

Frontal Attack. A uniform attack along the whole, or a considerable portion of the hostile front. An attack directed against the hostile front.

Frontier. The limits, confines, or boundaries of a country. That part of a country which fronts or faces another country. An outwork or minor defense.

Frontier Commands. The divisions of the land and sea frontiers of a nation for the purpose of command in war and for the preparation of the necessary plans in time of peace.

Fuel Distance. The usual distance which a motorized vehicle can travel on the fuel contained in its tank.

Fugelman, or Fugleman. A man placed in front of a body of soldiers, whose motions they follow. The leader of a file.

Full Distance. Distance between elements in close order, such that if line be formed from column the elements will have their prescribed intervals.

Full Dress. The uniform worn on occasions of ceremony.

Full Interval. Interval between elements in close order, such that if column be formed from line the elements will have their prescribed distances.

Full Pack. A pack containing all the prescribed equipment.

Full Pay and Allowances. Includes base pay and longevity increases thereof, any additions such as pay as aide de camp, pay for aviation duty, etc., and allowances, such as pay for mounts, and rental and subsistence allowances.

Full Retreat. An expeditious and complete retreat before a victorious enemy.

Full Revetment. A revetment which protects the entire face of a slope. A revetment which extends from the bottom to the fire crest of a trench.

Full Sight. An aim in which the entire front sight is seen through the notch in the rear sight.

Full Step. A step of the full prescribed length, 30 inches in quick time, 36 inches in double time.

Fulminate of Mercury. The most violent of the high explosives, commonly used in small quantities as a detonator or igniter for less sensitive explosives, in caps, fuzes and primers. It is now being largely replaced by more stable priming compositions.

Fulmination. The explosion of a fulminate. A violent explosion. The act of exploding.

Function. The office or duty assigned to or belonging to an individual group or unit. The purpose of any organization or machine. In mathematics, anything whose value depends on the value of something else; e.g., the ballistic coefficient is a function of the weight of the projectile. To operate properly or to perform duty satisfactorily.

Functional. Based upon the nature of the duty or operation to be performed; e.g., the functional divisions of the general staff.

Functioning. Action or performance, especially of a machine. Mechanical operation. Working correctly or satisfactorily.

Fundamentals of Training. The fundamentals of training are: (1) Individual and collective discipline; (2) Individual physical development; (3) Leadership; (4) Individual and collective morale; (5) Individual proficiency; (6) Teamplay; (7) Company proficiency; (8) Battalion and regimental training; (9) Combined training.

Funeral Escort. A military escort at a funeral.

Funeral Honors. The ceremonies and last honors paid to a deceased officer or soldier.

Funk Hole. A shelter dug in the side of a deep narrow trench. A dugout. A place of safety.

Furl. To wrap, roll, or fold, as a flag, around its staff.

Furlough. A leave of absence granted to an enlisted man for a stated period.

Fuselage. The body, of approximately streamline form, to which the wings and tail unit of an airplane are attached.

Future Position. The predicted position of a moving target, especially an aerial target, at the end of the time of flight.

Fuze. A device, usually with explosive or combustible elements, which detonates or ignites the charge of a shell, bomb, grenade, or other projectile or demolition charge, at the time and under the circumstances desired. A device inserted in a projectile to cause the explosion of the bursting charge on impact, or at a certain time during the flight. Fuzes are classified as percussion, time, mechanical, combination, detonating point, base, delayed action (long and short), non-delay, super-quick, safety, and instantaneous. See type of fuze in question.

Fuze Error. The variation in fuze range from standard as determined for a particular lot of ammunition.

Fuze Lighter. A device used to ignite a safety fuze in blasting and demolitions.

Fuze Prediction Angle. The vertical angle between the line of position at the instant of observation and the line of future position.

Fuze Prediction Time. The time of flight to the future position plus the dead time, during which the travel of the target subtends the fuze prediction angle.

Fuze Range. A fictitious range, expressed in graduations or reference numbers on a time fuze, at which the fuze must be set in order to produce a burst at a certain height above the target. The fuze setting necessary to produce a burst at a given point.

Fuze-range Pattern. In antiaircraft fire, a pattern of simultaneous bursts which results from specified fuze settings of normal, less than normal, and greater than normal.

Fuze Setter. A device used for setting the desired time for bursting on a time fuze.

G

Gabion. A cylindrical basket with open ends, made of brush or metal ribbon woven on pickets, and used as a revetment in constructing field works. To strengthen by gabions.

Gabion Form. A circular wooden template, with notches in its circumference, used to separate the pickets during the making of a gabion.

Gabion Knife. A heavy clasp knife used for cutting and trimming brush in making brush revetments.

Gabion Revetment. A revetment made of gabions, usually employed in siege operations for protection of batteries, etc.

Gabionade. A breastwork constructed with gabions. A traverse made with gabions between guns or on their flanks, to protect them from enfilading fire.

Gait. The rate of movement of a mounted man or organization. The pace of a horse, as the walk, trot, or gallop.

Gallery. A casemate in a scarp or counterscarp. A covered passage, sometimes open on one side, in a fortification. A horizonal passage in a mine or dugout, or other subterranean work. An indoor range for rifle or pistol practice.

Gallery Practice. Firing at miniature targets at short ranges with small caliber rifles or pistols, usually in a gallery.

Gap. An opening which implies a breach, as through an obstacle. An interval between the flanks of two adjacent organizations which is greater than that which is proper or desired. The distance separating two adjacent wings of a multiplane. A low point in a range of hills. A pass.

Garand Rifle. The semi-automatic rifle adopted for use in the United States Army; it has a caliber of .30 inch, weighs 8.56 pounds and uses a clip containing 10 cartridges.

Garlands. Garnishings of burlap, or other artificial or natural materials, mounted on a fishnet; used in camouflage.

Garniture. The dressing of a camouflage net. A general term for all equipage.

Garrison. A body of troops stationed in a military post, fortress or fortified place. A group or unit which occupies and defends a tactical area, as a combat group, strong point, or center of resistance. A strong place in which troops are stationed for its protection. To place troops in. To be stationed in.

Garrison Artillery. The artillery pertaining to a garrisoned fort or place.

Garrison Belt. A leather belt worn by soldiers of the United States Army in garrison.

Garrison Flag. The national flag with 38 feet fly and 20 feet hoist, furnished only to important posts and displayed only on important occasions.

Garrison Gin. See Gin.

Garrison Prisoners. All prisoners at a post who are not general prisoners.

Garrison Ration. The ration prescribed in time of peace for all persons entitled to a ration, except when another ration is specified.

Gas. A chemical substance, either finely comminuted solid, liquid or gaseous, disseminated to produce a poisonous or irritant atmosphere. To poison or asphyxiate by the use of gas.

Gas Alarms. Devices for giving the alarm in case of gas attacks.

Gas Alert. A signal sounded whenever the presence of gas is known or suspected, whereupon all men don their gas masks.

Gas Alert Zone. An area, parallel to the front, in which effective enemy gas concentrations may be expected, and within which the gas mask will be worn in the alert position, i.e., so that it can be quickly donned.

Gas Attacks. Attacks in which a heavy concentration of gas is placed on a portion of the enemy's line.

Gas Bomb. A bomb containing a bursting charge of high explosive and a gas of some nature.

Gas Chamber. A room in which gas is released for the purpose of testing gas masks, and for gas training.

Gas Check. A device used in breechloading guns to prevent the escape of gas at the breech. An obturator. In weapons using fixed or semi-fixed ammunition the cartridge case provides the gas check.

Gas Curtain. A blanket closing the entrance to a shelter or building to exclude gas.

Gas Danger Zone.   The area between the gas alert zone and a line some miles in rear, in which the gas mask will be carried at all times.

Gas Defense.   All means and measures, individual and collective, employed to protect troops against the effects of gas.

Gas Discipline.   Observation and enforcement of all orders and regulations as to individual and collective measures of gas defense.

Gas Mask.   An apparatus which purifies the air which the soldier breathes and protects his eyes and face when he is in an atmosphere contaminated with toxic or irritating gases, vapors, or smokes.

Gas Obstacle.   A concentration of a persistent chemical agent laid down on a definite area in order to prevent troops from advancing over that terrain.

Gas Positions.   Alternate positions for use by artillery when the original positions are rendered untenable by gas.

Gas Proofing.   Measures for excluding toxic gases from military establishments, as by curtains and filters for entering air.

Gas Sentry.   A man trained to detect the presence of gas and posted to give the alarm.

Gas Shell.   A projectile, discharged from a cannon, that releases a toxic chemical agent upon bursting.

Gas Situation Map.   A map showing the location of all known gas obstacles and other gassed areas in the immediate vicinity, including those established by the enemy and friendly troops.

Gas Warfare.   The use of poisonous and asphyxiating gases in war.

Gases.   Chemical agents used in warfare to incapacitate personnel. Agents used primarily to produce a physiological effect due to their ordinary chemical action. A gas which may cause death is called a lethal agent, while a gas which, under field conditions, does not cause death or serious injury, is an irritant agent. The principal gases used in warfare are tear gas (lacrimator), suffocating gas, mustard gas (vesicant), nauseating gas (sternutators). Casualties resulting from gas attacks are few, only 1 in 14 affected die, and the remainder generally recover completely. A colloquial term for chemical agents.

Gas-proof Shelter.   Any shelter or inclosed space from which gas can be excluded.

Gassed.   The condition of having suffered from or been overcome by the fumes of gas.

Gather.   The toe-in or slight convergence of the front wheels of a self-propelled vehicle, which assists in steering. To tighten the reins and press the legs gently against a saddle horse as a preparatory signal for any movement.

Gatling Gun.   An American machine gun, consisting of a cluster of barrels arranged in a circle about an axis, which are revolved, loaded, and fired by means of a crank.

Gauge, Gage.   The caliber of a gun. The distance between the insides of the rails of a track (standard gauge is 4′ 8½″). The diameter of a wire or thickness of a metal plate. Any device used to measure gauge, as a track or wire gauge. A scale, sometimes provided with a float, to measure the height of the tide or stage of a river. A guarantee or security. A cap or glove thrown on the ground as a challenge. A substance which regulates the time of setting of mortar or concrete. To measure, estimate or appraise.

Gauntlet or **Gantlet.**   An iron glove. A military glove. A military punishment which consisted in passing along the line of soldiers and receiving a blow from each man's gantlet.

Gear or **Gearing.**   Military accouterments, equipment, etc. Cogged wheels and racks whereby motion is transmitted, or its rate or direction is changed; extensively used in gun carriages, motor vehicles, and other machines.

Gendarmery.   A military body employed as police in some European countries.

General.   The next grade in rank above lieutenant general, and just below that of field marshal. The roll of drums which calls the troops together. The signal for striking tents.

General Court-martial.   A superior military court having jurisdiction over all persons subject to military law and all offenses defined by the articles of war, and which may adjudge punishment within the limits authorized by law or regulations. It consists of any number of officers not less than five.

General Depot. A depot affording accommodation for the supplies and operations of two or more supply branches.

General Headquarters. The headquarters of the commander-in-chief of the field forces.

General Headquarters Reserve. Troop units of types not habitually required as an organic part of an army, or in excess of the normal requirements of the armies. Such reserve units are assigned as needed.

General Headquarters Reserve Artillery. See G. H. Q. Reserve Artillery.

General Hospitals. Large hospitals located outside the probable combat zone.

General Mobilization Plan. See Mobilization Plan, General.

General Officer. Any officer above the grade of colonel.

General Orders. Orders which usually include matters of importance, directive in nature, general in application, and of permanent duration, not readily susceptible of immediate incorporation in established forms of regulations. Orders given to sentinels on all posts of an interior guard.

General Prisoners. Prisoners sentenced to dismissal or dishonorable discharge and to terms of confinement.

General Recruit Depot. A place where recruits on probation, or applicants for enlistment are assembled in large numbers for recruit instruction and assignment to units.

General Recruiting Stations. Places at which applicants for enlistment are examined and from which, if considered qualified for enlistment, they are forwarded to recruit depots.

General Reserve. A reserve retained under the control of the supreme commander of large forces.

General Service. A detachment of enlisted men for the performance of various duties not strictly military in character.

General Service Schools. Include the Army War College, the Army Industrial College, and a group of schools located at Fort Leavenworth, Kansas, consisting of the Command and General Staff School and Correspondence Schools; the general purpose of these schools is to teach higher command and staff duties, and industrial procurement and mobilization.

General Situation. A statement as in a tactical exercise or an order, etc., of the military conditions known or presumed to be known to both sides.

General Staff. A body of specially trained officers for the performance of staff duty in the War Department or with divisions and higher units. The sections of the general staff, corresponding to the functional duties of the commander, are as follows: G-1, personnel; G-2, intelligence; G-3, operations and training; G-4, supply; and with the War Department, a fifth section, war plans. The general staff is headed by a chief of staff who may be assisted by one or more deputy chiefs. Each section is headed by an assistant chief of staff, and may include any number of officers, enlisted men, clerks, etc.

General Staff with Troops. Selected general staff officers who render professional aid and assistance to their immediate division or higher commanders.

General Support. Fire, especially artillery fire, delivered in support of the entire unit (division, corps or army) of which the artillery is an organic part or to which it may be assigned. See also Direct Support.

Geneva Cross. A red Greek cross on a white background; used on a brassard, flag, or as a ground mark for the protection of medical personnel and establishments in time of war. Mohammedans use a red crescent for the same purpose. The Red Cross.

Geophone. A highly sensitive microphone used in military mining.

Geoscope. A form of periscope for use in a tank.

Getaway Man. The rear man of a patrol, or one on the side away from the enemy, who stays out of any fire fight and, if the patrol is overwhelmed, returns to report its fate and any information gained.

G.H.Q. General Headquarters.

G.H.Q. Reserve Artillery. A pool of artillery, particularly of heavy and special types, which can be assigned or attached to combatant units as needed.

Gilding Metal. An alloy of copper and zinc used for jackets for bullets.

Gin. An engine for lifting heavy weights, such as guns; it usually consists of a single pole or spar, secured in a position slightly inclined from the vertical by means of guy lines. The garrison gin is a breast derrick or A-frame with a pry pole, provided with a winch. A trap of any kind.

Glacis. A long, sloping mound of earth at the front of a fortification, which protects the scarp, and eliminates all dead space.

Glacis Slope. The exterior slope of the glacis.

Glanders-farcy. A highly contagious, common and incurable disease of horses and mules, characterized by a sticky discharge from the nostrils, and ulcers in the nostrils, under the jaw, and inside the legs.

Glide. A descent of an aircraft with reference to the air at a normal angle of attack and without engine power sufficient for level flight in still air, the propeller thrust being replaced by a component of gravity along the line of fight. The act or action of gliding. To descend at a normal angle of attack with little or no thrust.

Glider. An aircraft heavier than air, similar to an airplane, but without a power plant.

Globe Sight. A kind of front sight having a pin with a small ball on its end, or a disc with a hole in it.

Gold Medal Cot. A small, folding camp cot or bed.

Gondola. The car of an airship. A large, open freight car with deep sides.

Goose Step. The stiff legged parade step used in the German Army.

Gooseberry. A portable obstacle consisting of a globe or sphere of barbed wire.

Gorge. A deep or rugged ravine. The rear face, or open portion between the flanks, of a redan, bastion, or similar work. A protection for the throat.

Gorge Trench. A trench erected across the gorge of a fortification, usually provided with parapets on both sides to afford protection from or permit fire to the rear or to the interior of the work.

G. P. F. Grande Puissance Filloux, the high-powered Filloux. The French 155mm gun.

Grade. The rank of an enlisted man. Grades are designated by number, as 1st grade, 2nd grade, etc., or by title, as master sergeant, technical sergeant, staff sergeant, sergeant, corporal, private 1st class, and private. The slope of a road, railroad, ramp, etc. A grade of 1% is one that rises one foot vertically in each 100 feet of horizontal distance.

Gradient. A steep slope, as of a trench or embankment, expressed as a fraction in which the numerator is the vertical rise and the denominator is the corresponding horizontal run; e.g., 1/3, expressed as 1 on 3.

Grand Army of the Republic. An organization composed of the veterans of the American Civil War, who fought in the Northern Armies.

Grand Tactics. Tactics on a large scale. The tactics of large forces, or of forces including several or all of the combatant arms. The art of generalship or of independent command.

Grape Shot. A cluster of small iron balls, held together by plates at the top and bottom.

Graph. A diagram of any kind which conveys information graphically. To indicate by means of a graph.

Graphical Scale. A scale, drawn or printed on a map, in which ground units, such as miles or thousands of yards, are shown to the scale of the map.

Grapnel. A small anchor used in mooring boats or pontons for military bridges. A grappling device.

Gratuitous Issue. An issue of supplies or equiqment which is not charged against any allowance.

Gratuity. A money payment in addition to all authorized pay and allowances; as a payment made to a beneficiary of a deceased soldier.

Graves Registration Service. An organization charged with supervision of mortuary matters pertaining to the personnel of the army in time of war.

Graze. A point at which a projectile bursts or rebounds on impact. To pass close to the surface of the ground or water.

Graze Burst. A burst of a projectile on grazing the ground.

Grazing Point. The point at the near limit of dead space where the slope of the ground is equal to the slope of the trajectory.

Grazing Trajectory. The trajectory which strikes the grazing point.

Grenade. A small bomb which can be hurled a short distance by hand, or by means of a catapult or rifle. Incendiary hand grenades are of great antiquity, their use being known as far back as the siege of Saloniki in 104 A. D. The explosive grenade as a weapon of warfare dates from the 15th century, the earliest types being made of baked earth. Grenades made of wood, bronze, etc., were used in the 16th century. Hand grenades were employed about 1660 by special troops called grenadiers, and were in common use in the 17th and 18th centuries. About 1760 they disappeared from the battlefield, to be resurrected and used in the Crimean and American Civil Wars. High explosives grenades were used extensively in the Russo-Japanese and World Wars. See Hand Grenade; Rifle Grenade; Bomb.

Grenade Net. A net placed in front of or over a trench or other work as a protection against grenades.

Grenade Trap. See Bomb Trap.

Grenadier. Originally, a soldier with the special mission of throwing grenades. In modern times, a member of a regiment known as the Grenadier Guards. Lately revived and applied to a soldier equipped with a rifle grenade discharger.

Gribeauval System of Artillery. See Artillery.

Grid. A system of parallel and perpendicular lines dividing a map or sheet of paper into squares to facilitate rapid location of points. In the standard grid system (U. S.) the country is divided into zones and the central grid meridian or Y-line of each zone is coincident with the true meridian at that place. The X- or east and west lines are perpendicular to the central meridian. The grid squares are 1000 yards on a side on large scale maps, and 5000 yards on small scale maps.

Grid Azimuth. Azimuth measured from grid north.

Grid Declination. The angle, at any locality, between the true meridian and the north and south grid or Y-lines.

Grid Lines. Lines used to divide a map into squares. The lines running east and west are called the X-lines, those running north and south are called the Y-lines.

Grid North. The north indicated by the grid lines on a map. Lambert north.

Grip. The handle of a sword, saber, pistol, or revolver.

Grip Safety. A safety device which prevents the discharge of a pistol except when released by the grip of the hand.

Grooved Bullets. Bullets having grooves or cannelures. The bullets of practice dummy cartridges, so-marked for identification.

Grooves. The spiral channels within the bore of rifled guns. Channels cut around a projectile near its base to hold the cartridge case in fixed ammunition. See also Cannelure.

Ground. A connection of an electric current to the ground, or an accidental grounding of a circuit. On the ground or terrain, terrestrial as distinguished from aerial; e.g., ground reconnaissance. To bring to the ground or place on the ground.

Ground Angle. See Angle, Landing.

Ground Gear. The gear, or equipment, necessary for the landing and handling of an airship on the ground.

Ground Loop. See Loop, Ground.

Ground Observation. Observation of the enemy positions, activities, friendly artillery fire, etc., conducted from locations on the ground. Terrestrial observation.

Ground Officer. An air service officer whose duty it is to assign proper missions to aviators and receive their reports on their return.

Ground Organization. The ground facilities and non-flying personnel used in in the operation of aircraft.

Ground Reconnaissance. Reconnaissance conducted by agencies which operate exclusively on the ground.

Ground School. An air service school giving instruction in aerodynamics, map making, photography, etc.

Ground Scouts. Trained scouts who precede troops to ascertain whether the ground in front is passable and free of the enemy.

Ground Signal Cartridge. A cartridge containing a pyrotechnic signal, fired from the ground by means of a projector.

Ground Speed. The speed of an aircraft with reference to the ground. The air speed corrected for the velocity of the wind.

Ground-return. Utilization of the earth as part of an electric circuit by grounding the metallic circuit at two points.

Group. A number of men formed for a specific purpose. A provisional tactical unit formed of two or more batteries from different battalions for convenience in carrying out particular tactical missions. See also Combat Group; Gun Group; Mine Group.

Group, Airplane. Two or more squadrons, al of the same type of aviation.

Group A Supplies. Items of general supplies the requirements for which should be consolidated for the whole army and which can be advantageously purchased by one office; it includes only those items for which requirements can be definitely estimated for relatively long periods of time and whose prices normally remain stable for extended periods.

Group B Supplies. Items of general supplies the requirements for which should be consolidated for certain regional areas and which can be advantageously purchased by one office in such area; it includes only items for which requirements can be definitely estimated for at least a few months.

Group C Supplies. Items of general supplies not included in Group A or B.

Group of Armies. Two or more armies organized under a single commander.

Group of Mines. In the United States service, a group of 19 submarine mines, electrically controlled through a single distribution box. A number of mines placed close together.

Group of Parts. In a weapon or machine, the various pieces that make up some functional part of the machine; e.g., the bolt group, buffer group, etc.

Groupment. A provisional tactical unit of artillery temporarily formed from two or more artillery battalions (or groups) or larger tactical units. A tactical command within the harbor defense, containing two or more groups, the normal fields of fire of which cover the same or adjacent zones or areas; any provisional command.

Grousers. Cleats forming a part of or attached to the track plates of a track-laying vehicle to increase the tractive effect.

Guard. A body of men whose duty it is to insure that a force or place will not be surprised by the enemy. A man, or body of men, performing guard duty of any nature. A posture of readiness in fencing, bayonet exercise, etc. A member of a regiment of guards. A sentinel. To secure against attack or surprise in the immediate vicinity.

Guard Cartridge. A cartridge with reduced charge for use in the performance of guard duty.

Guard Detail. The men detailed for guard duty from an organization.

Guard Duty. The duty required of guards.

Guard, Interior. See Interior Guard.

Guard Mounting. The ceremony of installing the new guard and relieving the old.

Guard of Honor. A guard detailed to accompany distinguished persons.

Guard Rail. A curb or railing placed at the side of a bridge deck, edge of a platform, etc., as a measure of protection for passengers and vehicles (wheel guard, hub guard, handrail).

Guard Report. The daily report of the commander of the guard.

Guard Roster. A roster which shows the number of days that each member of an organization has been present and available for duty since the beginning of his last tour, and from which details for guard duty are made.

Guardhouse. A building occupied by the guard and in which prisoners are confined.

Guardroom. The room occupied by a guard during its tour of duty.

Guards. The household troops of a sovereign, especially charged with his protection. The term has been applied to certain elite regiments of various armies, which often constitute the final reserve.

Guardsman. An officer or soldier of a regiment of guards. A member of the National Guard.

Guerilla Warfare. An irregular method of warfare, usually carried on by independent bands.

Guerillas. Irregular troops.

Guide. An individual upon whom an organization or element regulates its march or alignment. One who indicates a route, as in unfamiliar country or for a night movement. To regulate upon in marching.

Guidon. A company emblem; a small flag carried at ceremonies and other times as prescribed by the commander. A small flag or streamer carried by mounted troops to indicate the direction of the guide, and to mark a line on which troops are to form. One who carries a guidon. (Guidon bearer.)

Gun. A term generally applied to all firearms, but in its more restricted and technical sense to a piece of ordnance with a relatively long barrel, fired from a carriage or fixed mount. As compared with a howitzer, a gun has a relatively long barrel, (30 to 50 calibers or more), high muzzle velocity, and a more or less limited maximum elevation, except antiaircraft guns.

The actual date of the introduction of cannon, and the country in which they first appeared, have been the subject of much research, but no definite conclusion has been reached. Some writers attribute their invention to the Chinese or to the Arabs, but the weight of opinion gives the credit to Bernard Schwartz, a German monk (1313) generally believed to have been the first to discover the propulsive power of gunpowder. The earliest guns were small and vase-shaped. From these guns cannon were evolved. The primitive cannon were made of wood lined with sheet iron or copper, and bound externally with numerous iron bands. They were very small, the largest in 1339 weighing 46 pounds. But by the end of the 14th century they had become of huge dimensions and fired stone shot weighing as much as 450 pounds. The hand gun was derived from the cannon in the latter part of the 14th century, and the improvement in this weapon brought about a gradual reduction in the weight of cannon.

The evolution of guns may be divided into four periods. The first period (1313-1520), during which stone shot were principally employed. Stone balls came into general use for the larger pieces during the 14th century as they were cheaper and lighter than iron balls. The guns during this period were largely of wrought iron; they were constructed of rods or bars beaten or welded together lengthwise, and reinforced by iron rings. Breechloading guns date from the end of the 14th century, when the gun was made in two sections, the breech section being wedged to the muzzle section for firing. But the danger of explosion and the inconvenience caused by the powder gases was so great that the breechloading system was discarded until about the middle of the 19th century.

The second period (1520-1854), during which cast-iron shot were generally employed. Cast-iron shot came into use about 1520 and the guns gradually became smaller and the calibers more uniform. In the 17th century cast-iron muzzle-loading guns came into general use. In 1739 progress was made by boring guns from solid castings, which resulted in still greater uniformity of caliber. Rifling and a built-up construction involving shrinkage were introduced. But the progress in gun construction during this period was remarkably slow. The manufacture of gunpowder improved and the deviation of projectiles decreased; the guns were strengthened to meet this progress, but the main principles of gun construction remained unaltered.

The third period (1854-1885) witnessed the introduction of a satisfactory breech mechanism by Armstrong and the general use of rifled guns and elongated projectiles. The fourth period (1885-.............) is the era of high velocity guns. The introduction of nitrocellulose powder made possible longer ranges and necessitated longer and stronger guns. This era is notable for the steady advance in caliber and hence the weight of projectiles; also for the development of rapid-fire guns.

It is by no means certain when wheeled carriages were first used. They are mentioned as having been used in 1376, but the size of the gun is not known. They were probably developed to overcome the difficulties of transportation and recoil. The remarkable fact about the ancient carriage is that its general design is the same as that used today. About the middle of the 15th century trunnions became an integral part of the gun.

The earliest guns were not provided with sights or other means for directing them, but this was not important as the range was little more than a hundred yards. Direction was easily obtained by looking along the length of the gun and moving the trail, and elevation was fixed by inserting a wedge under the breech. The quadrant for elevation was invented by Tartaglia about 1545, and a dispart sight was used in 1610. A fixed rear sight was used early in the 19th century and in 1829 a tangent sight was introduced into the British navy. The telescopic sight was invented about the middle of the 19th century.

Gun Barrel. The tube of a gun which serves to confine the gases of the propelling charge and to give direction to the projectile.

Gun Battery. Two or more cannon of the same kind, organized as a fire unit, with all the personnel, equipment, transport, etc. A firing battery, as distinguished from a headquarters, service, searchlight battery, etc. An emplacement for cannon.

Gun Breech. That part of a gun back of or at the rear end of the barrel.

Gun Carriage. A carriage designed to bear and transport guns, and to afford a stable mount for firing.

Gun Chamber. That part of a gun which receives the charge.

Gun Chart. A chart on which is noted complete firing data for the gun in question. A graphical firing table.

Gun Circle. A base ring.

Gun Crew, Detachment, Squad. A group of men assigned to the service of a single piece.

Gun Commander. An enlisted man who commands a gun and its crew.

Gun Defense. The particular class of defense provided by guns, as distinguished from other elements of a defense.

Gun Deflection Board. A mechanical device for computing the algebraic sum of direction corrections for travel of the target during time of flight and for wind and drift; or azimuths corrected for wind and drift.

Gun Differences. The differences, due to displacement, between the ranges or directions from the several guns of a battery to the target and from the directing point or gun to the target. See Convergence, Distribution, Divergence Difference.

Gun Displacement. The horizontal distance from the vertical axis of any gun in a battery to the directing gun or point.

Gun Group. A coast defense tactical unit composed of several (gun, howitzer, or mortar) batteries, mobile or fixed, whose fields of fire cover the same general water area, with the personnel and installations for the employment of the group as a unit.

Gun Hitting Volume. A solid volume representing approximately the space around a towed target within which a shellburst would produce effective results. A hypothetical target.

Gun Levers. The two large arms on a disappearing carriage, supporting the gun trunnions at one end and the counterweight at the other, being attached near their middle points to the top carriage by the gun lever trunnions or axle.

Gun Metal. Bronze used in the manufacture of cannon. Metal having a blue or black finish, similar to that used for firearms.

Gun Parallax. See Parallax.

Gun Pits. Excavations or sunken emplacements for the protection of cannon.

Gun Platform. A substantial base upon which a gun carriage rests.

Gun Pointer. A member of a gun crew who points or lays a gun in direction.

Gun Shelters. Bullet proof blinds or shields arranged to mask the mouths of embrasures when the guns are not in battery.

Gun Shield. A fixed or movable armor plate placed on a gun carriage to protect the gunners and the mechanism.

Gun Sight. See Sight.

Gun Sling. A sling for lifting a gun from its carriage. A combination strap used on the rifle for carrying purposes and as an aid in aiming.

Gun Tackle. A tackle rigging consisting of two single blocks, one attached to the load and one to the anchorage. It has a mechanical advantage of 3.

Gunfire. The firing of a gun. The use of artillery, etc., as weapons of war.

Gunlock. The firing mechanism of a small arm.

Gunner. One who operates or assists in the operation of a gun. According to proficiency gunners are rated as 2nd class, 1st class, and expert.

Gunner's Rule for Overhead Fire. A method for determining troop safety for overhead fire by use of the sight leaf; applied to ranges not over 900 yards. See also Leader's Rule.

Gunnery. The art of constructing and using guns and projectiles. The art and science of firing guns. The practical use of cannon.

Gunpowder. The compound or mixture used to propel projectiles from cannon and small arms; specifically, a mixture of charcoal, sulphur and niter. It is likely that gunpowder is a development rather than a discovery, as its invention cannot be traced to any one person. Incendiary compounds were known to the Chinese, Greeks and Arabs in very early times. However there is no good evidence that any explosive resembling gunpowder was discovered prior to the 13th century. By many, Roger Bacon is given credit for having enunciated the ingredients of gunpowder in 1248, but he apparently was not aware of its projecting power; this is generally credited to Bernard Schwartz, a German monk, who accidentally discovered it about 1313. Its use became general at the beginning of the 16th century.

Until the end of the 16th century it was used in the form of fine powder or dust. To overcome the difficulty in loading from the muzzle the powder was then given a granular form. With the same end in view attempts were made to develop a breechloader, but without success as no effective gas check was devised. No marked improvement was made until 1860, when General Rodman, U. S. A., discovered the principle of progressive combustion of powder, and that the rate of combustion, and consequently the pressure exerted in the gun, could be controlled by the use of larger grains of greater density. As a result of his investigations powder was thereafter made in grains of different sizes, each suited to the gun for which intended. General Rodman also advocated the idea of perforated grains. A further control of the velocity of combustion was obtained in 1880 by the substitution of an underburnt charcoal for the black charcoal previously used. A still further advance was accomplished in 1886 when smokeless powders made their first appearance. These powders are chemical compounds, and not mechanical mixtures like the charcoal powders. Smokeless powders have almost entirely replaced black and brown powders as propelling charges in guns. Flashless powder has been developed but is not yet in general use.

Gun-recess. A trench, at the side of the barbette, for the concealment of a gun when not in action.

Gunshot. The effective distance to which a projectile can be thrown by a gun. The range of a gun. The point-blank range of a gun.

Guns of Position. Heavy guns not designed or intended for rapid movement.

Gyroplane. A type of rotor plane, whose support in the air is chiefly derived from airfoils rotated about an approximately vertical axis.

Gyroscope. A wheel with a comparatively heavy rim, which rotates at high speed, the effect of which is a tendency to maintain the position of its axis of rotation against disturbing forces.

Gyroscopic Compass. A compass operating on the gyroscopic principle.

H

Hachures. A method of shading used on maps to indicate ground forms; now generally superseded by contours.

Hair Trigger. A trigger which requires only very light pressure to release its firing pin or hammer.

Hairbrush Grenade. A grenade in the shape of a hair brush, grasped by its handle.

Halberd. An ancient long-handled weapon used for cutting and thrusting. The gisarme, or one of its varieties.

Half Bent. The first notch in the tumbler of a gunlock. The half cock notch.

Half-closed Work. A fortification having a parapet on the side of the enemy only, its gorge being closed by an obstacle.

Half Cock. The position of the cock of a gun when held by the first notch.

Half Face. A 45 degree face to the right or left.

Half Mast or Staff. The display of a flag with the middle of the flag at the middle point of the staff, or, in case of a flagstaff with crosstrees or guy cables, at the middle point of the staff above the crosstrees; used as a symbol of mourning. When displayed at halfstaff the flag is first raised to the top and lowered to the halfstaff position; on lowering it is first raised to the peak of the staff.

Half-track Vehicle. A vehicle of the caterpillar type, having also a pair of wheels in front.

Hall Rifle. The first successful breechloading rifle, designed by Captain J. H. Hall, Ordnance Dept., U.S.A. (1811)

Halt. A stop in marching. To stop marching. To cause to stop marching.

Halt Order. An order issued when a command approaches its camping place, giving instructions for the encampment and necessary sanitary measures of the command. A halt and outpost order.

Halving and Coincidence Adjustment. An adjustment of the coincidence type of range finder whereby the proper proportions of images appear in the erect and inverted fields and coincidence is secured on objects at known ranges or at infinity.

Halyards, or Halliards. The ropes used in raising and lowering flags on poles or staffs.

Hammer. That part of a gunlock which strikes the percussion cap or firing pin. To push an attack with vigor.

Hand Arms. Individual weapons, carried and operated by hand, including both cutting and thrusting and missile weapons.

Hand Cannon. Various types of small, crude cannon, appearing as early as 1400 A. D., adapted for use as individual weapons by fitting with wooden stocks of various forms, which rested against the breast or shoulder, under the arm or over the shoulder of the operator. The cannon was thus the prototype of the musket.

Hand Firearms. Weapons fired from one hand, as pistols and revolvers.

Hand Fuze Setter. A rapid, portable fuze setter used by the field artillery.

Hand Grenade. A small grenade designed to be thrown by hand.

Hand Guard. A wooden covering for the barrel of a rifle which protects the hand of the rifleman from the heat of the barrel. A part of the hilt or handle of a sword, dagger, lance or other hand weapon, which protects the hand of the user.

Hand Gun, Hand Culverin, Escopette. An old name for a small gun used in the 15th and 16th centuries. Any gun carried and fired in the hand. The hand gun was derived from cannon and came into use in 1446 to fill the need for a weapon for the individual. At first it consisted of a short iron tube, prolonged behind the breech into a rod which was used to manipulate the gun and which was tucked under the arm when the gun was fired. Later a straight wooden stock was added, which rested upon the shoulder; the short curved stock, which rested against the breast, followed this. The gun was fired by the application of a match to a touchhole at the top of the barrel. The harquebus was a development of this gun. Any gun carried by the individual soldier in medieval times.

Hand Litter. A litter or stretcher carried by either 2 or 4 men.

Hand Salute. The prescribed salute rendered with the hand when the rifle is not carried.

Hand-to-hand Fighting. Immediate personal combat with the enemy.

Hang Fire. The failure of a weapon to fire, or delay in firing, due to any cause.

Hangar, or Hanger. A structure for the housing of airplanes.

Harass. To annoy and disturb the enemy by continual fire, raids, frequent small attacks, etc.

Harassing Agents. Chemical agents designed to reduce the efficiency of hostile troops by compelling them to wear gas masks, with consequent reduction of their mobility and power of endurance. Chemical agents whose physiological action is limited to producing an intolerable irritation.

Harbor Defense District. See Coast Artillery District.

Harbor Defense Artillery. Includes all seacoast artillery, fixed, tractor drawn, and railway, employed as a part of a harbor defense and organized primarily for defense against hostile naval operations. Submarine mines and obstructions usually supplement the artillery defense.

Harbor Defenses. The strong points of coast defense. A fort or forts with all armament, personnel and accessories, including controlled mines and aircraft, provided for the defense of a harbor.

Hard Bread. A hard cracker issued as a component of the ration.

Harmonizing Rifles or Rifle Sights. Adjusting the sights of a number of rifles so that when fired at short range, as on a 1000 inch target, any shot will strike at a point a fixed distance above the point aimed at.

Harquebus. A hand firearm, the prototype of the modern military rifle, which appeared about the middle of the 15th century. It superseded the perfected crossbow, which it resembled in some details. It was originally a matchlock, but was subsequently fitted with a wheellock, and finally with a flintlock. It was at first so heavy as to require a forked rest in firing, but lighter types were later used. The introduction of this weapon greatly increased the fire power of the infantry. It was effective up to about 200 yards. It was superseded by the musket in the latter part of the 16th century. The harquebus was variously known as Arbalist; Arcobugio; Archobugio; Arquebus; Crochert; Hack; Hackbush; Hackbut; Hagenbushe; Hagbut; Haquebut; Hookbut.

Harry. To make a hostile invasion or raid upon with destruction or seizure of property. To harass, ravage, pillage, or lay waste.

Hasta. The light lance of the Roman legion, which varied from three to five and a half feet in length.

Hasty Intrenchments. Informal and incomplete intrenchments quickly constructed, often in the presence of the enemy, when there is no reason or opportunity for more elaborate defenses. Hasty intrenchments are often the first phase of deliberate intrenchments. See also Deliberate Intrenchments.

Hasty Organization. Organization of the ground when in contact with the enemy with only a short period of time available for completion. Informal organization.

Hasty Sling. A gunsling adjustment used to steady the rifle while firing, by passing it under and behind the left arm. See also Loop Sling.

Hasty Trenches. Trenches constructed on the battlefield during mobile situations, under enemy fire or in the near presence of the enemy.

Hat Cord. A cord of differing colors to represent the different branches of the service, worn on the service hat. A cord, braided with gold thread, worn by officers.

Haversack. A case, usually of cloth, in which the soldier carries rations, toilet articles, and small articles of clothing on the march or in the field. See also Knapsack.

Hawkins Rifle. A celebrated rifle made in St. Louis about 1800.

Head. The leading element of a column, in whatever direction the column may be facing or moving.

Head Cover. A shield or other cover, which protects the heads of troops from fire. See also Overhead Cover.

Head of an Army. The front of an army.

Head Wind. A wind blowing head-on or from the front, parallel to the line of flight or travel, or nearly so, which reduces the speed of an aircraft, vessel, or projectile.

Heading. The angular direction of the longitudinal axis of the aircraft with respect to true north; in other words, it is the course with the drift corrections applied; it is the true heading unless otherwise designated. The most advanced end of a sap or gallery, where excavation is in progress. The first part of an order.

Head-log. A log placed on chocks or mounds on the fire crest to protect the heads of riflemen who fire under it; a form of head cover.

Headphone. A telephone receiver held over the ear by a band, leaving the hand free. See also Headset.

Headquarters. The office or command post of the commanding officer and staff of an organization, from which orders are issued and administrative and tactical control are exercised.

Headquarters and Service Company. A single company in which the functions of a headquarters company and a service company are combined.

Headquarters and Staff. The personnel, including the commander, staff, etc., which functions at the headquarters of any command.

Headquarters Company, Battery, Troop. An administrative organization which is usually an element of each tactical organization of the size of a battalion or larger; it furnishes the personnel for and performs duties in connection with administration, intelligence, communications, and certain other functions.

Headset. A telephone in which both receiver and transmitter may be attached to the head, leaving the hands free.

Headspace. The adjustable space between the rear end of the barrel and the front end of the bolt of a machine gun.

Heavier-than-air. A term generally applied to aircraft which do not use a gas lighter than air for sustentation; the airplane, of course, being the only practical form.

Heavy Cavalry. Heavily armed cavalry composed of men and animals of large size.

Heavy Field Artillery. A term used to designate the heavier guns and howitzers of the field artillery; they include the 155 millimeter and 6 inch guns and all guns and howitzers of larger caliber.

Heavy Marching Order. The complete individual equipment for permanent and continuous field service, when carried on the person.

Heavy Shellproof Shelter. A shelter which affords protection against continuous bombardment by at least 8 inch shells.

Heavy Tank. A tank of over 35 tons in weight.

Hectograph. A hand apparatus for reproducing small maps in colors. See Duplicator.

Hedgehog. A portable obstacle consisting of a wooden frame of three crosspieces at right angles to each other, strung with barbed wire.

Heel. The corner of the butt of a rifle which is to the rear in the position of order arms. The part of a sword blade next to the hilt.

Heel-piece or **Heel-plate.** Armor for the heels. The plate on the butt of a rifle.

Height Finder. An instrument used to determine the altitude of aerial targets. There are several types, e.g., the coincidence type and the stereoscopic type.

Height of Burst. The vertical angle between the base of the objective and the point of burst as seen from the gun. A normal height of burst is one that gives the maximum effect at the target.

Height of Site. The altitude of a gun above the assumed datum plane, generally mean sea level, or mean low water.

Helicopter. See Gyroplane.

Heliogram. A message transmitted by heliograph.

Heliograph, or Heliotrope. An instrument for signalling by means of the sun's rays reflected by a mirror. To signal by means of a heliograph.

Helium. A non-inflammable gas used as a filler for aerostats.

Helmet. A defensive covering for the head.

Herald. An officer whose duty is to bear messages, challenges, or information. An official bearer of important news. To announce or proclaim.

H Hour. The hour designated for an attack to be launched, or for a movement to begin

Hierarchy. A body of officials, military or civil, arranged in orders and grades. The chain of command.

High Command. Includes all echelons in military command from the President of the United States down to commanders of corps areas and territorial departments in time of peace and to and including commanders of armies in time of war. Leaders of the highest rank, such as the commander-in-chief of all the forces of a nation, an army commander, etc.

High Explosive Shell. A shell with comparatively thin walls, containing a bursting charge of high explosive; used in practically all guns and mortars.

High Explosive Shrapnel. A shrapnel shell containing a matrix of high explosive in place of the inert matrix contained in ordinary shrapnel shells.

High Explosives. Explosives of great force, in which the combustion or explosive reaction is transmitted so rapidly that it is practically simultaneous throughout the mass. High explosives produce a shattering effect; hence they are not suitable as propelling charges in firearms. They are used to charge shells, bombs and grenades, to detonate other explosives, and for demolitions. See also Low Explosives.

High Morale. A condition which exists when troops respond readily and willingly to the will of their commander and manifest general satisfaction with the conditions under which they are serving.

High-burst Ranging. Adjustment of fire by observation of air bursts.

High-burst Transfer. A transfer of fire from a check point in the air. See Fire, Transfer of.

High-wing (Low-wing) Monoplane. A monoplane with the wing attached near the top (bottom) of the fuselage.

High-wire Entanglement. A barbed wire entanglement consisting of a series of parallel fences, about four feet high and usually about ten feet apart, connected by diagonal wires; a very formidable obstacle.

Hilt. The handle of a cutting or thrusting weapon.

Hippology. The study of the structure, disposition, endurance, diseases and care of the horse.

Historical Ride. A tactical exercise in which an historical incident, such as a battle or campaign, is studied on the ground on which it took place.

Hit. An impact actually on the target. To reach or strike an object aimed at.

Hitch. A knot, especially one used to attach a rope to some other object. A slang expression meaning a term of enlistment. To attach a team of animals to a vehicle.

Hitting Area. An area, symmetrical with respect to the center of impact, such that if a target is included therein, there will be a reasonable probability of hitting; arbitrarily taken as extending three probable errors on each side of the apparent center of impact, both in range and in direction.

Hitting Volume. See Gun Hitting Volume.

Hoist. The perpendicular height of a flag. A crab, windlass or winch, used to lift a weight vertically, especially one operated by mechanical power.

Hold. To maintain or retain possession of by force, as a position or area.

Holdfast. An anchorage for a rope or cable. A picket holdfast consists of a number of pickets inclined to the rear, the top of each being lashed to the bottom of the one next in rear. A deadman.

Holding. The skill required to squeeze the trigger of a small arm without deranging the aim.

Holding and Reconsignment Point. A rail or motor center with considerable capacity to which cars or trucks may be sent to be held until their destination is determined.

Holding Attack. An attack made by a portion of a force for the purpose of holding the enemy in position, preventing him from shifting his reserves or making other arrangements to meet an attack at a different point or from a different direction. In an envelopment of the enemy the holding attack is usually the frontal and secondary attack, while the enveloping attack is the principal or decisive attack.

Holding Force. A containing force. A force which makes a holding attack. Any element of the garrison of a defensive position, or any component thereof, whose principal mission is that of passive defense.

Hollow Square. A formation in which troops are drawn up in a square with a hollow space in the middle, for the colors, etc.

Holster. A leather or web case for a revolver or pistol, usually conforming in shape and carried on the belt or saddle. See Scabbard.

Home Guard. A local organization, usually of men not fit or required for active military service, formed for local defense or to maintain order in case the regular militia organization is called away.

Home Service. Service in the home country or locality, in contradistinction to active service against the enemy.

Home Station. The place designated for the permanent headquarters of a unit.

Honest and Faithful Board. A colloquial term for a board of officers that examines the records of officers placed in Class B to determine whether or not their service has been honest and faithful. If an officer's service is found to be honest and faithful he is placed on the retired list, otherwise he is discharged.

Honor. A decoration for distinguished service. A consciousness of worth and virtue in the individual. Consideration due or paid as a reward for distinguished service. To regard with deference or respect. To bestow marks of honor or esteem upon.

Honor Graduate. One of a small number of the highest graduates of a school.

Honorable Discharge. A discharge which honorably releases a soldier from the term of enlistment to which it applies.

Honorary Rank. Rank which confers title and precedence without command.

Honors. Marks of courtesy or respect paid to persons entitled to receive them, on arrival at and departure from a post, camp, or station where troops, a band or field music are present.

Honors of War. Considerations or honorable terms or conditions granted to a defeated enemy, such as marching out of camp or position with arms and colors flying, with all baggage, etc.

Hood. A foot covering attached to a stirrup. A movable protective covering for an engine or other machine. A flexible covering for the head and back of the neck, attached to an overcoat.

Hook, Arresting. A hook attached to an airplane which engages the arresting gear in landing.

Hook-up. A slang term for an assemblage of apparatus used in radio transmission or reception. Any combination of electrical instruments or other machinery for a particular purpose.

Hoop. A supporting cylinder screwed or shrunk on the tube of a gun.

Horizontal Base System. A method of locating targets by using azimuths (directions) from two base end stations. It is based on the geometrical theorem that a triangle can be constructed when one side and the two adjacent angles are known.

Horizontal Equivalent. In map reading, the horizontal distance between two adjacent contours for any given slope. Map distance.

Horizontal Jump. See Jump.

Horizontal Range. The distance to which a gun will project a shell on a horizontal plane. The horizontal projection of the line gun-target. The base of the vertical triangle in space, of which the vertical leg is the altitude and the hypothenuse is the line of position. See also Curved Range.

Horizontal Shot Group. A shot group on a horizontal target or surface.

Horizontal Stabilizer. See Stabilizer.

Horizontal Velocity. The horizontal component of the velocity at any point of the trajectory.

Horizontal Vrille. See Barrel Roll.

Horn. The extremity of a body of troops in crescent formation. A short lever attached to a control surface of an aircraft, to which the operating wire or rod is connected. The operating lever of a control surface of an aircraft. The pommel, or forward part of the bow of a saddle.

Hornwork. An outwork composed of two demi-bastions joined by a curtain, connected with the works in rear by long, almost parallel, wings.

Hors de Combat. Incapable of further action.

Horse. A common name for a body of cavalry. Mounted soldiery.

Horse and Foot. A force including both mounted and dismounted elements. The cavalry and infantry.

Horse Artillery. See Artillery, Horse.

Horse Grenadiers. Mounted grenadiers who fight either mounted or on foot.

Horse Guards. A regiment of English cavalry familiarly known as the Oxford Blues, the third cavalry regiment of the Household Brigade.

Horse Length. A term of measurement which, for convenience in estimating distance is considered to be 3 yards; actually it is about 8 feet.

Horse Pistol. A large pistol such as was formerly carried by a horseman.

Horse Trench. A trench for sheltering horses or mules.

Horse-drawn Artillery. See Artillery, Horse-drawn.

Horseholder. A trooper assigned to hold or guard a number of horses, as during a dismounted action of cavalry.

Horses, Mobile, Immobile. Led horses of a mounted force are said to be mobile when the number of horseholders is sufficient to move the horses readily from place to place, as one man to each four horses. When the number of horseholders is insufficient to do this, the horses are said to be immobile, and each dismounted trooper must return to his own mount.

Hospital. A place for the care of the sick and wounded.

Hospital Battalion. One of the battalions of the medical regiment, including a headquarters and 3 hospital companies, whose function is the establishment of hospital stations in the field.

Hospital Flag. A flag used to indicate the location of a hospital, dressing or ambulance station.

Hospital Ship. A ship equipped as a hospital.

Hospital Station. A field hospital.

Hospital Tent. A large tent intended for hospital purposes.

Hospital Trains. Trains used to transport patients from the combat zone to hospitals in the rear.

Hospitalization. Includes the personnel, shelter, beds, messes, operating rooms, and all other appurtenances necessary in providing adequate shelter, care and treatment for the sick and wounded. The procedure of placing in and caring for in a hospital.

Hospitalize. To place in a hospital for treatment.

Hostage. A person held as a pledge or guarantee for the performance of certain conditions or obligations. Anything given or held as a pledge.

Hostilities. A rupture in relations between people or nations, involving armed conflict. A state of war.

Hotchkiss Machine Gun. An early machine gun. A machine gun made at the Hotchkiss works in France, in several models and calibers; used in the French, Japanese and other armies.

Hour of Signature. The hour stated in the heading of an order.

Housewife. A small case containing needles, thread, buttons, thimble, pins, scissors, etc., carried by soldiers for convenient use in repairing clothing.

Hover. To remain in the air with little or no horizontal or vertical motion. To cruise at a speed just above stalling speed in order to fly as slowly as possible.

Howitzer. A comparatively short cannon with a relatively low muzzle velocity and curved trajectory; intermediate between the gun and mortar.

Howitzer Company. An infantry unit armed with small cannon. A cannon company.

Howitzer Company Weapons. Specifically, a light cannon of 37mm (about 1½ inches) caliber, and a small mortar of 81mm (about 3 inches) caliber. (Old.)

Hub. The hilt of a weapon. A stake placed in the ground to mark a point located by a survey. The central part of a wheel.

Hull, Airship. The main structure of a rigid airship consisting of a covered elongated framework which incloses the gas cells and supports the cars and equipment.

Hull, Seaplane. The float of a seaplane or flying boat which supports it when in contact with the water, and contains accommodations for crew and passengers. A combined float and fuselage.

Hundred Per Cent Zone. See Beaten Zone; Dispersion Diagram; Zone of Dispersion.

Hurdles. Woven rectangles of strong wicker or brush work used in revetting, flooring bridges, and for other purposes.

Hussar. Originally, a name applied to some of the cavalry of Hungary and Croatia; later applied to light cavalry regiments of modern Europe.

Hydraulic Gun Carriage. A gun carriage having hydraulic apparatus to check recoil.

Hydrography. The science of coastal, riparian and under-water surveying, especially in the interests of navigation.

Hydrophone. An instrument for listening to sounds transmitted through the water, as in the detection of the presence of submarines.

Hydroplane. A submerged surface, like a fin, which acts in the same manner as an airfoil in motion. It governs the movement of a submarine or lifts a surface craft partially out of the water, thus increasing its speed. A light motor boat, equipped with hydroplanes, and driven by water screws or aerial propellers. A term incorrectly applied to a seaplane. See Plane.

Hydro-pneumatic Gun Carriage. A carriage in which a combination of hydraulic and pneumatic apparatus is used to check recoil and return the gun to its firing position.

Hypothetical Target. See Gun Hitting Volume.

I

Identification Marks. Marks on aircraft to indicate country of origin, type, or make, etc.

Identification Panels. Panels of cloth or other easily handled material which are displayed by ground troops on signal from friendly command aircraft to indicate the position reached or occupied by a unit.

Identification Record. The identification record of a person includes his signature, finger prints, height, general description, date of birth, etc.

Identification Tag. A metal tag worn on the person by means of which the wearer can be identified if killed or wounded.

Identifications. Any means by which hostile units, personnel or equipment may be identified. These include prisoners, deserters, captured documents, materiel, and uniforms, including insignia or other distinctive marks.

Igniting Charge. A charge, usually of black powder, placed in contact with the propelling charge to insure prompt and complete inflammation of the latter; often included in the primer.

Ignition. The setting on fire of a part of the grain or charge in a gun or projectile, or of the explosive mixture in the cylinder of an internal combustion engine. Setting on fire.

Illuminating Lights. Those lights of a searchlight group which are employed normally for lighting up enemy aircraft that have been located by the pilot light. Searchlights used to illuminate a target in order that it may be tracked and fired upon.

Illuminating Shell. A shell discharged from a gun that, upon bursting in air, ignites and releases one or more flares, often supported by parachutes, for purposes of illumination. A star shell.

Immediate Action. The immediate and automatic application of a probable remedy for a stoppage of fire of an automatic weapon. In immediate action, field guns are unlimbered and fire is opened as promptly as possible upon indicated targets.

Immelman Turn, Normal. A maneuver made by completing the first half of a normal loop; from the inverted position at the top of the loop, half-rolling the airplane to the level position, thus obtaining 180 degrees change in direction, simultaneously with a gain in altitude.

Immobilization. The tying down of a force to a particular place or position. Prevention of movement.

Immobilize. To tie down. Te deprive of mobility.

Immunization. A treatment, usually an inoculation, whereby a person or animal is rendered immune to a particular communicable disease. Prophylaxis.

Impact. The strike of a projectile on a target or the terrain. A blow or stroke of a body in motion against another body. An impact on the water is called a splash.

Impact, Center of. See Center of Impact.

Impress. To seize persons or property for public service. To compel a person to enter the military service.

Impressment. The seizure of persons, supplies, or property for military use.

In Abatage. The position of a gun when the wheels rest on the brake shoes.

In Action. A cannon or machine gun is said to be in action when mounted on its carriage or tripod, and firing, or loaded and ready to fire.

In Battery. Guns in position and in readiness for firing.

In Garrison. Doing duty in a fort or garrison.

In Kind. Actual, as rations in kind.

In Line. A formation with the units abreast.

In Line of Duty. See Line of Duty.

In Mass. In close column, line, or other formation, with less than normal intervals and distances. See Mass.

In Observation. Guns unlimbered and all preparations made for opening fire at the desired moment upon existing or expected targets. Detached and posted to observe the movements of the enemy, or a designated area.

In Place Halt. The command to stay the execution of a movement in drill for the correction of errors.

In Position. Guns in position and ready to fire, and the necessary systems of observation and communication established.

In Principle. A term which implies the principles and practices, based on common sense and experience, which are correct and normal in any par ticular procedure or situation, and are to be taken as a guide. For example, in principle, fire trenches should be placed just below the military crest, on a forward slope. They will be so placed unless there is a good reason, in a particular situation, why they cannot or should not. In theory. As a rule.

n Readiness. See Position in Readiness.

n Reserve. In rear of the front line acting as reserve to front line units.

n Series. A number of electrical resistances, such as lamps, instruments, or cells, in a circuit are in series when they are on a single circuit, the entire current passing in succession through each; i.e., when the negative pole of each is connected to the positive pole of the next.

n the Air. A term signifying that the flank of a force is not protected by natural obstacles or a supporting force, and is vulnerable to attack.

n the Clear. Not in code. A term meaning that a dispatch is written or sent in common language. The placing of an electric line so that it is free from obstructions.

n the Execution of His Office. To be engaged in any act or service required or authorized to be done by statute, regulation, the legal order of a superior, or military usage or necessity.

nactive. Not immediately available for active duty. In reserve for certain contingencies. Authorized but not actually organized.

nactive List. A list of personnel which may be called to duty in certain cases or under certain restrictions. A retired list.

nactive National Guard. Units allotted to the several states but not authorized for organization or equipment. Federally recognized inactive personnel. The reserve of the National Guard.

ncendiary, Incendiary Agent. A chemical which causes destruction, primarily of materiel, by ignition or combustion. Incendiaries include spontaneously inflammable materials, such as phosphorus, and liquids such as phosphorus dissolved in carbon disulphide; metallic oxides such as thermit; oxidizing combustibles such as magnesium; and inflammable substances such as pitch, oil and resin. They are applied by means of incendiary arrows, bullets, flame-throwers, projectors, drums, grenades, shells and bombs. Their use is not limited to the battlefield. Incendiaries have been used in battle for thousands of years, but the World War gave a great impetus, due to the longer ranges at which they could be applied, by the artillery and air service.

ncendiary Arrow, Grenade. An arrow or grenade conveying an incendiary.

ncendiary Shell. A shell containing a bursting charge and incendiary.

ncidental Protection. Protection incidentally enjoyed as a result of being near antiaircraft or other defensive weapons, or troops, posted for other purposes.

ncised Wounds. Wounds made by cutting and thrusting weapons.

nclination of the Trajectory. The angle between the tangent to the trajectory and the horizontal at any point.

ncline. A diagonal movement for the purpose of gaining ground simultaneously to the front and a flank. An inclined surface or ramp. An entrance to a cave shelter. To make a change of direction or movement to front and flank.

nclined Sights. A condition which exists when a piece is canted.

nclinometer. An instument for measuring inclination or slope, as of the ground or of an aircraft with reference to the ground. A clinometer. A dipping compass.

nclosure. A paper, not an integral part of but explanatory or accessory thereto, attached to a basic communication.

ncrease the Gait. To change from a slower to a faster gait, as from a walk to a trot.

ncrement. One of a series of additions or increases in strength or number. That which is added or gained. An additional propelling charge which increases the muzzle velocity and range of a projectile.

ncursion. The invasion of a sovereign state with hostile intentions. A temporary invasion. A raid.

ndemnification or Indemnity. An allowance for losses sustained on actual service. Repayment or compensation for damage suffered.

ndependent. Acting alone and without support. Not under the orders of a particular commander.

Independent Cavalry. Cavalry which precedes divisions and corps and acts under the orders of the army commander.

Independent Light. A searchlight used for a particular mission without reference to the use made of the other lights of the battery.

Independent Unit. A unit which is not a part of a larger unit at the time in question.

Indeterminate Error. In firing, an error, either systematic or non-systematic, of such nature that it is incapable of precise determination.

Index Error. The amount that a scale of any kind reads when it should read zero. The correction to be applied is numerically equal to the error and of opposite sign.

Indian File. Single file. The usual manner among Indians of traversing woods.

Indirect Laying. Laying a piece in direction by the use of a sight and an aiming point other than the target, or by the azimuth circle on the carriage, and in elevation by range drum or quadrant; the target is usually invisible to the gunner. See Laying; Pointing.

Indirect Laying Position. An artillery position from which direct laying is precluded because of a mask or cover, or for tactical reasons.

Individual Control. A method of fire control employed with machine guns in which fire is controlled by the individual gunner.

Individual Equipment. See Infantry Equipment.

Individual or **Miniature Intrenching Tools.** Small tools carried on the persons of soldiers for use in field fortification. They commonly include a shovel, pick mattock, hand-axe, and wire-cutting pliers, one tool being carried by each man.

Individual Practice. The firing on the range by individuals for instruction purposes and to determine their classification.

Individual Proficiency. The possession by each person in the military service of a knowledge of the duties and responsibilities of his grade, together with the determination to perform and exercise them under all circumstances. A standard or criterion of individual proficiency.

Individual Protection. Apparatus and measures for protection, as against chemical agents, etc., applicable primarily to the individual. Cover or shelter provided by an individual for his own use.

Individual Requirements. The supplies and equipment required to enable the individual to function as a soldier.

Individual Rolling Shield. A small mounted bombproof, capable of being controlled from inside, which protects a soldier in his advance toward the enemy.

Individual Trench. A trench or pit for the accommodation of a single individual.

Indoors. In the rendering of salutes and courtesies offices, hallways, kitchens, amusement rooms, bathrooms, dwellings, etc., are considered as indoors.

Indorsement. A statement on a basic communication made by any commander or office in the channel of communication.

Induced Detonation. Detonation of high explosive resulting from the explosion of another charge nearby. Sympathetic detonation.

Induct. To initiate or bring into the service by the customary ceremonies.

Induction. The process whereby a registrant under selective service is transformed from his status as a civilian to that of a member of the armed forces. Production of current in an electric circuit when it is moved in a magnetic field, the principle of the dynamo. Production of current in a coil of wire by variations in the current in another coil, the principle of the induction coil and static transformer. Temporary magnetization of soft iron when near a magnet or electric current.

Induction Station. The place to which draftees are sent by their draft board to have final physical examination, be sworn into the service, and where initial records are prepared. See also Reception Center and Reception Training Center.

Industrial Control Agency. An agency to which the President delegates authority to coordinate and control certain war activities of the executive departments of the government and of the industry of the United States.

Industrial Mobilization. The diversion from normal tasks of such part of the nation as may be necessary to insure the procurement in such quantities and at such times as needed of the material requirements of the armed forces in war.

dustrial Mobilization Plan. The program prepared in time of peace under the incentive of the Assistant Secretary of War in response to statutory requirement, to accomplish industrial mobilization in war.

fantry. The largest and most important combatant branch of an army, which marches and fights on foot, and whose principal weapons are the rifle and bayonet. The earliest soldiers in most countries were foot soldiers, if for no other reason than lack of suitable mounts. The organization and tactics of infantry in all ages have been largely determined by the weapons in vogue. When armed with the pike the infantry was formed in compact masses to insure discipline and solidarity, and counteract the weakness of the individual soldier. This form of tactics found its highest expression in the Greek phalanx and the Roman legion, which were as well trained and disciplined infantry as the world has ever seen. But after the overthrow of the Roman Empire, infantry as a decisive element in battle practically ceased to exist. During the Dark Ages the knights or cavalry became the principal combatant element. The infantry degenerated into a mere rabble, armed with a miscellany of weapons, not disciplined or uniformed. Often it did not appear at all on the field of battle, and it was never a decisive influence. But the prowess of the Swiss mountaineers and the exploits of the English bowmen in the 14th century restored the infantry to something approaching the dignity it had enjoyed under the Greeks and the Romans. By the middle of the 15th century infantry, with improved weapons and tactics, better discipline and training, was reestablished side by side with the cavalry.

The birth of modern infantry dates from the introduction of firearms, and its development and growth in importance have been in direct proportion to the improvement in its principal weapon, the musket and its successor, the modern military rifle. Musketry fire was slow, and for a long time after the introduction of the musket pikemen were retained to support and supplement the musketeers. As the range and rapidity of fire of the musket increased the pikemen gradually disappeared. Today the infantry soldier with his bayonet is the last representative of the ancient. pikeman. Coincident with this development there was a very gradual improvement in infantry tactics and organization. The number of ranks was reduced, and a more flexible organization, possessing a degree of maneuverability, was developed.

These improvements were gradual, as was the improvement in the musket, but such great commanders as Gustavus Adolphus and Frederick the Great recognized that in mobility and fire action lay the real strength of infantry, and they brought both to a degree of perfection never before realized.

The next great advance, quite fortuitous and unstudied, occurred in the American Revolution. The American colonists were skilled in the use of firearms, but quite unused to discipline, orderly formation or drill. These untrained men were thrown forward in thin lines of skirmishers to harass and delay the enemy. Eventually, skirmishing was regularly employed and became a recognized element of infantry tactics. During the Napoleonic Wars these methods were utilized and greatly improved. Small columns were used for maneuver, and deployed lines for firing.

In the American Civil War the infantry tactics of today were foreshadowed. The principal developments were the use of successive lines of skirmishers, entirely replacing the old precise lines of battle, the advance by rushes, and the use of hasty intrenchments, or field fortifications.

The World War ushered in such modifications of infantry tactics as were necessitated by the greatly increased fire power of all weapons, better observation and means of communication. The principal features were the increase in the volume and accuracy of artillery and machine gun fire, resulting in the intensified use of field fortifications, deployment in great depth with strong reserves, covering fire for the infantry in both attack and defense, and the development of the counter attack as an element of the defense. New weapons made their appearance and old ones were greatly improved.

Infantry Cannon. Small guns which form part of the armament of infantr'

Infantry Equipment. The individual outfit of the infantryman or foo' soldier. The term properly comprises all dead weight on the soldier's bod' but is usually construed to include all except his weapon and the clothing o his person. At various times it has been carried in a knapsack, haversac' or a roll, or combination of these carriers; the whole being called a pac' In all modern armies the pack is supported between the shoulders by strap' in the manner that an Indian squaw carries her papoose. The more in' portant items of the equipment are the shelter (half) tent, bedding, ove' coat or raincoat, extra clothing and shoes, mess kit, toilet articles, ration' intrenching tool, etc. In addition the modern soldier must carry a gas mas' and steel helmet, filled canteen, first aid kit and ammunition. The la' three items are, in our service, supported by a waist (cartridge) belt wi' suspenders, and bandoleers. On the march or on entering battle the soldi' discards part of his equipment, and takes on extra ammunition. T' infantry equipment is scientifically designed to be as efficient and as lig' as possible. The total load on the soldier should not exceed about ⅓ h' weight, or 50 lbs. for an average man; and 45% or 67 lbs. is a maximu' that should never be exceeded. Actually the total of essential equipme' considerably exceeds the lower figure, and constant studies are made wi' a view to reducing its weight. At present the solution is to carry part ' the equipment on the combat train.

Infantry Liaison Plane. An observation airplane whose principal missio' for the time being, is the maintenance of liaison between the infant' front lines and higher headquarters.

Infantry Tactics. The art and science of maneuvering infantry and en' ploying it in cooperation with other arms.

Infantry Weapons. The various weapons with which the infantry is arme' In the United States Army they include at present: The M1903 rif' (Springfield); The M1 (Garand) semiautomatic rifle; the M1 carbine; t' bayonet; the pistol; the hand grenade; the .30 caliber machine gun, lig' and heavy; the Browning automatic rifle; the .50 caliber machine gun; t' 60mm and 81mm light mortars; the 37mm gun and the 37mm antitank gu'

Infantryman. An infantry soldier.

Inferior. Junior in rank. A subordinate.

Inferior Court-Martial. Any court below a general court-martial.

Infiltrate. To pass, or cause to pass, troops in relatively small numbe' through gaps in the enemy's position or in his field of fire. To advan' troops into or nearer a hostile territory or position by sending forwa' single men or small groups o lines of men at widely separated interval'

Infiltration. The act of infiltrating a hostile position, as a trench or woo'

Infinity Bar. An adjusting lath for a range finder.

Inflammation. The spread of ignition from point to point of the grai' or from grain to grain of a charge of explosive.

Information. See Military Information.

Information Patrol. A reconnoitering patrol.

Information, Sources of. See Sources of Military Information.

Initial. Marking the commencement. Beginning.

Initial Aiming Point. A definite point or object at as great a distance a' possible, on which guns are laid initially, and from which angles are mea' ured in determining direction data for machine gun fire by indirect layin' methods.

Initial Data. Data not corrected as a result of actual firing.

Initial Dispositions. The dispositions made at the beginning of an engag' ment

Initial Firing Position. The position designated from which attackin' troops will open fire on the enemy.

Initial Line or Point. A designated line or point at which the various el' ments of a command will arrive at the proper times to take their prescribe' places in a marching column without interfering with other elements.

Initial Location. The location prescribed for the various elements of ' command at the beginning of an operation or movement.

Initial Requirements. See Requirements, Initial.

Initial Velocity. The velocity at which a projectile leaves the muzzle of ' gun, expressed in feet per second.

Initiate. To begin. To set in motion.

Initiative. Energy or aptitude which enables a commander to conceive and carry out a plan, or to act with energy and foresight when a favorable opportunity offers. The right or power to act. In a military sense, the initiative means the power to lead and dictate the course of a campaign or battle. A combatant who, by vigorous aggressive action forces his opponent to abandon his own plans and devote all his efforts to meeting the moves of the attacker, is said to hold the initiative. Possession of the initiative is essential to success in war.

Initiators. High explosives which initiate the explosion of larger charges; they include priming compositions and detonators. They should be stored and transported separately from other explosives.

Inner Flank. The flank nearer the point of rest, or farther from the enemy.

Inroad. A sudden and desultory incursion. Encroachment.

Insignia. Distinguishing marks of authority, office, rank or honor, branch of the service. Badges. Emblems.

Insignia of Rank. Distinguishing marks of rank or grade.

Inspected and Condemned. See Condemned Property.

Inspection. A strict or critical examination. An examination and report as to the condition of accounts, property, state of training, etc.

Inspections and Standards. A term which implies the procedure in the inspection of a military unit with a view to determining its state of efficiency or readiness for service, with the standards or criteria by which proficiency in each detail is measured.

Inspector. A regularly detailed examining officer. One who inspects.

Inspector General's Department. That branch of the service which makes inspections and reports on all matters concerning the efficiency of the troops, their conduct and discipline, condition of uniforms, supplies, and expenditure of public funds for military purposes.

Installation. Any set-up or establishment. An entire plant, including its accessories.

Instantaneous Fuze. A fuze that bursts on impact with any object. A tube filled with a compound which burns at a very rapid rate; used in connection with a safety fuze for detonating demolition charges.

Instruction Practice. The prescribed firing on the range which precedes record practice and which is devoted to the instruction of the soldier.

Instructions. Military directions, specifically those conveyed in some manner other than by formal orders.

Instrument Flying. The art of navigating an aircraft solely by the use of instruments; sometimes called blind flying.

Instrument Landing. The landing of an aircraft by aid of its instruments, without seeing the terrain. A blind landing.

Insubordination. Deliberate disobedience to lawful orders or authority.

Insurgent. A person in a state of insurrection against his government or civil authority. A rebel, not generally recognized as a belligerent, who rises in opposition to civil or political authority. Insubordinate. Rebellious.

Insurrect. To rise up. To rebel against constituted authority.

Insurrection. A rising of people in arms against their government.

Integrity of Tactical Units. A military principle that tactical units should not be broken up unnecessarily; thus if six men are required, use a squad; if two squads are needed, use a section, etc.

Intelligence. See Military Intelligence.

Intelligence Maps. Special maps on which items of intelligence or certain classes of intelligence are shown graphically and by notation. Maps prepared for intelligence purposes.

Intelligence Net. Concentric cordons of observers to give warning of the approach of hostile aircraft toward a defended area. The various means of securing information within any command, considered as a whole.

Intelligence Officer. A staff officer of a brigade or lower unit, having charge of the collection, study and dissemination of military information. His duties correspond to those of the intelligence section (G-2) of the general staff.

Intelligence Section. A part of a headquarters or headquarters company charged with the collection, study, interpretation and distribution of mil-

itary information. A section of the general staff or other correspondi
staff, charged with such duty. An intelligence department or division.

Intelligence Situation Map. See Intelligence Maps, Situation Maps.

Inter-Allied. Of or pertaining to a number of allies.

Interbranch Procurement. The procurement of supplies by one supp
branch of the War Department from another such branch.

Intercept. The art of copying and recording messages between two or mo
intercommunicating agencies; usually applied to enemy radio transmi
sion. To copy and record enemy messages.

Intercept Station. A station that records enemy radio messages to obta
information; or friendly messages for the purpose of supervision.

Interceptor Plane. See Aviation, Classification of.

Interdict. To prevent or hinder the use of an area or route by the applic
tion of chemicals or fire, or both.

Interdiction Barrage. A barrage applied to a particular line, area, or poi
to prevent its use by the enemy.

Interference. Disturbance of reception due to strays, undesired signals,
other causes. That which produces the disturbance.

Interior Ballistics. That branch of the science of ballistics which deals wit
the movement of a projectile within the bore of a gun.

Interior Crest. The crest of the interior slope of the parapet of a fortific
tion. The fire crest of a trench.

Interior Economy. The interior regulation or management of an organiza
tion.

Interior Guards. Guards used in camp or garrison to preserve order, pr
tect property, and enforce police regulations.

Interior Lines. A term used to signify that the lines or routes available t
a combatant are such that he can concentrate on any line more quickl
than the enemy. See also Exterior Lines.

Interior Slope. The inner front slope of a rampart, breastwork or trench.

Interior Works. Works constructed within the parapet of the main work
of a fortification. They include cavaliers, redoubts, defensible barrack
citadel, etc.

Intermediate Armament. Coast defense guns of medium (5 and 6 inch) cali
ber.

Intermediate Depot. A general or branch depot, designated as such by th
commander of the theater of operations, and located in the intermediat
section of the communications zone, for the storage of balanced stocks fo
warded from base depots or procured in the intermediate section.

Intermediate Objective. The objective whose attainment precedes and i
usually essential to the attainment of an ultimate objective.

Intermediate Position. A position, usually three or four hours march fron
the assault position, where a tank unit is to make final preparation for en
tering action. An additional or supplementary defensive position, lyin
between two other positions.

Intermediate Section. That portion of the communications zone lying be
tween the advance and base sections.

Intern. To confine or restrict combatant troops that have taken refuge o
neutral territory. To confine an alien enemy.

International Law. The principles and rules of action which are acknowl
edged by civilized states as controlling their mutual relations.

International Morse Code. The general service code of the United States.

International Salute. A salute of 21 guns to a national flag.

Interpolation. The process of determining an intermediate value, as be
tween two graduations of a scale, or two tabular values, the elevation o
ground between contours, etc.

Interpretation of Information. An analysis of information to determine its
probable significance in the existing conditions.

Interrupted Screw. A device employed in the breech of certain cannon, con
sisting of alternate threaded and slotted sectors in the breech and breech
block, whereby the block may be engaged or disengaged by turning throug
a small arc. The French or Canet type of breechblock.

Interval. An open space between military units or elements on the same
line. The interval between men in ranks is 4 inches and is measured from
elbow to elbow; between companies, squads, etc., it is measured from the

left elbow of the left man or guide of the group on the right, to the right elbow of the right man or guide of the group on the left.

ntolerable Concentration. A concentration of a chemical agent which renders unprotected personnel helpless; usually applied to irritant chemicals only.

ntrench. To construct intrenchments. To protect a position or body of troops by constructing intrenchments. To fortify.

ntrenched. Provided with intrenchments, fortified.

ntrenched Camp. A large area, often comprising a city or town, surrounded by mutually supporting defensive works. A battle area fortified for defense.

ntrenching Tools. Tools used in field fortification. They include commercial type, pioneer, and individual (miniature) tools.

ntrenchment. An earthwork consisting of at least a parapet and a trench or pit. Any defense or protection provided by excavation. The construction of trenches.

nvade. To enter a country with hostile intent. To make an invasion.

nvasion. A hostile entrance into the territory of another nation. The incursion of an army for conquest or plunder.

nventory. A physical check of property. A verified list or schedule of property.

nventory and Inspection Report. A form on which unserviceable property is listed for inspection and recommendation as to disposal.

nvest. To surround so as to prevent ingress or egress.

nvestment. The act or state of investing a place or fortress. See Line of Investment.

nward Traverse. See Fire, Sweeping, Inward Traverse.

ron Cross. A German decoration in the shape of a Maltese Cross, awarded for gallantry.

rregular Operations. Operations by or against poorly organized or irregular forces.

rregular Troops. Troops which do not form part of the regular forces of a state or nation, usually raised in time of war only.

rregularity. A minor violation of orders, regulations or customs of the service. An offense less than a crime or misdemeanor.

rritant Candle. A smoke candle which produces a cloud of irritant (not screening) smoke.

rritant Gases. Chemical agents which cause sneezing and irritation of the eyes, nose and throat even in low concentrations, nausea and certain nervous phenomena. Lung irritants, sternutators, and lachrymators.

rritating Concentration. A concentration of chemical vapor sufficient to produce an irritant effect upon a man without injuring his body functions or seriously impairing his working efficiency.

Island Traverse. A detached traverse in a trench.

Isolation. Separation of a person suffering from a communicable disease from other persons, to prevent spread of the disease.

Issue. A delivery of supplies. Specifically, the delivery of supplies of any kind by a supply department to responsible persons authorized to receive them on behalf of their organizations. The supplies so delivered. To send out officially or publicly, as orders or communiques. To emerge or sally forth, as from a defile or fortress.

Items, Contributory. Components, materials or equipment used in the fabrication of primary items. Contributory items divide themselves into the following classification: (1) Those which must be purchased by a supply arm or service; (2) Those which must be purchased by prime contractors from sub-contractors.

Items, Essential. An item that is necessary to the combat efficiency of troops and which is unobtainable from civil stocks or production in time and quantity required.

Items, Finished. Items of military requirement in form for immediate use.

Items, Primary. Military supplies in the form delivered to the using arms or services.

Itinerary. A plan for a march, patrol, etc., sometimes accompanied by a map or sketch, and including information as to the route to be followed, conditions to be encountered, etc.

J

Jab. An upward thrust with the bayonet at close quarters. To thrust quickly. To stab. To punch.

Jacket. A short military coat. A hard metal covering surrounding the lead core of a rifle bullet. The principal forging shrunk on the breech end of the tube of a cannon.

Jacob's Ladder. A short rope ladder with wooden treads used in fortifications in the absence of ramps or steps.

Jam. A stoppage of a machine, as an automatic weapon, from any cause. Malfunctioning of a weapon. To cause interference in radio transmission. To render radio transmission unintelligible by sending out other radio signals or messages in an interfering manner.

Javelin. A short, light spear, thrown by hand, used before the introduction of firearms; it had a range of 30-40 paces. A thrusting weapon with a staff. A close, defensive formation of a squadron or group of airplanes.

Jerkin. A leather military coat. A buff military coat having a light collar.

Joint Army and Navy Exercises. Field maneuvers in which both the army and the navy participate.

Joint Board. An advisory board consisting of the heads of certain staff departments of the army and navy, which considers matters relating to joint action by the army and navy, and joint army and navy policy relative to national defense.

Joint Operations. See Combined Operations.

Joint Plan. A war plan whose purpose is to establish the basis and prepare the necessary plans for joint action of the army and navy in a given situation.

Journal. A brief summary of messages sent or received at a headquarters or message center, intelligence section etc., kept chronologically. That part of a shaft or axle which turns in a bearing.

Judge Advocate. An officer who conducts a trial by court-martial.

Judge Advocate General's Department. The legal branch of the United States Army.

Judgment Mine. A land mine which can be fired at any desired time, from any distant point, by means of an electric current.

Jump Curves. Curves determined from experimental firing, showing the lateral and vertical jump at all ranges. These corrections are incorporated in the firing tables.

Jump of a Gun. The angle between the axis of the bore when the gun is laid and as the projectile leaves the muzzle. The amount that the direction of the axis of the bore changes while the projectile is moving through the bore. Jump has a horizontal or lateral and a vertical component, which are allowed for in the firing tables.

Jump-off. The start of a prearranged attack against the enemy, usually on a wide front. To leave one's front lines with a view to attacking the enemy.

Jump-off Line. The line from which an attack starts or is to start. A line of departure in attack.

Junction Box. A casting used to make a watertight connection between two sections of submarine cable.

Junior. One having lower military rank. Lower in rank.

Junior Division of the Reserve Officers Training Corps. Students in high schools and private schools of corresponding grade, who receive military instruction in accordance with law.

Jurisdiction of a Court-martial. The power of a court-martial to try and determine cases legally referred to it and, in the case of a finding of guilty, to award punishment for the offense within the limits prescribed. Jurisdiction includes the person to be tried, the crime charged, and the locality in which the crime is alleged to have been committed.

K

K. The ratio of the adjusted or gun range to any point to the map range to the same point; used in making transfers to new targets.

Kampf. A battle, fight, combat, engagement, or action. (Ger.)

Kaserne. Barracks. (Ger.)

Keel. The assemblage of members at the bottom of the hull of a rigid or semirigid airship, which provides special strength to resist hogging and sagging and also serves to distribute the effect of concentrated loads along the hull.

Keep. The central and principal tower of medieval fortifications and castles, where the final defense was made. An interior, central stronghold. A fortress.

Kepi. A close fitting round cap with a curved visor. A flat topped cap with a horizontal visor. A forage cap.

Kettledrum. A drum made in the form of a kettle with parchment stretched over the mouth.

Key Enlisted Men. Certain men, necessary to an organization, requiring expert knowledge of some nature in order to function properly in the duties they are required to perform; e.g., clerk, cook, electrician, etc.

Key Phrase. An arbitrary phrase used as the base for a simple cipher alphabet.

Key Points. Points whose possession gives control of a position or area. Points that control the communications, or that afford observation either into a position or over the foreground, and the capture or possession of which is indispensable to the successful progress of an attack.

Key Position. A position which, by being strengthened, can be made a formidable point of defense. A point or position whose capture or retention is considered important.

Keyholing. Said of bullets that do not revolve about their longer axes, but tumble and pass through the target in various positions, often sidewise.

Khaki. A brownish cotton uniform cloth. A uniform made of khaki.

Kick. The recoil of a hand firearm on discharge.

Kilometer. One thousand meters, about 0.62 English statute miles, or 1112 yards.

Kit. The personal necessaries of a soldier that he is required to have in his possession. The necessaries of a soldier packed in a very small compass. An assemblage of tools or implements for a special purpose, as a flag kit.

Kitchen Police. Enlisted men detailed to help the cooks wash dishes, set the table, etc., in the organization mess.

Knapsack. A bag of canvas, or other material, designed to contain certain of the soldier's necessaries, and carried on his back. See also Haversack.

Kneeling Position. A firing position in which the soldier kneels on his right knee.

Kneeling Trench. A trench designed to shelter a rifleman and permit fire in the kneeling position.

Knife Rest. An obstacle consisting of a frame of wood or steel in the shape of a carving knife rest or a sawbuck, on which barbed wire is strung. A wire strung cheval-de-frise.

Knight. Originally, a man at arms who attended his feudal lord or sovereign on horseback in time of war; now, a title conferred as a mark of a sovereign's esteem or in recognition of services rendered.

Knot. A kind of epaulet or braided ornament. A thong on the hilt of a saber or sword whereby it may be attached to the wrist. A nautical mile, approximately 1.15 statute miles. A loop or fastening in a rope.

Krag-Jorgensen Rifle. A bolt type repeating rifle used in the armies of Denmark and Norway. It has a magazine capacity of 5 cartridges. Formerly used in the United States Army.

Kris or Creese. A kind of knife or bolo, the universal weapon of the Malay archipelago. Weapon of the Moros of Mindanao and Sulu.

Krieg. War (Ger.)

K-transfer. A transfer of fire based on the ratio of the adjusted or gun range to any point to the map range to the same point (K).

L

Labor Troops. Organizations that regularly furnish labor details to work under the supervision of technical troops; as engineer labor battalions, pioneer infantry.

Lacrimators, Lachrymators. Chemical agents which exert an intense irritant action on the eyes, and cause so profuse a flow of tears and so much discomfort that vision becomes impossible. Commonly known as tear gas. The principal tear gas is chloracetophenone.

Lacrimator or Lachrymose Shells. Shells charged with lacrimators (tear gas).

Lambert North. See Y-line.

Lambert System. The system of rectangular coordinates now employed on most military maps.

Lampert Bridge. A floating footbridge using canvas pontons.

Lance. A weapon consisting of a long shaft with a steel blade. A kind of spear carried by a horseman. A pointed staff, to which is attached the color or guidon of a mounted or motorized unit.

Lance Corporal, Sergeant. An acting corporal or sergeant.

Land Defenses. Fortifications employed in land warfare.

Land Forces. Troops intended and trained for land service only. Generally applied to the entire available land forces of a nation.

Land Forces of the United States. The organized land forces consist of the army of the United States; the unorganized forces include the unorganized militia, i.e., all persons who are liable by law for military service. See Army of the United States.

Land Grant. A grant of public lands to railroads, educational institutions etc. Land grant roads are required to furnish certain transportation to the government. Land grant colleges are required to maintain courses of military instruction under the Morrill Act, passed in 1862.

Land Mine. See Mine, Land.

Landing. The act of terminating flight in which the aircraft is made to descend, lose flying speed, establish contact with the ground or water, and finally to come to rest.

Landing Attack. A forced landing of troops on a shore, or an attack immediately following a landing.

Landing, Blind. A landing of an aircraft in low visibility.

Landing Field. Any area of land or water designated for the take-off and landing of aircraft.

Landing, Forced. A landing forced on a pilot because of engine failure, structural failure, lack of fuel, hostile fire, weather, etc.

Landing Gear. The understructure which supports the weight of an aircraft when in contact with the land or water and which usually contains a mechanism for absorbing the shock of landing.

Landing Speed. The minimum speed of an airplane at the instant of contact with the landing surface in a normal landing.

Landing Strip. A landing runway laid parallel to an important highway, or other place, for emergency use of airplanes. The Marston system of metal paving utilizes perforated steel sheets which can be hooked together to form a temporary pavement for a landing field.

Landmark. A prominent, easily identified feature of the terrain which may be used as a reference point in defining the boundaries of a sector, establishing a direction of march, designating a target, etc.

Landplane. An airplane designed to rise from and alight on land.

Lands. The raised portions of the bore of a rifled weapon between grooves.

Landscape Sketching. Sketching in perspective, as distinguished from topographical or cartographical sketching.

Landscape Target. A picture or representation of a landscape used for instruction in marksmanship and musketry.

Lane. A body of soldiers formed in two ranks, facing one another. A belt of land leading to the front, as a zone of action.

Languet. A small metal apron on the hilt of a sword, which overhangs the scabbard. The ear of a sword.

Lantaka. A small, swivel-mounted cannon, in various sizes, used by the Malay tribes.

Lanyard. A strong cord with a hook at one end, used in firing cannon. A cord used to attach a pistol to the wrist.

Large Scale Maps. See Maps, Classification of.

Large Units. Divisions and larger units.

Laryngaphone. A telephone in which the transmitter is strapped against the larynx instead of being held in front of the mouth; used in tanks.

Lashing. Any fastening made by means of a rope, as in a ponton or spar bridge, shears or tripod.

Latch. An ancient crossbow having a trigger like a latch.

Lateral. To one side. Measured along the front. Measured perpendicular to the plane of fire, or the line gun-target or directing point-target, as a deviation. As used in antiaircraft fire, the term means measured at right angles to the plane of position, as for an error or deviation; it corresponds to lateral, as applied to terrestrial targets.

Lateral Communications. Communications between different commands or the elements of a command which are approximately abreast of each other. Trenches constructed to facilitate lateral communication between strong points, etc.

Lateral Deflection Angle. In antiaircraft fire, the horizontal angle between the vertical plane through the present position and the vertical plane containing the axis of the bore when the gun is laid. The algebraic sum of the principal lateral deflection angle and the lateral pointing correction.

Lateral Deflection Setting. The lateral deflection scale setting of the sight corresponding to the lateral deflection angle.

Lateral Deviation. The lateral distance between the plane of direction and the point of impact of a projectile. The lateral divergence of a point or center of impact from the target.

Lateral Error. The lateral, horizontal divergence of a point of impact or burst from the center of impact.

Lateral Fork. See Fork.

Lateral Observation. Observation of fire from a point considerably displaced from the line of fire.

Lateral Pointing Correction. That part of the lateral deflection angle due to causes other than the travel of the target.

Latrine. A pit or trench dug for the deposit of excreta. A cesspool.

Law Member. That member of a general court-martial, designated as such by the appointing authority, whose duty is to rule on questions of law arising during the conduct of a case, in addition to performing his other duties as a member of the court.

Law of Hostile Occupation. See Military Government.

Law of Nations. International law.

Laws of War or **Arms.** Recognized rules governing the conduct of war among civilized nations.

Lawful Belligerent. Comprises not only members of regularly organized armies, but also militia and volunteer corps who: (1) Are commanded by a person responsible for his subordinates; (2) Have or wear a distinctive emblem or uniform recognizable at a distance; (3) Carry arms openly; and (4) Conduct their operations in accordance with the laws and usages of war.

Lay. To aim a gun, especially in cases where the target is not visible to the gunner. To impose, as a duty. To lay down one's arms.

Laying. The process of aiming or pointing a gun without directing the sight upon the target. See Case III Pointing; Indirect Laying.

Lead. To conduct or act as a leader. To command. To precede. To have precedence. To possess the initiative. To aim at a point on the course and in front of a rapidly moving target in order that the target and projectile may meet. The distance or deflection allowed in aiming at such a target. A target length, especially the length of an aircraft, used as a unit in measuring lead.

Lead Table. A table giving the approximate leads for antiaircraft fire at all ranges.

Lead Team, Horses or **Mules.** The leading pair of a 4-, 6-, or 8-line team of draft animals.

Leader. One having authority. A commander, usually of a smaller unit (squad, section, platoon).

Leader's Rule for Overhead Fire. A method for determining troop safety for overhead fire by use of the inverted sight leaf in the field glass. Applied to ranges over 900 yards. See also Gunner's Rule.

Leadership. The ability to lead or the qualifications of a leader. The possession by a leader of those qualities of character and professional attainments, and readiness to exercise responsibility, that inspire confidence and loyalty throughout his command.

Leading. The acts of a commander in exercising control of his unit by personal direction. Guiding. The farthest advanced. In the lead.

Leading Edge. The foremost edge of an airfoil or propeller blade. Also called entering edge.

Leading File. The first two men (front and rear rank men) who march in forming a column of files. The man at the head of a column of files.

Leading Guide. The foremost guide of a column.

Leading Troops. The advance guard of a force marching away from the enemy. (Old.)

Leaf Sight. A form of rear sight consisting of a hinged metal leaf or leaves.

League. A confederacy. An agreement. A compact. To unite in a league.

Lean-to. An improvised shelter affording protection against high-angle fire. A shelter formed by boards or other material leaning against a wall or steep slope.

Leapfrog. To advance the elements of a command in the attack by passing them successively through or by the other elements, which cover the advance of the moving elements.

Leave of Absence. Permission granted an officer to absent himself from his station or duty for a specified time.

Lebel Rifle. A bolt-type rifle which has a caliber of .315 inches, a magazine capacity of 5 cartridges, and weighs 9.21 pounds; used in the armies of France, Belgium, Greece, Rumania, and Yugoslavia. It was the first small calibered rifle and the first in which smokeless powder was used.

Led Horses. The mounts of a cavalry unit fighting on foot. Horses without riders linked to another animal or led by hand.

Lee-Enfield Rifle. A bolt-type rifle which has a caliber of .303 inches, a magazine capacity of 5 cartridges, and weighs 8.65 pounds. It is used in the armies of Great Britain, Canada, Egypt, and India.

Leeway. Lateral movement of a vessel or aircraft, due to the wind or component of the wind at right angles to its course.

Left. The left extremity or element, or the terrain to the left of a line or body of troops.

Left Wing. The portion or element to the left of the middle point or element of a force deployed for battle.

Legend. The key printed on a map or chart giving the explanation of the conventional signs used thereon to show natural features or works of man. An explanatory note on a map.

Legion. A term applied rather indefinitely to many military organizations in our own and other countries at various times. A legion usually consisted of a relatively large number and of various arms, constituting a complete fighting unit. An army. A multitude.

Legion of Honor. A military order instituted in France in 1802 by Napoleon 1.

Legion, Roman. A military organization of varying numbers at different periods of Roman history. About 300 B. C. its heavy infantry consisted of three classes. The hastati, or first line, comprised men between twenty-five and thirty years of age; the principes, or second line, were from thirty to forty; and the triarii, or third line, were veterans from forty to forty-five years of age. The velites, or light troops, consisted chiefly of young men from seventeen to twenty-five years of age. The tactical unit of each of the three lines was the maniple, consisting of two centuries. The maniple of the hastati and the principes comprised 120 men, twelve men front by ten men deep; while the maniple of the triarii consisted of 60 men on a frontage of six men. The cavalry was divided into ten turmae of thirty horsemen each, arranged in files three deep. When the ten turmae were formed in line they composed an ala, or wing. One maniple each of hastati, principes, and triarii, 120 velites, and one turmae of cavalry, constituted a cohort, a tactical unit, with a minimum strength of 450 men. There were ten cohorts in a legion. The normal strength of the legion was 4200 foot and 300 horse. The term legion often meant one Roman legion and one allied legion. The usual consular army consisted of two Roman and two allied legions, numbering from 18000 to 20000 men, of whom 1800 were cavalry. Two hundred of the cavalry, with 840 foot troops, made up a body in each legion known as the extraordinarii, who formed a sort of reserve. The army usually drew up in line of battle with the two Roman legions in the center and the two allied legions on their right and left. The cavalry was either on the flanks or in the front or rear. Each legionnaire occupied a frontage and depth of about five feet, which was reduced to resist cavalry or to form a testudo. The distances between lines of maniples was about 100 yards. The velites, at the beginning of an engagement, were posted in front of the hastati; when driven back, they formed in the line of the triarii. The principal weapon was the gladius, or short sword; in addition the hastati and principes carried a heavy lance or pilum and a lighter lance or hasta; the triarii had a pike from ten to fourteen feet long and several darts; the velites carried a sword and several darts.

Lengthen. To extend in length, as the step.

Lengthening Out. Increase in the length of a column of troops on the march, due to failure to maintain proper distances.

Lensatic Compass. A type of compass, much used in machine gunnery, so named because it is provided with a lens or magnifying glass for reading graduations.

Lethal Concentration. A concentration of a chemical agent in the field which will produce death upon exposure for a definite period of time.

Letters of Instruction. A method by which the plans of the superior commanders are communicated and which regulate movements and operations over large areas and for considerable periods of time.

Level. Horizontal. Perpendicular to the direction in which gravity acts. Any instrument or device used to establish the level. To establish the horizontal. To make level. See Level, Hand, Surveyor's, etc.; Spirit Level.

Level Point. The point on the descending branch of the trajectory which is at the same altitude as the muzzle of the piece. The point of fall.

Levelling. The process of adjusting a gun and mount, or any fire control or observing instrument, so that all vertical and horizontal angles will be measured or applied in truly vertical and horizontal planes. Operations with a surveyor's level. The process of making level.

Levelling Screws. Vertical screws on the mounting of an instrument, base of a gun, etc., for bringing it to a level.

Level-off. To make the flight of an airplane horizontal after a climb, glide, or dive.

Levy. The compulsory raising of troops. A demand for funds or supplies. To exact or impose by authority.

Levy en Masse. The spontaneous act of the inhabitants of territory threatened by the enemy in taking up arms in self defense, or in response to a call by proper authority.

Lewis Chart. A chart for the rapid and accurate graphical solution of triangles, used in antiaircraft gunnery.

Lewis Light Machine Gun. A light machine gun used in the British and Russian Armies. It has a caliber of .303 inches, weighs 27 pounds, and has a magazine capacity of 47 cartridges.

Liaison. The linking together of the different units, so as to secure proper coordination and cooperation in campaign or battle; accomplished by the interchange of liaison officers and detachments of enlisted men between the different units, and by the use of all available means of communication. Close touch maintained between units or formations by agents or officers to interchange information and insure cooperation.

Liaison Agents. Officers or enlisted men sent from the headquarters of one unit to that of another for the purpose of transmitting information between the headquarters of the two units concerned.

Liaison Detachment. A detachment sent from one headquarters or activity to another with a view to maintaining liaison therewith.

Liaison Officer. An officer sent from the headquarters of one unit to the headquarters of another unit for the purpose of maintaining liaison between the headquarters of the two units concerned.

Lieutenant. A commissioned officer below a captain in grade. In the United States Army there are two grades, first and second lieutenant. The grades additional, third, provisional and brevet lieutenant have also been recognized.

Lieutenant Colonel. An officer next in rank above a major and below a colonel.

Lieutenant General. An officer next in rank above a major general and below a general.

Life of a Piece. The number of rounds a gun will fire before becoming unserviceable or inaccurate.

Lifting Jacks. Specifically, the jacks used to raise a mobile gun and mount clear of its traveling bogies, to level it during emplacement for firing, and to return the gun to its traveling position.

Light Artillery. See Artillery, Light.

Light Horse. Mounted soldiers who are lightly armed and equipped, such as chasseurs, hussars, mounted riflemen, etc. Light cavalry.

Light Infantry. Lightly armed and equipped foot soldiers, selected and trained for rapid movement.

Light, Landing. A light carried by an aircraft to illuminate the ground while landing.

Light Marching Order. A soldier equipped with arms, ammunition, canteen and haversack is said to be in light marching order.

Light Pack. A colloquial term applied to a pack made up for practice marches, etc., when it is not desired to carry the full pack.

Light Prison. A moderate form of punishment.

Light Railway. A narrow gauge railway.

Light Shellproof Shelter. A shelter which affords protection against continuous bombardment by all shells up to and including 6 inch caliber.

Light Shelter. A shelter which affords protection against direct hits and, in some cases, against a continued bombardment by 3 inch shells. Temporary shelter for supports and reserves until better can be constructed.

Light Tank. A tank weighing between 5 and 16 tons. Anything weighing less than 5 tons is classed as very light. Very light or light tanks can be transported on carriers (trucks).

Lighter-than-air. Used with respect to aircraft which are supported by means of a gas lighter than air, such as hydrogen or helium.

Lights, Identification. A group of lights, clear and colored, carried on the rear part of an airplane for identification at night.

Limb. A graduated arc or circle of an angle-measuring instrument. The edge of a disc or circle.

Limber. A two-wheeled vehicle to which a piece or caisson is attached for transport. It usually also mounts an ammunition chest. To attach a piece or caisson to its limber.

Limber and Caisson Type of Vehicle. A type of vehicle consisting of two sections, each having one axle, the limber being attached to a team or tractor and the caisson attached as a trailer to the limber. The type of vehicle used by light artillery.

Limber Pits. Excavations for the protection of the gun limbers, each large enough to cover a limber and two horses.

Limber Up. To limber.

Limber-chest. A box on the limber in which tools and ammunition are carried.

Limbered. With the trails of the pieces and caissons attached to their respective limbers.

Limit of Visibility. See Visibility.

Limitations. The limits of the powers and capabilities of any person, group, branch, or weapon. See also Powers.

Limited Service Men. Men who fail to meet the requirements for general military service, but whose services may, nevertheless, be used advantageously.

Limiting Elevation. The greatest elevation permitted by the sights or elevating mechanism of a particular gun.

Limiting Points. Designated points where the several lines in a defensive position shall cross the unit sector boundaries, used to insure coordination between the works of adjacent units.

Limiting Ranges. The minimum and maximum ranges of any weapon under any particular circumstances.

Limits of Arrest. The confines prescribed for an officer or soldier in arrest

Limits of Traverse. The maximum arc over which a gun carriage may be traversed.

Line. A formation in which the various elements are abreast of each other; when the elements are in column the formation is called a line of columns. A straight or nearly straight line determined by the positions of individuals, units or establishments. The combatant branches of the army considered as a whole, as distinguished from the staff. A rope or twine. A railway. Poles and wires carrying electric current as for a telephone, telegraph or electric power. That part of an electric circuit exterior to the source of power. A rein. To take position in line or to form in line.

Line, Mooring. A line attached near the bow of an aircraft for securing it to the ground, buoy, anchor, or to a mooring mast. A line securing a ponton or other vessel to its anchor.

Line of Aim. The line established by the sights and the point aimed at.

Line of Battle. The arrangements or dispositions of an army when drawn up for battle. The disposition of troops for battle.

Line of Circumvallation. A more or less continuous line of defenses by which a besieged place is surrounded. Its purpose is to confine the garrison within its works, to prevent its communication with the outside, to protect the besiegers against sallies by the garrison, and to serve as the line of departure for the operations of the siege. The line of investment.

Line of Close Columns. An infantry formation in line of masses with companies in close line of platoons in column of squads.

Line of Close Lines. An infantry formation in line of masses with rifle companies in close line of platoons in column of squads.

Line of Collimation. The central line or longitudinal axis of a cylinder, as a telescope, or other observing instrument; it passes through or should pass through the optical centers of the lenses of the telescope.

Line of Columns. A number of columns with heads abreast or in line.

Line or Lines of Communication. All the routes, land, water and air, which connect a military force in the field with its base of operations, and along which supplies and reinforcements move to the front. The construction and maintenance of lines of communication are the functions of the engineers.

Line of Contravallation or Countervallation. A more or less continuous line of defenses around a besieged place, exterior to the line of circumvallation. Its purpose is to protect the besiegers against attacks from the outside, as by a relieving force.

Line of Departure. A line designated for the initial forward movement in an attack in order to secure coordination between units and arms. The tangent to the trajectory at the muzzle; the axis of the bore prolonged at the instant when the projectile leaves the muzzle.

Line of Duty. Performance of authorized or prescribed duty. A medical term signifying that a sickness or injury of a soldier is due to no fault or neglect on his part.

Line of Elevation. The axis of the bore prolonged, when the piece is laid for firing.

Line of Fall. The tangent to the trajectory at the point of fall or level point.

Line of Future Position. The line joining the muzzle of a gun and the future position of an aerial target (at the end of the time of flight).

Line of Impact. The tangent to the trajectory at the point of impact.

Line of Investment. See Line of Circumvallation.

Line of Least Resistance. The shortest distance from a subterranean charge of explosive to the surface of the ground.

Line of March. The regular disposition of the component elements of a force for the march. The arrangement of troops for marching. The course or direction taken by a body of troops on the march. The route prescribed for a march.

Line of Observation. The line occupied by the most advanced elements of the outpost position, whose principal mission is observation. The line from a position finder to a target at the instant of a recorded observation.

Line of Operations. The line or direction by which an army advances from its base into the theater of war; it may be single, double, or multiple. The routes, taken collectively, over which an army moves in campaign.

Line of Platoons. An infantry formation with rifle companies in line or platoons in column of squads at normal distances.

Line of Position. A straight line joining a point of origin and a point in space, as a gun and a target. See also Line of Site.

Line of Present Position. The line joining the muzzle of a gun and an aerial target at the instant of firing.

Line of Resistance. The line, in any defensive position, on which the principal defense of the position is made; it is usually the forward line of the combat echelons. The line of resistance of the principal defensive or battle position of a defensive system is called the main line of resistance, and is the reference line which, to a great extent, governs the location of all elements of the defensive system. The firing line of a defensive position.

Line of Retreat. The general direction of movement to the rear of a command, considered as a whole, in a retreat. The routes of retreat.

Line of Section Columns. Section columns in line at full deploying intervals (Old.)

Line of Sections. An infantry formation with rifle companies or platoons in line of sections in column of squads at normal intervals.

Line of Sight. A line of sighting or vision. To avoid confusion with Line of Site, this term should not be used.

Line of Sighting. The line of vision of a sight, angle-measuring instrument or observing instrument when directed on a definite point. The straight line connecting the front and rear sights of a rifle and the target when the piece is aimed. Any line of vision, either with the naked eye or an optical instrument.

Line of Site. A straight line joining the muzzle of a gun, or an observer and the target. In antiaircraft gunnery, a straight line joining a point of origin (gun, observer, position-finding instrument) with a point in space. This term is the same as Line of Position, which latter term is preferred because it avoids the confusion between Sight and Site.

Line of Squad Columns. Squad columns in line at full deploying intervals.

Line of Thrust. In an aircraft, the line along which the power or thrust of the propelling engine is applied. In a single-motored plane the line of thrust is the prolongation of the axis of the propeller, it lies in the plane of symmetry and passes through or very close to the center of gravity, coinciding with the longitudinal axis. In a symmetrical multi-motored plane the lines of thrust of all motors are parallel, and the resultant of them all lies in the plane of symmetry, and generally, but not always, passes through or near the center of gravity.

Line of Withdrawal. The general direction of movement to the rear of the subordinate elements of a command in a withdrawal.

Line Officer. An officer belonging to a combatant branch of the service.

Line Shot. An impact or burst on the line observer-target.

Line Troops. Troops which engage in actual combat with the enemy, viz: infantry, cavalry, artillery, signal corps, engineers and air service.

Lineal. Pertaining to length, or to the line of promotion of the army. Lineal

Lineal Promotion. Promotion by seniority according to lineal rank.

Lineal Rank. The relative rank of an officer. But see also Promotion List.

Linear Height of Burst. The distance in yards or meters of the burst above the plane of site or the base of the target.

Linear Tactics. A form of battle tactics in which the force is deployed on a wide front with little depth in order that all the firepower available may be brought to bear on the enemy from the very beginning of the attack.

Linear-speed Method. A method of prediction in antiaircraft fire based upon continuous measurement of the ground speed of a target by the data computer.

Liner. An inner tube, installed in modern cannon, that may be replaced when worn out.

Line-route Map. A map which shows the routes of wire circuits, with all telephone and telegraph instruments, switchboards, etc.

Lines of a Defensive Position. The lines or axes, approximately parallel to the front, along which the successive echelons of the defense are grouped. In a typical battle position they include, from front to rear, the line of observation or local security, the line of resistance, the support line, the battalion reserve line, and the regimental reserve line. These lines are actually lines of defended tactical areas, and may or may not be marked by trenches. A defensive position consists not of successive lines, but of areas properly distributed to control the terrain and afford mutual support. See Battle Position, Defensive Position.

Lines of Action. The possible plans open to a commander in a particular position or situation.

Lines of Defense, Strategical. The natural military frontiers of a state, or favorable lines within the frontiers, on which the general defense is or may be conducted.

Lines of Information. Channels for the receipt or transmission of information.

Lining In. Laying a piece for direction by lining the axis of the tube on the target by eye.

Link. That which connects. One of the sections of a chain. A bar or plate with a hole for a link pin at each end. A length of 7.92 inches. A snap to connect the bridles of two horses.

Link Horses. A method of fastening horses together by linking their bridles.

Liquid Fire. Flaming liquid ejected from an apparatus, usually carried on the back of the operator, and known in the German Army as flammenwerfer.

Listening Gallery. A subterranean gallery driven to the front for the purpose of detecting enemy activity by means of listening instruments.

Listening Post. A concealed or sheltered position established in advance of a defensive line for the purpose of early detection of the enemy's intentions and movements.

Listening-in. The adjustment or connection of a receiving apparatus, radio, telegraph, or telephone, so as to intercept messages being sent to another station. A method of espionage employed in warfare.

Litter. A stretcher for carrying sick and wounded.

Litter Bearer. One who helps to carry a litter.

Litter Squad. The men assigned to carry a litter, either two or four.

Live Ammunition. Ammunition charged with explosives as distinguished from dummies. Ball cartridges.

Live Axle. A vehicle axle to which the power of a motor is transmitted.

Live Load. The load which a bridge or other structure carries, not including its own weight, which is the dead load.

Live Roller. A roller which rests directly upon the supporting surface instead of bearings at its ends, and which moves with the object it supports; a free roller.

Livens Projector. A crude form of chemical mortar, usually installed in large numbers, to establish a gas cloud of high concentration by simultaneous discharge.

Load. A single charge of powder and a single projectile as combined for firing. To charge a firearm. To fill a shell.

Loading Platform. The surface upon which the cannoneers stand while loading the piece.

Loading Tray. A tray inserted in the breech of a cannon to facilitate loading, and prevent injury to the breech or the projectile. A shot tray.

Local Attractions. Any local conditions which derange the readings of a compass.

Local Counterattacks. Counterattacks on a small scale by local supports or reserves.

Local Mobilization. The initial assembly of units at their home stations, during which they will, in so far as is possible, be expanded to the prescribed war strength, equipped, supplied, and trained. Mobilization at or near home stations.

Local Reserves. The reserves of smaller units, used for local reinforcement or counterattack.

Local Security. The protection secured through the use of security detachments maintained by component elements of a larger force, independent of the security detachments of the force as a whole.

Locate. To designate the site or place of. To establish. To station or place. To determine the position of a point in surveying or sketching, or to fix its position on a map or on the ground.

Location. Situation. Place. Locality. Act of locating. A location survey as for a road.

Lock. That part of a firearm which explodes the charge, especially in the older forms of weapons, e.g., flintlock, matchlock, etc. A firing mechanism for a weapon.

Lodgment. The occupation and holding of a position. An intrenchment thrown up in a captured position. The hollow or cavity in the under par of the bore of a gun where the projectile rests when rammed home.

Logistics. That branch of military art that comprises everything relating to the movement and supply of troops. See also Art of War.

Logistics Plan. A plan that covers the details of concentration, supply, evac uation and administration of a war plan.

Long Bow. An ancient bow, about 5 feet long, which shot an arrow 45 inche long as far as 600 yards, but its effective range was about 200 yards an it could discharge 12 or more arrows per minute. It made its appearanc late in the 12th century, A.D., and was primarily an English weapon. I succeeded the crossbow in England in the 13th century and, in spite o the introduction of the harquebus, remained one of the principal weapon of the English soldier until about 1590. See also Bow; Crossbow.

Long Fuze. A super-quick point fuze, which has a long neck.

Long Range. See Ranges, Classification of.

Long Roll. A prolonged roll of the drums.

Longe. A long rope used to guide or lead an animal in training or exercis The use of the longe. A place for training or exercising animals.

Longeron. A principal longitudinal supporting member of the framing o an airplane fuselage or nacelle, usually continuous across a number of point of support.

Longevity. Length of service, especially as affecting pay.

Longevity Pay. Additional pay for length of service. Fogies.

Longitudinal. Lengthwise. Measured along the range, or parallel to th gun-target line, or directing point-target line, as a deviation.

Longitudinal Axis of an Aircraft. The axis that runs from front to rea in an aircraft, and is horizontal during level flight. See Axes of an Aircraf

Longitudinal Deviation. The horizontal distance from the point or center impact to the target or adjusting point, measured parallel to the line gu target.

Longitudinal Error. The divergence of a point of impact from the center o impact, measured parallel to the line gun-target.

Lookout. An observer, sentry or watchman, usually for a special purpose.

Lookout, Air. A ground observer whose duty is to watch for hostile aircraf An air scout.

Lookout Posts. Protected positions, usually located in the front line trenche where good observation may be had of the enemy's position. Advanced o servation posts.

Loop. A maneuver executed in such a manner that the airplane follows closed curve approximately in a vertical plane. A bight of a rope. A loo antenna.

Loop, Ground. Turning an airplane quickly about the vertical axis while ru ning on the ground, by means of hard rudder action. An uncontrollab violent turn of an airplane on the ground, while taxying or during take-o or landing.

Loop, Inverted Normal. A loop starting from inverted flight and passir successively through a dive, normal flight, climb, and back to inverted fligh

Loop, Inverted Outside. An outside loop starting from inverted flight a passing successively through a climb, normal flight, dive, and back to i verted flight.

Loop, Normal. A loop starting from normal flight and passing successive through a climb, inverted flight, dive, and back to normal flight.

Loop, Outside. A loop starting from normal flight and passing successive through a dive, inverted flight, climb, and back to normal flight, the pil being on the outside of the flight path.

Loop Sling. A gunsling adjustment used to steady a rifle while firing, t left arm passing through a loop in the sling. See also Hasty Sling.

Loop the Loop. To perform a loop in an airplane.

Loop, Upside Down. A loop started from a position of inverted flight a ending in the same position, having followed the same path as that made by an outside loop.

oopholes.   Openings in a parapet, wall or head cover through which fire may be delivered.   Small embrasures.

oose Pieces.   Rifles, automatic rifles and light machine guns not used in making stacks, leaned against the stacks on their completion.

osses.   The killed, wounded, captured or missing persons belonging to an organization, and property captured or destroyed by the enemy.   Casualties.

ot.   A lot of ammunition manufactured or received at one time, all of which is supposed to possess practically the same ballistic properties.   A group or collection of articles or materials of the same kind, assumed to be interchangeable.

ow Explosives.   Explosives of relatively slow action in which the chemical reaction is a rapid but measurable combustion or deflagration.   The effect of low explosives is in the nature of a push, rather than a blow, and they are hence suitable for propelling charges in firearms, but when properly tamped may also be used for blasting when a rending rather than a shattering effect is desired.   See also High Explosives.

ow Position.   A firing position of the machine gun in which the tripod is spread wide so that the gun is close to the ground.

ow-wire Entanglement.   A wire entanglement about 18 inches high, similar to the double apron fence.

uff Tackle.   A tackle rigging similar to a gun tackle but consisting of a single and double block.   A double luff consists of two double blocks.   A luff-on-luff is rigged by attaching a block of one luff tackle to the fall of another.

uftwaffe.   The German air force.

unette.   An outwork or detached work of fortification consisting of two faces forming a salient angle, and two flanks, with an open or partially closed gorge.   An eye or ring on the end of the trail of a gun or pole of a trailer, which drops over the pintle of the towing vehicle.

ung Injurants or Irritants.   Gases that particularly attack and injure the bronchial tubes and lungs, often with fatal results.   Chlorine, phosgene, and diphosgene are examples of this class.

unge.   A thrust in fencing or bayonet exercise in which the arm is fully extended and the body thrown forward by advancing one foot.

ying or Prone Trenches.   Hasty trenches affording protection to and permitting fire by riflemen in the prone position.   Skirmishers' trenches.

ying Position.   The prone position for rifle firing.

M

Machete. A large knife, resembling a cutlass, used both as a tool and a weapon by the inhabitants of tropical America.

Machine Cannon. A machine gun, particularly one throwing projectiles larger than those used in small arms.

Machine Gun. A gun that fires small arms ammunition on the automatic principle. The line of demarcation between the machine gun and the automatic rifle is not sharp, but the former is usually distinguished by having some form of rigid mount, as a tripod or wheels and trail, which sustains the force of recoil, and by means of which the direction of fire may be clamped. Machine guns are operated either by the recoil or by the pressure of the powder gases; they are fed either from a belt, clip, or magazine; and are cooled either by a water jacket around the barrel or by radiation from metal fins (air-cooled). According to size and weight they are classified as light and heavy.

The development of machine guns has rather closely paralleled that of the rifle. An early form was the "carte with gonnes" or ribaudequin, which appeared early in the 15th century. It was simply a number of harquebuses mounted side by side and fired simultaneously by a common combustion box or pan. Late in the 15th century this idea was further developed as the orgue or organ gun, consisting of several barrels which could be fired successively. Adaptations of the organ gun were used as late as the World War. A harquebus of the 16th century had a single barrel and four chambers, and may be regarded as the prototype of the modern machine gun. Early in the 18th century a single barreled gun with a revolving cylinder of 7-9 chambers, mounted on a tripod, made its appearance. This was followed by a compound gun of several barrels, fired by a flintlock.

The first real machine gun was the Gatling gun, which appeared in 1862 and was first used in the American Civil War. It consisted of a number of barrels, usually ten, each having its own lock, the whole revolving about a central axis. It was operated by a crank. The Montigny mitrailleuse was adopted in the French Army just before the Franco-Prussian War. It consisted of 37 barrels in a casing, revolving about a central axis. This was soon followed by the Gardner, Palmcranz and Nordenfeldt. In 1884, Maxim, an American, produced the first automatic gun, in which the force of recoil of the barrel was utilized to perform the operations of extraction, feeding and firing. This was followed by various guns utilizing the pressure of the powder gases to perform these operations, including the Colt and Hotchkiss.

Machine guns were first employed in battle on a large scale in the Russo-Japanese War. In the World War their use was so extensive as to usher in a new era in battle tactics. The following are some of the guns now in use in various armies: Beardmore; Bergman; Berthier; Browning; Chauchat; Chautellerault; Colt; Darne; Eriksen; Farquhar; Fedorov; Fiat; Fusil Furrer; Gardner; Gast; Hotchkiss; Lewis; Madsen; Maxim; Nambu; Nordenfeldt; Parabellum; Praga; Revelli; Schwarzlose; SIA; St. Etienne; Thomson; Vickers. The Browning machine gun, which is typical of these weapons, has a cyclic rate of over 500 shots per minute, and a practical rate of about 250.

Machine Gun Barrage. A barrage of machine gun fire. Machine gun barrages may be stationary or moving, and are employed both in attack and defense, usually in coordination with artillery fires.

Machine Gun Defense. A planned defense of a position by the coordinated fire of machine guns, suitably located. Defense of a definite area against aerial attack by antiaircraft machine gun fire.

Machine Gun Marksman, Sharpshooter, Expert. Grades of classification proficiency for machine gunners.

Machine Gun Nest. One or more machine guns in firing position, emplaced or concealed.

Machine Gun Synchronizer. A device which synchronizes the fire of an airplane machine gun so that the bullets pass between the blades of the rotating propeller.

Machine Rifle. An automatic rifle. See Browning Machine Rifle.

Mafstab. The scale of a map. (Ger.)

Magazine. A chamber in or attached to a gun, designed to hold a number of cartridges to be fed automatically, or otherwise, to the piece. A case for holding a number of cartridges. A stock or store of ammunition or other supplies, or the place of storage for such.

Magazine Filler. A funnel used to fill the magazine of an automatic weapon.

Magazine Gun. A breechloading gun having a magazine capable of holding a number of cartridges. A repeater.

Magazine Pockets. Pockets, usually attached to a belt, in which cartridge magazines are carried.

Magistral Line. The trace or outline of a work. The line of intersection of the face and top of the scarp wall of a fortification.

Magnetic Azimuth. Azimuth measured from the magnetic meridian as an origin. See Azimuth.

Magnetic Declination. The angle between the true and magnetic meridians at any locality; it is said to be east when the magnetic meridian points east of the true meridian.

Magnetic Meridian. A magnetic north and south line.

Magnetic North. The direction at any point to the magnetic north pole.

Magneto Exploder. A small magneto or dynamo in a box, operated by a rack and pinion, which gives a momentary current of high voltage; used in detonating blasting and demolition charges. A blasting machine or plunge battery.

Magnitude Method of Adjustment. A method of adjustment used when the actual magnitudes or amounts, as well as the senses of the deviations can be determined.

Mail. A metal network or flexible fabric of interlocked metal rings used as defensive armor. A piece of mail armor. Any defensive covering. To don a coat of mail or armor. To arm defensively.

Main Attack or Effort. The principal attack, in which the greatest possible offensive power is brought to bear with a view to bringing about a decision. The decisive or principal attack which is intended or expected to decide the issue of battle.

Main Body. The principal part of a command. A command less all detachments.

Main Guard. The principal guard from which the necessary details for minor or special guards are made.

Main Line of Resistance. A line near the front of a battle position designated to coordinate the defensive fires of all units and supporting arms.

Maintenance, Motor Transport. The maintenance of motor transport is performed in four echelons, as follows: **First Echelon.** Drivers maintenance, being simple operations that can be performed by the drivers with the tools and materials on their vehicles. **Second Echelon.** Maintenance, other than First Echelon, performed by the using arms or services in their local shops. **Third Echelon.** Maintenance performed in the field by mobile shops of the supply services (Ordnance and Quartermaster). **Fourth Echelon.** Maintenance performed at properly equipped shops in the rear areas by the supply services.

Maintenance Section. An organization charged with the repair and maintenance of weapons and other equipment.

Major. An officer next in rank above a captain and below a lieutenant colonel. Greater in number, quantity, or extent. Greater or higher in dignity, rank, or importance. Principal or most important.

Major Armament. Fixed seacoast armament comprising guns of 8-inch or greater caliber and 12-inch mortars.

Major General. An officer next in rank above a brigadier general and below a lieutenant general.

Major Operation. An important operation on a large scale.

Majority. The rank and status of a major.

Malfunctioning. Improper or faulty functioning, as of a gun or machine.

Malinger. To feign or protract sickness, incapacity or fatigue in order to avoid duty or service.

Man. To furnish with men. To garrison. To provide with an operating crew or detachment.

Manchu Law. A law which provides that no line officer of the United States Army may be detached from duty with troops for more than four years out of any six; so-called from the Manchus of China, who were dethroned and exiled.

Mandatory. Positive, allowing no discretionary judgment.

Mandatory Sentence. A sentence determined in kind and amount by statut in which case no different punishment may be lawfully substituted.

Maneuver. A movement designed to place troops, materiel, or fire in favor able strategic or tactical locations with respect to the enemy. A tactica exercise executed on the ground or map in simulation of war and involvin two opposite sides, though one side may be outlined, represented, or imagin ary. A tactical or acrobatic evolution performed by aicaft. To execut maneuvers.

Maneuver March. A march conducted during a campaign, when not in actu contact with the enemy, for the purpose of improving the strategical or ta tical situation.

Maneuver of Fire Power. The application or transfer of fire to a favorabl or more favorable target.

Maneuver Tactics. The movements by which troops are brought to an disposed upon a field of battle. The connecting link between strategy an tactics.

Maneuver Unit. Any tactical unit having component elements which may b maneuvered separately in battle.

Maneuverability. The capability of maneuver of a body of troops. Tha quality of an aircraft which makes it possible for the pilot to change it attitude rapidly. The facility with which an airplane can fly around a poin

Maneuvering Blocks. Short pieces of wood used for blocking and cribbin in mechanical maneuvers.

Manifesto. A public declaration of intentions or motives.

Manipulation Exercises. Exercises in which a machine gun or other weapo is manipulated, as in traversing or searching, or the bolt of a rifle i manipulated, without actually firing, etc.

Manning Table. A chart which shows the permanent assignment of eac officer and enlisted man for duty in the fire section of coast artillery bat teries.

Mannlicher Rifle. The rifle used in the armies of Austria, Bulgaria, Ecuado Greece, Netherlands, Rumania, and Yugoslavia. Its caliber varies fror .256 inches to .315 inches, it has a magazine capacity of 5 cartridges, an weighs about 8 pounds.

Manta. A portable bulwark or shelter. A mantelet. A waterproof canva used to cover pack loads.

Mantlet or **Mantelet.** A screen or shield for an embrasure. Portable over head cover for besiegers. A bullet-proof observation post.

Manual. An instructional pamphlet which prescribes certain matters in de tail. A handbook.

Manual of Arms. Prescribed and formal movements with the rifle at drill and ceremonies, and the changes from one position to another at comman The positions include (at present): order; right (left) shoulder; port present; trail; and inspection arms; parade rest; stand at ease; fix an unfix bayonets; stack and take arms; rifle salute.

Manual of the Color. Prescribed movements of the color at drills and cere monies.

Manual of the Guidon. The prescribed positions in which the guidon is hel or carried, and the manner of changing from one position to another.

Manual of the Sword or **Saber.** Prescribed positions and movements with th sword or saber. They include: draw; order; carry; port; present; retur saber; parade rest.

Map. A graphic conventionalized representation, to scale, of a portion o the earth's surface. A graphic representation of any portion of the earth surface drawn to scale in which the amount of land is greater than th amount of water. To sketch. To make a map of. See also Chart.

Map, Air-navigation. Air-navigation maps embrace strips of ground be tween airports about 80 miles in width; for the use of air navigators.

Map Declination. The angle between the grid and magnetic meridians at an point.

Map Distance. The horizontal distance between contours corresponding to given slope. Any distance measured on a map. See Horizontal Equivalen

Map Exercise. A tactical exercise (commonly called a one-sided maneuver) which is similar to a map maneuver, except that the players are assigned to one side only, the other side being represented by the director.

Map, Fire-Control, Training. A complete topographic map suitable for troop training, map problems, detailed work, artillery fire control, and defense of tactically important areas.

Map, Geographic. A small-scale map of the type usually found in an atlas. It covers a large area and generally shows only such detail as locations of cities, rivers, railway lines, territorial boundaries, important mountain ranges, etc.

Map in Colors. A topographical map in which certain classes of features are distinguished by characteristic colors. The convention for the standard 4-color map is as follows: black, all cultural features, grid lines, lettering and scales; green, vegetation; blue, water; brown, ground forms (contours).

Map Maneuver. A tactical exercise in which a military operation, with opposing sides, is conducted on a map, the troops usually being represented by markers, constructed to the scale of the map, which are moved to represent the maneuvering of the troops on the ground.

Map Measurer. A small device, consisting of a gearing, dial and indicator, for measuring distance on a map.

Map, Military. A map conveying information useful for military purposes, or specially prepared for military use.

Map Problem. A written tactical exercise, requiring a written solution, in which a military situation is stated and solved on a map, which is the only guide as to terrain, distance, character of roads, etc.

Map Projection. See Projection.

Map Range. The range to any point as scaled or computed from a map.

Map Reading. The art of interpreting the information conveyed by a map. The ability to recognize the conventional signs on a map so as to visualize the nature and configuration of the terrain or area represented, and to solve problems by the aid of a map.

Map Reference. The designation of the map or maps referred to or required by a field order.

Map Reproduction. The various processes by which maps are reproduced in small or large numbers. The means employed in the military service include blue prints, blue line prints, black and white prints, ozalid process, hectograph and duplicator, and lithography.

Map Scale. The ratio between distances shown on the map and the corresponding distances on the ground. It is expressed as a ratio or representative fraction, in words and figures, and graphically by a scale printed on the map.

Map, Special. A map which conveys special information, or to which special information has been added.

Map, Standard. A geographic or topographic map, ordinarily made in time of peace as an element of preparedness, or for the economic development of the country. Refers especially to the printed maps issued by various government bureaus.

Map, Strategic. A geographic map on a relatively small scale, showing rail lines, maintained water and air ways, terminals, principal roads of military importance, boundaries, important cities, mountain ranges, etc., used for stategic purposes.

Map, Tactical. A complete topographic map of sufficient scale, detail and accuracy to facilitate the study and conduct of tactical operations.

Map, Topographical. A map which portrays all the natural and artificial features of the terrain, especially the ground forms.

Map Transfer. A transfer of fire in which the relation between the target and the check point is determined from a map.

Maps, Classification of. Maps are classified as standard or special. According to scale they are classified as: Small (1:1,000,000—1:7,000,000). Intermediate (1:200,000—1:500,000). Medium (1:50,000—1:125,000). Large (1:20,000 or larger). According to use they are classified as general, strategic, tactical, battle, aeronautical. According to method of reproduction they are classified as lithographic, fluid duplicator, contact prints, mimeograph, hectograph.

Maps, Information to Appear On. Every map or sketch should show on its face, usually with the title, the following information, or as much thereof as may be necessary in the particular case: (1) The name of the locality. (2) The date. (3) The scale. (4) The contour interval. (5) The direction of north. (6) The grid. (7) Latitude and longitude. (8) Explanation of any special symbols used. (9) How the map was prepared (original survey, compilation, sketch, etc.). (10) The person or persons responsible for the map.

March. The movement by which a body of troops moves from one place to another. The signal or command for troops to move. The distance passed over in a given time or from one point to another. A territorial border or frontier. To advance in a steady, regular manner.

March Column. Comprises all the elements marching on a single route under the control of a single commander, including security units.

March Conditions. The conditions which obtain as to dispositions and security while in march formation.

March Discipline. The observance and enforcement of the rules of good marching, especially as relates to the position of units in the column and the position and conduct of individuals and vehicles.

March Dispositions. The arrangement of troops for a march, based on the proximity of the enemy, weather, available roads, and other conditions.

March Graph. A graph showing the progress of all columns on the march, which enables the commander to tell where the head or tail of any organization should be at any time.

March in Readiness for Battle. A march executed prior to contact with the enemy in which the command moves on a wide front, partially developed, in order to reduce the time required to deploy for battle.

March in Review. The march of troops in front of a reviewing officer during certain ceremonies.

March Off. To leave a place of ceremony or other duty, as to march off guard.

March On. A term used in an order when doubt exists as to whether the unit will reach the destination mentioned. To march toward. To begin the performance of a prescribed duty, as to march on guard.

March Order. A field order including detailed instructions for a march.

March Outpost. A temporary outpost established for the protection of the command during a halt, or while regular outposts are being established.

March Table. A table, accompanying a march order, containing the names of organizations, location, time of starting, routes to be followed, destination, and other pertinent instructions.

March To. A term used in an order when it is reasonably certain that the destination mentioned will be reached.

March Unit. A subdivision of a marching column which moves and halts at the command or signal of its commander.

Marching Fire. Assault fire.

Marching Flank. The flank of a body of troops farthest from the pivot when executing a movement in drill.

Marines. Troops which serve at naval stations and on board ships of war. Members of the U. S. Marine Corps. Sea soldiers.

Mark. The target or point aimed at. A particular model of a weapon, machine, or article of equipment, as Mark V tank. To indicate by discs or spotters the location of hits on a target, or a point aimed at.

Mark Time. A marching step in which the feet are alternately raised about 2 inches and lowered without gaining ground.

Marker. A symbol, letter, or figure on the ground, visible from aircraft, by means of which the operators are able to determine their position. A small flag to indicate position, limit, change of direction, etc., at ceremonies. An individual, flag or stake placed to mark a position or route. Bits of colored paper, strings of beads, etc., used in map exercises to represent troops or positions.

Marking Discs. Colored discs mounted on a staff, used by a man in the target pit to indicate the position and value of hits on the target.

Marksman. One skilled in shooting. A degree of classification in rifle shooting just below that of sharpshooter.

Marksmanship. The ability to shoot a firearm. The skill of a marksman.

Marquee, Marquis, or Markee. A large tent, as for a high ranking officer. A fly erected immediately in front of a tent.

Marshal. A field marshal. A grade above that of general. To arrange in order. To assemble, as for battle.

Martial. Pertaining to war or to an armed force. Suited to war. Warlike.

Martial Law. Military authority substituted for civil government in the home country or any district thereof, either by proclamation or as a military necessity when the civil government is temporarily unable to exercise control. It is distinguished from military law in that it applies to the civilian population, and from military government or the law of hostile occupation in that it is exercised in the home country. It is not statutory in character, but arises, in every case, out of strict necessity.

Martinet. A strict disciplinarian. In general, one who lays stress on a rigid adherence to the details of discipline, or to its forms and fixed methods.

Martini-Henry Rifle. A breechloading rifle adopted by the British Army in 1871.

Mask. Any natural or artificial obstruction which affords shelter from, or interferes with view or fire; usually an intervening hill, woods, etc. Friendly troops located between a gun and its target may constitute a mask. A force of troops or a fortress is masked when a force of the enemy holds it in check while some hostile maneuver is being carried out.

Masked. Hidden or protected from view of the enemy or from his fire.

Masked Battery. A battery screened or protected by a natural or artificial mask.

Masked Sap. A covered sap which affords protection against shrapnel or shell fragments.

Masking. The act of detaching a force to prevent an enemy garrison or force from interfering with the operations of an army. The act of masking.

Mass. A formation in which the troops are arranged with reduced intervals or distances, or both. A concentration of troops. A mask. To form a mass.

Mass of Maneuver. The element employed to outflank the enemy while he is held in front by the pivot of maneuver. In combat involving long fronts it may be the force employed to effect a penetration, the enemy's attention having previously been diverted elsewhere.

Mast. A flagpole or mooring post. See also Pylon.

Master Electrician. An enlisted noncommissioned specialist having a knowledge of electricity.

Master Gunner. A gunner of the highest rating.

Master Lines. The stream and ridge lines of a terrain, which constitute its skeleton.

Master Schedule. A training schedule in the form of a schematic chart or table showing the subjects of training, the estimated total number of hours to be devoted to each, and the allocation of these hours by weeks.

Mastery or Command of the Air. Means that a nation is in a position to prevent the enemy from flying, while retaining for itself freedom of air action.

Matchlock. A medieval small arm invented toward the end of the 14th century, having a long hammer, pivoted in the stock, which held a piece of slow match. When the trigger was pulled it pressed down the burning match until it touched the powder in the pan and thus fired the weapon. It was the first attempt at automatic ignition and was the principal weapon of the infantry soldier until superseded by the flintlock in the latter part of the 17th century. See also Flintlock; Wheellock.

Materialize. To locate accurately and mark on the terrain points selected from a map, or established by a survey; a surveying operation incident to conduct of fire.

Materials, Critical Raw. Raw materials essential to national defense, the domestic resources of which have been developed to supply the normal needs of the industries of the country but which will require stimulated production and controlled distribution in time of war.

Materials, Raw. Materials, in the form in which they enter a given factory or plant for fabrication into components or finished items.

Materials, Strategic Raw. Raw materials essential to the national defense for the supply of which in normal times dependence is placed in whole or in large part on sources outside the continental limits of a state.

Materiel. One of the two main subdivisions of military force, personnel being the other. All material things, such as instruments, weapons, machines and engines of war, designed for the use of the army, as well as transport, supplies, equipment, and stores. Materiel is not synonymous with material, but is sometimes used to include the latter.

Materiel Effects. The effect on a standard trajectory of materiel conditions not standard.

Mauser Rifle. A bolt-type rifle of various models, with calibers varying from .276 to .315 inches and weights varying from 8.8 pounds to 9.7 pounds It is used by the armies of Germany, Argentina, Belgium, Brazil, Chile, Columbia, Czechoslovakia, Lithuania, Mexico, Paraguay, Persia, Peru, Portugal, Spain, Sweden, Uruguay, Turkey, Venezuela, and Yugoslavia First used in 1871 and converted to a repeater in 1889.

Maxim Machine Gun. A machine gun in both light and heavy types, used in the German Army. It has a caliber of .315 inches.

Maxims of War. The established principles governing the conduct of war especially the immutable principles. The principles laid down by Napoleon.

Maximum Ordinate. The vertical height of the summit of a trajectory above the muzzle of the gun.

Maximum Range. The extreme range of a projectile or weapon. See Ranges.

McClellan Saddle. The standard army saddle in the United States service. It has a relatively high pommel and cantle, and hooded stirrups.

M Day. The first full day of open preparation for mobilization. The date of the War Department order directing mobilization.

Mean. Arithmetical or algebraic average.

Mean Absolute Error. The arithmetical average of the absolute errors of a series of shots.

Mean Deviation. The arithmetical mean of the deviations of a series of shots.

Mean Error. The arithmetical average of the values of similar kinds of errors, regardless of sign.

Mean Height of Burst. The height of the center of burst. The average of the heights of burst of a series of shots.

Mean Lateral Deviation. The arithmetical mean of the lateral deviations of a series of shots.

Mean Lateral Error. The arithmetical mean of the lateral errors of a series of shots.

Mean Longitudinal Deviation. The arithmetical mean of the longitudinal deviations of a series of shots.

Mean Longitudinal Error. The arithmetical average of the longitudinal errors of a series of shots.

Mean Range. The mean of the several ranges of projectiles fired with the same data; the range to the center of impact.

Mean Trajectory. The trajectory passing through the center of impact.

Measured Angle. The horizontal angle, measured at the observation post, between the target and the aiming point, base point or other datum point.

Mechanic. An enlisted specialist in line companies who is able to make minor repairs to the company equipment.

Mechanical Advantage. The power of a tackle rigging or combination. Thus, if a man weighing 150 pounds can by means of a tackle just lift a weight of 1500 pounds, the mechanical advantage of the tackle is 10. Friction is disregarded in determining mechanical advantage.

Mechanical Functioning. The functioning of any weapon which is wholly or partly automatic.

Mechanical Maneuvers. The procedure of handling heavy weights; specifically, mounting and dismounting large cannon.

Mechanical Operation. The mechanical operation, by an individual or team, of any weapon or machine; specifically, of a non-automatic weapon, or non-automatic functions requiring judgment on the part of the operator.

Mechanical Pilot. See Automatic Pilot.

Mechanical Time Fuze. A time fuze operated by a clockwork.

Mechanical Training. Instruction in the mechanics of operation of weapons or machines.

Mechanization. The application of mechanics directly to the combat soldier on the battlefield. The use of mechanical means or methods to assist and protect the human element in actual conflict with the enemy.

Mechanized Cavalry. Cavalry equipped with armed, and armored self-propelled motor vehicles.

Mechanized Force. A fighting force that moves by motor power. It includes or consists of armored elements, and is capable of relatively independent action.

Mechanized Unit. A unit which moves and fights in motor vehicles, the bulk of which are armed and armored, and self-contained as to crew and weapons.

Medal of Honor. A military medal awarded by the President of the United States in the name of Congress to an officer or enlisted man who, in action involving actual conflict with the enemy, distinguishes himself conspicuously by gallantry and intrepidity at the risk of his own life above and beyond the call of duty.

Mediation. A friendly intervention of a state or power in a dispute or hostilities between other states.

Medical Battalion, Regiment. The medical organization which forms an organic part of each division, corps or army. Its functions are the collection, temporary care and evacuation of sick and wounded; and sanitation and medical supply to the unit of which it is a part.

Medical Corps. The medical and surgical branch of the medical department.

Medical Department. That branch of the service which is charged with the care and treatment of sick and wounded men and animals, the control of disease and sanitation, and the furnishing of all necessary medical supplies. It includes the medical corps, the dental corps, the veterinary corps, the medical administrative corps, the army nurse corps, and contract surgeons.

Medical Detachment. Specifically, an organization of medical troops assigned for the immediate service of a combat unit (battalion or regiment).

Medical Laboratory. An organization which performs laboratory service in the field, including epidemiological investigations, supervision of sanitation, examination of water, etc.

Medical Squadron. A medical unit for service with a cavalry division. It includes a headquarters, collecting troop, ambulance troop, hospital troop, and veterinary troop. See also Medical Regiment.

Medium Field Artillery. See Artillery, Medium.

Medium Scale Maps. See Maps, Classification of.

Medium Tank. A tank weighing from 16 to 35 tons, too heavy or bulky to be readily transported by tank carrier.

Meeting Engagement. A meeting of two hostile forces not deployed for battle. A rencontre.

Melee. A hand-to-hand conflict, specifically one following a cavalry charge.

Melinite. A high explosive, similar to lyddite, chiefly picric acid, used in France.

Memoranda. Matter that is directive, advisory, or informative in nature, and may be either temporary or permanent in duration. Official papers issued in lieu of bulletins or circulars.

Memorandum Receipt. A receipt given for government equipment or supplies by the person responsible for their proper care and use.

Mess. A number of officers or enlisted men who take their meals together. The place where meals are taken. To take meals in a mess.

Mess Council. A council of administration composed of the commanders of all companies belonging to a mess.

Mess Kit. The individual mess equipment carried by the soldier. Cooking utensils and mess furniture.

Mess Sergeant. A sergeant in charge of a company mess under the supervision of the company commander or mess officer.

Mess Tin. An individual metal pan or plate having a bail which is hinged to its side and which serves both as a handle and to hold down the cover of the vessel when closed; carried by every soldier.

Message Center. The agency at a headquarters or command post charged with the receipt, transmission, and delivery of all communications pertaining to the headquarters or command post.

Messages. All communications, in plain language or code, transmitted by means of signal communication.

Messengers. Persons used to carry written or verbal messages or orders;

they include runners, mounted messengers (horse and motorcycle), and, under some conditions, airplane messengers. See Couriers.

Metal. The effective power or weight of the projectiles thrown by any number of guns. Broken stone, etc., used for surfacing roads.

Metal Fouling. Fouling of the bore of a gun caused by metal stripped from the jacket or rotating band of a projectile.

Metal Fouling Solution. A solution used to remove the metal fouling.

Metal Jacketed Bullet. A bullet having a jacket of hard metal, usually cupro-nickel, inclosing a lead core. A full metal patched bullet.

Metallic Ammunition. Fixed ammunition put up in metallic cases.

Meteorological. Pertaining to atmospheric conditions, especially as affecting artillery fire and the operations of aircraft.

Meteorological Datum Plane. The level assumed as the basis for the data as to atmospheric conditions. The altitude of the meteorological station.

Meteorological or Metro Message. A tabulation of meteorological (atmospheric) conditions for the use of artillery in determining firing data.

Methods of Instruction. Military instruction commonly includes four steps, viz.: explanation; demonstration; practice; test.

Metric System. The French, or C.G.S., system of measures. Its fundamental unit is the meter, which equals approximately 39.37 inches. The prefixes deka-, hecto-, kilo-, and myria-, indicate 10, 100, 1000, and 10000; deci-, centi-, and milli-, indicate 1/10, 1/100, and 1/1000. The metric system is incommensurable with the British system, except for units which are universally employed; e.g., hour, degree, second, electrical units, etc.

Micrometer. An instrument for setting off or measuring small values; specifically, of the subdivisions of a scale. A graduated tangent or slow motion screw.

Mid Range. See Ranges, Classification of.

Middle Ages. The period in European history from the fall of the Roman Empire, near the end of the 5th century A. D., to the fall of the Byzantine Empire in the 15th century A. D., during which time the present states of Europe were developing. This period includes also the Age of Chivalry and the Dark Ages.

Mil. A unit of angular measurement used in gunnery. A true mil is the angle measured by an arc whose length is 1/1000 of its radius. As the true mil is incommensurable with the circumference, the mil in general use is 1/6400 of the circumference.

Mil Formula. $M = 1000W \div R$, in which M is the angular width of a target or other object in mils; W is the width or frontage of the target, and R is the range from the observer to the target, both in the same unit. A formula used in gunnery.

Mil Scale and Inverted Sight Leaf. Scales engraved on the reticle of a field glass for measuring small angles and determining troop safety in overhead fire, etc.

Mileage. A prescribed money allowance of so much per mile for the expenses of authorized travel. The aggregate distance in miles traveled.

Mileage Tables. Tables showing the distances by the shortest usually travelled routes between important points.

Militant. Engaged in warfare. Warlike. Aggressive.

Militarism. Military spirit, policy, or government. Placing emphasis on the military spirit and military preparedness.

Militarist. One who believes in and advocates a policy of militarism. One devoted to military pursuits.

Military. The spirit of soldiers. The whole body of soldiers. Anything belonging to or pertaining to the profession of a soldier, to arms, or to war.

Military Area. A primary subdivision of a corps area, created to facilitate the decentralization of certain specified functions.

Military Art. The art of war. It is properly divided into five main branches, viz., strategy, tactics, logistics, communications, and engineering.

Military Attaches. Duly accredited military observers from foreign nations, or those sent to foreign nations. They are usually attached to embassies.

Military Aviators. Formerly, officers who served three years creditably as Junior Military Aviators and received the advanced rating. Aviators in military service.

Military Bridge. A temporary bridge rapidly constructed by military methods. The types of bridges most commonly employed for military purposes include ponton or floating, trestle (including spar), pile, truss, girder and suspension bridges.

Military Channels. The prescribed routes of military correspondence and official intercourse. The succession of offices or headquarters through which a communication passes between addressor and addressee.

Military Command. The authority which is exercised in the military service by virtue of rank and assignment. The place, area, or troops commanded.

Military Commission. A criminal war court resorted to in certain cases under the recognized laws of war that are outside the jurisdiction of courts-martial; as the trial of spies in certain cases.

Military Courtesy. Invariable politeness in official and personal relations, including deference to superiors and respect and consideration for subordinates, which is enjoined upon all persons in the military service. The customs which govern personal intercourse between those in the military service. Military courtesy has many outward manifestations and forms, of which the most common is the military salute.

Military Crest. The line nearest the crest of a ridge or hill from which all or nearly all of the ground toward the enemy and within range may be seen and reached by fire. On a concave slope the military crest is at or very near the topographical crest; on a convex slope the military crest is in front of and below the topographical crest. Except in delaying actions or for relatively long range fire, trenches are generally located at or below the military crest.

Military Decorations. Decorations awarded or bestowed for distinguished service.

Military Demolitions. Demolitions by fire, explosives, or other means, of structures or materiel, executed as a matter of military necessity.

Military Department. A territorial subdivision of a country for military purposes.

Military Deportment. Conduct appropriate to one in the military service.

Military Diary. A detailed contemporaneous account of the operations of a unit during the period covered, and such items of military interest as the state of the weather and roads, health of command, etc. It is required to be kept during actual or threatened hostilities by headquarters of units in the theater thereof and during maneuvers by troops engaged therein. A war diary.

Military Discipline. A spirit or state of mind in a properly trained organization that causes each and every man to put forth his best effort to carry out the will of the leader. The habit of cheerful and intelligent obedience to proper authority. The system of disciplinary training used in the army. Obedience to and enforcement of orders and regulations. Punishment administered in the interest of discipline, especially company punishment. To give disciplinary training or apply any measure of discipline.

Military Education. The process of training or imparting military knowledge through prescribed courses of study, discipline and practical training. The training of individuals to fit them for military duty and command. The knowledge so gained. See Military Training.

Military Engineering. The art of utilizing the materials and forces available to meet the tactical needs of the moment, or to provide for the security, effectiveness, health and comfort of troops in the field, by works of engineering. An adaptation of civil engineering to military needs. The earliest engineers were military engineers.

Military Geography. A study of the probable influence of the topography of a country upon contemplated operations of war. Geographical features in their relation to military operations.

Military Government. A government under military control established in enemy territory, which comes within the domain of international law; its rules are the rules of war. The law of hostile occupation. It differs from military law in that it applies to civilians, and from martial law in that it is exercised in a hostile or conquered territory.

Military Hierarchy. A hierarchy determined and constituted within its sphere of action by grades of rank created by law, and by other laws or orders regulating the exercise of rank and command.

Military Hygiene. The science of the care of troops from a hygienic standpoint.

Military Information. Information bearing on any or all matters of military importance, collected from any or all sources, at any or all times in peace or in war. See also Sources of Military Information.

Military Intelligence. Military intelligence may be defined as the winnowings of military information. Information is drawn from all sources, weighed, collated and evaluated. It is then critically analyzed for the tactical and strategical conclusions that may logically be drawn from it. The resulting intelligence is disseminated to all concerned, and is utilized in the preparation of all plans for military operations.

Military Jurisdiction. Jurisdiction defined and conferred by statute or derived from the common law of war, exercised by military tribunals.

Military Justice. The justice which prevails in the army and which is administered by military tribunals in accordance with the Articles of War.

Military Language. The terminology of the military profession, which should be used in all orders and other military documents and communications.

Military Law. The rules of action and conduct imposed by a state upon persons in its military service, with a view to the establishment and maintenance of military discipline. See also Martial Law; Military Government.

Military Maps. Maps prepared or used for military purposes.

Military Mining. The tactical use of underground works for the attack or defense of fortified localities. See also Countermining.

Military Necessity. The necessity which attends military operations and justifies ignoring private or personal rights and resorting to all measures which are indispensable for securing the object desired, and which are not forbidden by modern laws and customs of war.

Military Occupation. The occupation of foreign or hostile territory by the military forces. The conditions existing during the occupation of hostile territory by military forces.

Military Order. An authoritative direction to a military command. A military society or brotherhood.

Military Order of Foreign Wars. An order instituted in 1894 by the veterans and descendants of the veterans of the various foreign wars in which the United States has engaged.

Military Order of the Carabao. A military order composed of those officers of the United States military and naval forces who served in the Philippine Islands during the period 1898 to 1913.

Military Order of the Loyal Legion. An order instituted in 1865 by officers of the military and naval forces of the United States who participated in the Civil War.

Military Organization. See Organization.

Military Ornaments. See Ornaments.

Military Pits. An obstacle consisting of holes dug in the ground, sometimes covered with camouflage or a thin layer of natural material, and having sharpened stakes upright in the bottoms; trous-de-loup.

Military Police. A class of troops charged with the enforcement of all police regulations in the theater of operations and in other places and areas occupied by troops.

Military Policy. The traditional policy or practice of a nation in the organization of its military resources for defense or aggression, and in preparation for and prosecution of war. See also Army of the United States.

Military Position. Any site which may be occupied for tactical purposes.

Military Post. A place where troops are stationed or stores are kept.

Military Principles. The immutable principles governing the operations of war. See Principles of War; Tactical Principles; Combat Principles.

Military Prisons. Prisons for the incarceration of prisoners of war, or for the safe keeping and punishment of offenders against military law.

Military Punishment. The execution of a sentence pronounced by a court-martial, or without trial under the 104th Article of War.

Military Railways. All railways of every nature which are constructed, maintained, or operated for the general service of the army, whether by the personnel of the army or by civilians under their direction.

Military Rank. That character or quality bestowed on persons in the military service which marks their station and confers eligibility to exercise command or authority in the military service within the limits prescribed by law. It is divided into degrees or grades which mark the relative positions and powers of the different classes of persons possessing it.

Military Reconnaissance. An examination of an area of country made by or under the protection of an armed force. Any reconnaissance for military purposes.

Military Regulations. Rules and regulations for the government of a military force. Army Regulations.

Military Reservation. Land set aside for military purposes. The confines of any military post or station.

Military Resources. Men, material and money available for military use.

Military Roads. Roads of a more or less temporary character constructed for use during a campaign. Roads constructed in time of peace with a view to military use.

Military Sanitation. The establishment and maintenance of conditions essential to the health and physical efficiency of a military command.

Military Schools. Schools for the training of military personnel in the tactics and technique of their own branch, and of the arms and services in combination. Service schools of the army. Civil educational institutions organized on a military basis like the United States Military Academy.

Military Science. The scientific principles which govern the conduct of war, and their application to specific situations.

Military Science and Tactics. A term specifically applied to the military instruction in civil educational institutions (R.O.T.C.).

Military Service. The performance of military duty. The army as a whole.

Military Shelter. Structures necessary for the shelter of men, animals, equipment, and supplies of the military service.

Military Sketch. A hasty topographical map of a limited area of terrain or a route of travel and the terrain adjacent thereto, showing in detail the existing features of military importance with sufficient accuracy to meet special tactical requirements.

Military Sketching. The making of crude maps by rapid military methods.

Military Staff. The military staff of a commander, which may include a personal staff, general staff and special (administrative and technical) staff, which see.

Military Station. A place for the rendezvous or quartering of troops. A military post.

Military Stores. Supplies of every nature pertaining to an army.

Military Substitute. A substitute for service furnished by one who is unwilling to serve in person.

Military System. Specific rules and regulations for the government of an army. The means and methods employed in maintaining an army. The military policy and organization of a state.

Military Tactics. See Tactics.

Military Terminology. The technical language of the military profession.

Military Topography. The art of military sketching, map-making, and map reading, including the making and reading of aerial photographs.

Military Training. Military training, in its broadest sense, embraces national military education and military training proper. National military education comprises the inculcation in every citizen of a knowledge of our government, our military history, policy and organization, and an appreciation of his military obligations to the nation. Military training proper consists of the technical and tactical instruction of the authorized and potential military forces of the nation in order to develop an efficient fighting force, prepared for the particular type of war in which our country may be expected to engage.

Military Training Camps Association. An association organized by graduates of the military training camps of the World War period, which fosters and promotes practical outdoor military training for young men.

Military Tribunals. The agencies through which military jurisdiction is exercised, including: (1) Military commissions and provost courts, for the trial of offenders against the laws of war and under martial law; (2) Courts-martial, for the trial of offenders against military laws; (3) Courts of inquiry, for the examination of transactions of, or accusations or imputations against officers and soldiers.

Militia. A term orginally applied to the more or less irregular troops employed or maintained by the American colonies. The existence of the militia was recognized in the Constitution and Congress was given power to organize and control it. The Militia Act of 1792 laid down the doctrine that every able-bodied male citizen was liable for military service, but this service was to be rendered to the state and not to the nation. This Act remained the basic law concerning the militia for 111 years, or until 1903. Meantime state troops had increased in numbers and efficiency, but in all our wars had been found unsatisfactory for national service because of lack of federal control of organization, training, appointment of officers, etc. In 1903 the Dick Bill established a measure of federal control over the militia, and created a Bureau of Militia Affairs in the War Department. In subsequent acts, especially the National Defense Act of 1916 and 1920, federal control was further extended. The organized militia of the several states was designated as the National Guard, which when called into service, becomes in all respects a part of the Army of the United States. The unorganized militia continues to include all able-bodied male citizens between the ages of 18 and 45 years, who are subject to draft in the event of war. The militia as now defined by law, includes the National Guard, the Naval Militia, and the Unorganized Militia. See also National Guard.

Militia Bureau. See Militia; National Guard Bureau.

Mills Cartridge Belt. The woven fabric belt used in the United States Army.

Mimeograph. A method for reproducing maps and documents by means of stencils; or the prints thus produced. To make a mimeograph.

Mine. A charge of explosive placed upon or below the surface of the ground or water, which may be exploded at will or by contact with a vehicle, vessel, etc.; designed to inflict damage upon enemy personnel and materiel. A subterranean passage used to attack or counterattack the enemy by the use of charges of explosives. To place land or water mines. To engage in subterranean warfare.

Mine Battery. In coast defense a mine battery consists of the personnel assigned to the controlled mine defenses, whose mission is the installation, operation, and maintenance of the controlled mine fields.

Mine, Buoyant. A submarine mine which floats underneath the surface of the water, between the level of the armor belt and the keel of the vessel to be destroyed.

Mine Case. The metallic container of a submarine mine, made in the following shapes: spherical, plain sphero-cylindrical, and corrugated spherocylindrical.

Mine Casemate. A casemate room or protected building on shore where the electrical apparatus for controlling submarine mines is installed.

Mine Chamber. The cavity in a mine which contains the explosive charge.

Mine, Chemical. A mine whose detonator is ignited by the chemical combination of two substances which are brought together. See also Chemical Land Mine.

Mine Command. Such portions of the mine defenses, including mine planters and auxiliary boats, as may be necessary, the batteries for the protection of the mine field, and the assigned personnel.

Mine, Common. A subterranean mine that produces a two-lined crater, or one whose diameter is twice the line of least resistance.

Mine, Contact. An electric mine having an automatic circuit-closer, either in a buoyant mine, or in a buoy above a ground mine, which detonates the explosive charge when struck by a vessel.

Mine, Controlled. A land or submarine mine which may be fired from a distant point by electricity.

Mine Defense. All mine personnel and materiel pertaining to the defense of a harbor, together with all battery commands assigned to protect the mine field.

Mine, Electric. A mine whose detonator is ignited by a current of electricity.

Mine Field. A water area in which submarine mines are systematically planted.

Mine, Free. A mine which is not moored or anchored to the bottom, and which detonates automatically on contact with a vessel.

Mine, Ground. A submarine mine which is placed upon the bottom of a channel; it is not effective in depths greater than 40 feet.

Mine Group. A tactical and technical unit for the employment of submarine mines, including personnel, armament, equipment, structures and vessels.

Mine, Judgment. An electric mine which is exploded at will by the closing of a circuit by an operator.

Mine, Land. A charge, usually of high explosive, placed on the surface of the ground or buried in a shallow pit, and detonated at will or by the pressure of vehicles or troops passing over it; sometimes discharging a mass of broken stone. See Fougasse; Chemical Land Mine.

Mine Loading Room. A place where submarine mines are loaded.

Mine, Mechanical. A mine whose detonator is ignited by some mechanical device, as a firing pin.

Mine Planter. A vessel which places submarine mines.

Mine Planting. Planting submarine mines and controlling apparatus in the water.

Mine Planting Flotilla. A tactical unit within the mine group, consisting of one mine planter, one distribution box boat, and three mine yawls, together with the personnel to man these boats.

Mine Plotting Board. A plotting board used to determine the course and probable future position of a vessel with reference to submarine mines.

Mine Predictor. A device used in determining the time of travel of a vessel from any observed point to the position of a submarine mine.

Mine, Regulated. An electric mine which signals when struck and may then be exploded by an operator if desired.

Mine, Self-acting. A mechanical, chemical, or electric mine, in which the entire control and detonating apparatus is in the mine; it explodes when struck.

Mine, Submarine. A case containing a sufficient charge of explosive to rupture the hull of a vessel, with some form of detonator; mines are classed as buoyant and ground, contact, judgment, self-acting, regulated, controlled, mechanical, electric, chemical, and free. Specifically, a mine which floats below the surface of the water.

Mine Sweeping Gear, Tank. A heavy roller, attached to a spar extending in front of a tank for the purpose of exploding any mines in its path.

Mine Tank. A tank in which submarine mines are tested.

Mine, Tank. A land mine designed and placed for the destruction of tanks. It may be exploded by judgment (electrically), or by the pressure of a heavy vehicle passing over it.

Mine Thrower. A minenwerfer. Any of the numerous trench mortars used during the World War. Any device used to throw projectiles containing relatively large charges of high explosives.

Mine, Undercharged. A subterranean mine that produces a crater whose diameter is less than twice the line of least resistance.

Mine Warfare. Subterranean warfare conducted by mining and countermining; it was known to the ancient Romans.

Minenwerfer. A trench bomb-throwing weapon used by the Germans in the World War.

Mines, Group of. A group of 19 submarine mines placed about 100 feet apart, and electrically controlled through a single distribution box.

Minie Ball. A conical bullet with a cavity in its base, which expands and forces the metal into the grooves of the rifling. Used with early muzzle loading rifles.

Minimum Range or Elevation. The least elevation at any position which will cause projectiles to clear a mask intervening between a gun and its objectives. The least range at which any weapon may be fired.

Minimum Specifications. Statements of essential specific abilities and the lowest acceptable standards for a given grade or rating in the military service.

Minimum Standard of Proficiency. A prescribed standard which must be attained in order that an individual or organization may be rated as efficient.

Mining. All operations incident to the placing of charges of explosives underground and exploding them at the time desired for the purpose of destroying men, materiel, or works in their vicinity. The operations involved in subterranean attack and defense.

Minor Armament. Fixed armament in coast defenses consisting of the smaller guns.

Minor Tactics. The tactics of small forces, or of those consisting entirely of one arm.

Minor Warfare. Warfare conducted on a relatively small scale, which embraces both regular and irregular operations.

Minute Guns. Guns fired at frequent intervals, usually as a signal of distress or on the interment of an officer or personage of high rank.

Mirage. An atmospheric optical illusion presenting the appearance of waves in the atmosphere. Heat waves observed to rise from the ground on warm days, especially during firing on the range, which interfere with correct aim.

Misfire. Failure to fire, as of a defective cartridge or charge, engine, etc.

Missile or Missive. A weapon thrown or designed to be thrown. A projectile of any kind, as an arrow, stone, or bullet. Capable of being thrown.

Missile Weapon. A weapon that discharges a projectile.

Mission. A specific task or duty assigned to an individual or unit, or deduced from a knowledge of the plans of the immediate superior. The mission is the criterion for all plans and procedure.

Mistakes. Those personal errors, as in gunnery, sketching, etc., which may be avoided by proper care. Blunders. See also Errors.

Mitrailleuse. A French breechloading machine gun consisting of a number of barrels fired simultaneously or successively, first used in the Franco-Prussian War. A machine gun.

Mixed Force. A provisional force including several different arms.

Mixed Salvo. A salvo in which some of the shots fall short of and the others over the target. A straddle.

Moat. A deep and wide ditch around the wall or rampart of a fortification, often containing water. A wet ditch of a fortification.

Mobile Armament. Seacoast guns, howitzers and mortars on movable mounts, capable of being readily moved or transferred and of comparatively rapid emplacement.

Mobile Army. An army organized primarily for offensive operations and possessing the highest degree of mobility.

Mobile Artillery. See Artillery, Mobile.

Mobile Defense. A class of defense in which the troops are not distributed uniformly along the entire front but occupy mutually supporting defensive areas, the tactical continuity of the front being maintained by the counter attacks of local and, when necessary, of general reserves. An active defense.

Mobile Fire Commands. Mobile batteries of coast artillery.

Mobile Hospital. A field hospital, provided with tentage and transportation, which enables it to move and establish itself in another locality on short notice; it may include surgical, evacuation, and convalescent hospitals.

Mobile Reserves. Troops held mobile in favorable positions for probable employment as reinforcement or in counterattack. Reserve supplies held on trucks or cars for prompt movement. See also Rolling Reserves.

Mobile Searchlights. Searchlights capable of being moved for considerable distances on roads or across country.

Mobile Troops. Troops capable of moving promptly and rapidly.

Mobile Warfare. A warfare of movement, as distinguished from stabilized or trench warfare. Open warfare.

Mobilization. The change from peace to war footing. The organization and assembling of troops and all other resources of a nation for war.

Mobilization Area. An area in which a number of mobilization camps, cantonments, or mobilization points are grouped for administrative or training purposes.

Mobilization Camp. A camp or cantonment to which units come from their home stations to complete the second stage of mobilization, i.e., mobilization concentration.

Mobilization Center. A place or locality in the zone of the interior, designated as the point where certain units will mobilize or complete their mobilization.

Mobilization Concentration. The concentration in selected areas of units which are unable to complete their mobilization at home stations, due to lack of training, supply or organizing facilities, for the completion of these requirements.

Mobilization Order. An order which directs the execution of a plan of mobilization.

Mobilization Plan, General. A plan for the mobilization of the army for national defense. It does not envisage any specific enemy.

Mobilization Plan, Special. A plan for the mobilization of the army to meet a specific situation.

Mobilization Point. The home station, post, camp, cantonment, or other definite place at which a unit completes its local mobilization.

Mobilization Rate. The rate at which troops may be or are to be placed in readiness to enter the theater of operations.

Moderately Persistent Gases. Gases which give off vapor rather slowly and continue to contaminate the air in the neighborhood from three hours to one day.

Modified Bracketing Method of Adjustment. Adjustment of the range to an aerial target based on stereoscopic spottings, using the fork (4% of the altitude) as the unit, and limiting changes to 3 forks.

Molotov Breadbasket. A container, designed to be dropped from aircraft, which scatters small incendiary bombs over a considerable area.

Mono-block Gun. A gun manufactured by the internal pressure method. See Autofrettage.

Monocord Switchboard. A telephone switchboard which requires only one cord in connecting one subscriber to another. Commonly used in the field.

Monoplane. An airplane with but one main supporting surface, sometimes divided into two parts by the fuselage.

Mooring Mast. A mast or tower at the top of which there is a fitting to which the bow of an airship may be secured.

Mooring Point. A specially designed and strengthened point in the structure of an airship from which the mooring attachments are led.

Moot Court. A mock court for instruction purposes such as is held by students of law.

Mopping Up. The act of searching an area or position that has been passed over by friendly troops in the attack and of killing or capturing any enemy found.

Moral. The lesson taught or designed to be taught. Pertaining to the state of mind, especially of an organization or of troops as a whole, as distinguished from physical.

Morale. The mental state which renders a man or body of men capable of bravery and endurance in the face of the enemy, or under trying conditions. The psychological condition or mental state of an individual or body of troops, especially in relation to performance of duty.

Morning Gun. A gun fired every morning at permanent posts or stations at sunrise or reveille, at which time the flag is raised on the staff.

Morning Parade. The daily parade of troops held at reveille.

Morning Report. A report rendered daily to superior headquarters showing the duty status of the individuals belonging to an organization.

Morrill Act of 1862. See Land Grant; Reserve Officers Training Corps.

Morse Code. A system of signaling based on the Morse telegraphic code.

Mortar. A cannon having a relatively short barrel and a low muzzle velocity fired at high elevations. A mixture of sand and lime or cement, used for laying brick, etc.

Mosaic. A composite vertical air photograph made by piecing together adjacent, overlapping prints. If formed by matching detail at the margins of adjacent prints it is an uncontrolled mosaic. If the prints are placed in their correct positions by means of ground control, it is a controlled mosaic.

Motor Transport Maintenance. See Maintenance, Motor Transport.

Motorization. The process of equipping a military force with motor-propelled vehicles. Motorization may include the supply elements of all branches of the service, the use of motor transport to increase strategical mobility, and the employment of motorized vehicles and weapons in battle.

Motorize. To equip with motor-propelled vehicles.

Motorized. Equipped wholly with motor propelled transportation.

Motorized Artillery. Artillery drawn by tractors or other motor propelled means.

Motorized Force. A motor-transported force that fights dismounted.

Motorized Unit. A unit equipped, either organically or provisionally, with sufficient motor vehicles to carry all its personnel and materiel.

Moulinet. A circular swing of the sword or saber. A machine used for bending a crossbow by winding it up.

Mount. A riding animal. A gun carriage. To get upon, as for riding. To cause to mount. To ascend. To place a gun on its carriage.

Mountain Artillery. See Artillery, Mountain.

Mountain Battery. A battery of mountain artillery.

Mountain Warfare. Warfare conducted in mountainous terrain, with the modifications of normal procedure dictated by the situation.

Mounted Combat. Combat by cavalry actually mounted, in which the horse, saber and pistol are the weapons of the trooper. The saber is no longer used by cavalry in the United States service.

Mounted Defilade. A defilade which affords protection to mounted men.

Mounted Infantry. Infantry mounted on animals as a means of moving rapidly, but which fights on foot.

Mounted Messenger. A motorcycle, bicycle, or horse messenger.

Mounted Point. A small detachment of mounted men forming the leading element of an advance, or rear element of a rear guard composed primarily of foot troops.

Mounted Sketch. A road sketch made by a mounted man.

Mounted Units. In the United States service the following units are considered to be mounted and entitled to carry a standard: (1) Air Corps groups; (2) Cavalry regiments and separate squadrons; (3) Coast Artillery regiments (tractor and antiaircraft); (4) Engineer squadrons in Cavalry divisions and Engineer camouflage battalions; (5) Field Artillery regiments; (6) Field Artillery ammunition trains; (7) Infantry Tank regiments; (8) Medical regiments; (9) Quartermaster trains and repair battalions; (10) Signal battalions.

Movable Armament. Coast defense guns capable of being moved from place to place.

Movement. A component part of a maneuver or evolution. The regular and orderly motion of a body of troops for any particular purpose. A march.

Movement Order. An order issued by a corps or army headquarters generally to direct movements of corps and army troops, especially such movements as are not involved directly in a combat mission.

Moving Pivot. The arc of a small circle on which the pivots of a column of troops successively turn in making a change of direction, as column right. The man on the pivot flank of any rank in making such a turn.

Moving Target. A target capable of motion, or of being moved. A fleeting target.

Multicharge Gun. A gun having a number of auxiliary powder chambers along its bore, which are discharged as the projectile passes, thus increasing its muzzle velocity without a marked increase in the maximum powder pressure in the bore.

Multiplane. An airplane with two or more main supporting surfaces placed one above the other.

Munitions. All materials used in war. Military stores of all kinds. Ammunition.

Munitions Distributing Point. A point where munitions are distributed.

Munitions Officer. An officer of any unit who is charged with supervision of the supply of ammunition.

Musette Bag. A bag, with a strap to pass over a shoulder, for the carrying of small articles.

Mushroom or **Soft-pointed Bullet.** See Dumdum Bullet.

Mushroom Head. The enlarged front end of an obturator, which receives the pressure of the powder gases.

Musket or **Mousquet.** A smoothbore small arm of which there were various types. The musket proper, invented about 1540, was much heavier and more powerful than the harquebus, which it superseded. Like the early harquebus it was fired from a rest. The musket was a matchlock, weapons with other forms of locks being distinguished as snaphances, wheellocks, flintlocks, or firelocks, etc. On the abandonment of the matchlock firing mechanism about 1690, the term musket was for the time discontinued in favor of fusil or flintlock, which thenceforward reigned supreme until the introduction of a practicable percussion firing mechanism about 1830-1840. But the term musket has survived the weapon it originally designated and is now generally applied to individual military firearms between the harquebus and the modern rifle. See also Flintlock; Harquebus; Matchlock; Wheellock.

Musketry. The application and control of the collective fire of rifle units. That branch of training which treats of the ballistic qualities of rifle fire and the effect thereon of changing meteorological conditions, range, and terrain, the variations in the human element in handling the rifle, the functions of fire control and direction, and the principles and methods of training and handling a fire unit so as to insure a maximum of properly distributed hits in a minimum time. The principles of musketry pertain to the rifle, automatic rifle, semi-automatic rifle and light machine gun. (Old.)

Mustard Gas. A toxic chemical agent in the form of an oily liquid. having violent irritant and blistering effects. A powerful vesicant. The gas mask protects the face and respiratory passages against mustard gas, but additional gas-proof clothing is necessary to prevent burns.

Muster. A review of troops in order to take account of their numbers, etc. To enroll in the service.

Muster Roll. A list or register of all the officers and men in a military organization.

Muster Troops into the Service. To inspect and administer the oath of service to a unit accepted into the service.

Muster Troops out of the Service. To discharge a body of troops from the service.

Mutiny. Collective insubordination of two or more persons, usually involving collusion and conspiracy, in resisting lawful military authority. Insurrections against constituted authority. Forcible resistance to lawful authority on the part of subordinates. A resistance by force to recognized authority.

Mutual Fire Support. Reciprocal assistance by fire, on the part of adjacent elements.

Mutual Support. The support rendered one another by adjacent elements in case of need. involving fire or movement or both.

Muzzle. The forward end of a gun.

Muzzle Flash. See Flash.

Muzzle Loader. The name applied to any gun that is loaded at the muzzle.

Muzzle Velocity. The velocity of the projectile at the muzzle of the gun. See Initial Velocity.

N

Nacelle. An inclosed shelter for personnel or for a power plant of an airplane. A nacelle is usually shorter than a fuselage and does not carry the tail unit.

Napoleon Gun. A 12 pounder bronze gun, introduced in 1856 as a medium weight gun to replace the light and heavy caliber guns then in use.

Narrow Sheaf. A battery sheaf of fire narrower than a parallel sheaf.

National Anthem. The musical piece which is recognized as expressing the patriotic sentiment of a nation; the Star Spangled Banner is designated as the national anthem of the United States.

National Army. That part of the Army of the United States which was obtained during the World War by selective service, the officers coming largely from the Officers' Reserve Corps and officer training camps.

National Defense. The defense of a state or nation against attack. The measures taken in preparation for such defense.

National Defense Act. The Act of Congress of June 3, 1916, with subsequent amendments, concerning the Army of the United States and other matters relative to the national defense. The current basic law embodying the military policy of the United States.

National Defense League. A league founded in 1913, to foster all measures that will better prepare the United States for national defense.

National Flag. The flag which represents the living country and is considered as a living thing. The flag designated as such by proper authority.

National Guard. That portion of the militia of the several states and territories and the District of Columbia, which is regularly enlisted and commisioned, properly organized, armed and equipped, and federally recognized, under the provisions of the National Defense Act. The National Guard is subject to supervision and control by the War Department, and is supported in part by the federal government. See Militia.

National Guard Bureau. A bureau of the war department exercising supervision and control of all affairs pertaining to the National Guard. The chief of the bureau is a national guard officer.

National Guard of the United States. Those federally recognized National Guard units and organizations, and the officers, warrant officers, and enlisted members of the National Guard of the several states, who shall have been appointed, enlisted and appointed, or enlisted, in the National Guard of the United States.

National Rifle Association. An association for the promotion of rifle practice throughout the United States.

National Salute. A salute of 21 guns rendered to a national flag, to a president or ex-president, to a sovereign or chief magistrate of a foreign country, and to members of a reigning royal family.

Natural Fortifications. Such natural features of the terrain as are capable of being used to impede the enemy in his advance. A place is said to be naturally fortified when it is situated on the top of a steep hill or surrounded by impassable rivers, marshes, etc.

Natural Obstacle. Any geographical or topographical feature which hampers military maneuvers or operations, such as deserts, mountains, streams, swamps, forests, etc.

Navigation Head. A point on a navigable waterway corresponding to a railhead. The highest point to which a stream is naturally or artificially navigable.

Navigation Lights. A combination of lights on the under surface of an airplane for identification purposes.

Necklace. A charge of explosive strung like a necklace, as for cutting a tree.

Needle Gun. A breechloading, bolt-type rifle adopted in the Prussian Army in 1841. The earliest successful breechloader to be generally adopted.

Negative Information. Information which is not positive in character, such as that a certain locality is not occupied by the enemy at a certain time, but nothing more. Negative information may be of great value.

Negative Measures. The use of obstacles, both natural and artificial, demolitions, gas, etc., to hinder or block a hostile advance. Passive measures.

Neglect of Duty. The military offense involved in failure to perform any ordered or prescribed duty. Total omission or disregard of any prescribed service, or poor execution of same.

Net. The fabric formed by the ensemble of a number of posts or stations devoted to a particular duty; e.g., radio net, telephone net, observation net, intelligence net. See also Camouflage Net; Bomb Net.

Neutral. A person or nation that takes no part in a war in progress. Not involved in hostilities. Not taking part with any of the contending belligerents. It is customary for nations not engaging in a particular war to issue proclamations of neutrality in order that their neutral rights may be respected.

Neutrality. Neutrality, on the part of a state not a party to the war, consists in refraining from all participation in the war, and in exercising absolute impartiality in preventing, tolerating, and regulating certain acts on its own part, by its subjects and by the belligerents. The condition of being neutral.

Neutralize. To destroy or reduce the effectiveness of personnel or materiel by the application of fire or chemicals. To make insufficient. To nullify the effects of. To prevent the functioning of. See Fire, Neutralization.

Night Assault or Attack. An attack launched under cover of darkness with the intention of securing an objective or a decision before daylight.

Night Firing Device or Mark. An illuminated marker used as an aiming point for night firing.

Night Marches. Marches made under cover of darkness for any purpose.

No Man's Land. In modern warfare, the strip of land lying between the most advanced trenches or lines of the two opposing forces.

Nomenclature. A systematic classification of the names of the different parts of a weapon or apparatus. Technical designation.

Noncombatants. All persons connected with an army whose particular duty does not include combat service. The citizens of an area occupied by an army. In general, all persons not belonging to an army.

Noncommissioned Officer. An enlisted man holding a grade between that of first class private and warrant officer, by virtue of a warrant issued him by proper authority.

Non-delay Fuze. A percussion fuze that detonates instantly on impact.

Noneffective. Not fit or available for duty.

Non-mobile Troops. Troops unsuitable for active warfare and capable of only a passive defense.

Non-standard Conditions. Ballistic conditions which differ from the average or standard conditions on which firing tables are based, and for which corrections must be made; e.g., temperature, humidity, muzzle velocity.

Nonsystematic Errors. See Accidental Errors.

Normal. Perpendicular to. Usual or customary. An arbitrary value given to the actual zero for certain corrections in order to avoid the use of negative values.

Normal Barrage. A standing barrage laid in immediate defense of the sector which it supports. The barrage which is fired on prearranged signal without further orders. The barrage fired in case of a general attack by the enemy. It is coordinated with the defensive fires of the supported units in the defense.

Normal Corrector. The corrector setting which gives a normal uncorrected height of burst.

Normal Height of Burst. The height at which an air burst gives the maximum effect as shown by the firing tables, not corrected for non-standard conditions.

Normal Sight. For the rifle, a sight attained by bringing the top of the front sight and the target on a line with the center of the peep or notch of the rear sight.

Normal System of Maps. Maps of various scales in which the relation between scale and contour interval is such that the horizontal equivalent for any given slope is the same on all maps.

Normal Zone. A zone for artillery fire which coincides laterally with the division or corps sector or zone of action, and extends in rear of the enemy front line to a depth determined by the ranges of the guns employed. The zone within the objective zone for which an artillery unit is normally responsible and within which its normal fire is directed.

Nose. The forward end of an airplane. The point of a projectile. A short projecting ridge or spur.

Nose Dive. A steep descent of an aircraft, head-on.

Nose Spin. A nose dive in which an airplane rotates about its own longitudinal axis.

Nose-down. To depress the nose of an airplane in flight.

Nose-over. A colloquial term referring to the accidental turning over of an airplane on its nose when landing.

Nose-up. To elevate the nose of an airplane in flight.

Not in Line of Duty. A medical term which implies that a disease contracted or an injury suffered by a soldier is the result of his own neglect or misconduct.

O

Oak-leaf Cluster. A bronze decoration awarded to the holder of a medal of honor, distinguished service medal or distinguished service cross, for further service which justifies the award of any of these decorations, only one of either of which may be awarded to any person.

Obedience to Orders. An unequivocal and unhesitating performance of prescribed or ordered duty.

Objective. The immediate aim or object of military operations in the theater of war. A locality or terrain feature to be reached in the course of an attack, or during a movement. A hostile position or force which a command has been ordered to reach or overcome; the attainment of an objective usually completes a phase of operations. A result to be accomplished. A target or an area upon which fire is directed. The forward end of a telescope.

Objective of War. The purpose or ultimate objective of a state at war is to impose its will upon another state or people. This is usually best accomplished by destroying the power of resistance of the enemy state, and accordingly, as a rule, the hostile main army is the principal objective, all other objectives being secondary or intermediate. There will be cases, however, when the enemy will to resist can be more readily broken by the seizure of some locality or area of strategical importance, such as the capital city or a district from which important supplies are derived, or by an effective blockade, which may require the destruction of the hostile fleet, etc.

Objective Plane. The plane tangent to the ground or other material object at the objective point, or point of impact.

Objective Point. The point or result aimed at in the operations of an army. The point of impact, or point aimed at.

Objective Zone. The area which contains the objectives or targets which are or may be assigned to a fire unit, and whose ultimate limit is determined by the effective range and positions of the pieces concerned. See also Fire, Zone of.

Objects of Military Training. The objects of military training are: (1) To prepare the units authorized by legislation for prompt and efficient field service; (2) To produce trained officers, noncommissioned officers, specialists, and key men in such numbers as may be necessary for the authorized peace units and for carrying out complete and immediate mobilization in an emergency; (3) To prepare for war expansion; (4) To develop the art and science of war, and methods of training therein; and (5) In time of war, to produce in the minimum time, units and individuals trained to the minimum standards of proficiency in the necessary numbers.

Oblique. Diagonal to the front. At an angle of 45 degrees to. Half-oblique implies a change of direction or front of 22½ degrees.

Oblique March. A movement in which each man marches in a direction 45 degrees to the right or left of his original front.

Oblique Order of Battle. See Order of Battle, Oblique.

Oblique Photograph. An aerial photograph taken with the axis of the lens inclined from the vertical, giving a perspective of the terrain shown.

Observation. An organized method of gaining information visually without movement or physical contact with the enemy or object examined. Stationary surveillance of the enemy, terrain, effects of fire, progress of an attack, etc. Spotting.

Observation Aviation. See Aviation, Observation.

Observation Balloons. Captive balloons used principally for observation of artillery fire and enemy activities.

Observation Post or Station. A point selected for the observation and conduct of fire, for the observation of an area or sector, for the study of objectives, or for the purpose of securing information of the enemy and his activities. A position from which friendly and enemy troops can be seen and from which artillery and machine gun fire is controlled and corrected. In small units observation posts are combined with command posts when possible.

Observation, Spotting. An observation of artillery fire, for purpose of adjustment, from an airplane.

Observer. One detailed to make observations and report thereon. That occupant of an aircraft whose duty is observation.

Observer-pilot. A rating given to an officer who has successfully passed the tests for both aircraft pilot and observer.

Observing or **Observation Sector.** The field of view of an observer. An assigned sector of observation.

Obstacle. Any device or feature, either natural or artificial, used in field fortifications for the purpose of delaying the hostile advance. A natural terrain feature or artificial work which impedes the movements of the troops. Obstacles are classified as natural or artificial, tactical or protective, fixed or portable, etc.

Obstacle Ability. The ability of an armored vehicle to pass or cross obstacles.

Obstacles, Types of. Natural obstacles include any topographical or geographical features which interfere with the free movement of troops, weapons or transport. Artificial obstacles include abatis, slashings, trous-de-loup or military pits, caltrops, palisades, walls, trenches, chevaux-de-frise, mines, fougasses, inundations, barbed wire entanglements, etc. Barbed wire entanglements include portable types, such as concertina, ribard and plain spirals, knife-rests, hedgehogs, gooseberries, etc; and fixed types such as low wire, double apron fence, high wire, spider wire, etc. See Natural Obstacles; Wire Entanglements.

Obturator. A device used in guns firing separate loading ammunition for preventing the escape of gas from the breech of a gun.

Obus. Artillery shell (Fr.).

Occupation of Position. Movement into a prescribed or selected position.

Occupational Specialist. An enlisted man whose duty or assignment requires that he possess training of a special nature not included in the basic military training peculiar to the arm or service to which he belongs.

Occupied Areas. The areas in a defensive position actually occupied by troops and affording mutual support. Any portion of hostile territory occupied by the military forces of another belligerent.

Octagonal Trace. A trench trace of alternate fire bays and traverses, whose angles are those of a regular octagon.

Odometer. A device for measuring distance, attached to the wheel of a vehicle.

Oerlikon. A Swiss arms firm. The 20mm automatic cannon, Oerlikon, is used in the United States.

Off Duty. Having no prescribed duty to perform for the time being.

Off Limits. Areas or places which soldiers are forbidden to visit. The status of being off limits.

Offense. A state of attack. The act of attacking. Attack. A crime or offense committed.

Offensive. Of or pertaining to the attack.

Offensive and Defensive Operations. Operations whose object is not only to prevent the advance of the enemy, but also to attack him when a favorable opportunity offers. An active defense.

Offensive Fortifications. The various works employed in making an attack or in conducting a siege. Hasty intrenchments constructed during a pause in the attack, to hold a position gained, to protect the flanks, etc.

Offensive Operations. Operations which seek to seize and maintain the initiative and force an early and successful conclusion of hostilities. The role of an attacker or aggressor.

Offensive Tactics. The tactics of the attack.

Offensive War. A war involving invasion and carried on in the territory of the enemy.

Offensive Weapons. See Weapons.

Offensive-defensive. The act of taking the offensive in order to further a plan that is, in general, defensive in its nature, in accordance with the basic military principle that an active offensive is the best defense.

Officer. A commissioned officer. A person lawfully invested with military rank and authority by virtue of a commission issued him by or in the name of the sovereign or chief magistrate of a country.

Officer Candidate. An individual tentatively selected with a view to his appointment to a commissioned grade.

Officer Observer, Airplane, Balloon or **Airship.** An officer who has received a certificate of graduation from a school of aerial observation or a balloon school and who has subsequently passed a prescribed test in aerial gunnery.

Officer of the Day. An officer who has general charge of the interior guard and prisoners for a particular day or period of days.

Officer of the Guard. An officer detailed for duty with the guard. The senior officer so detailed is the commander of the guard.

Officer's Patrol. A patrol commanded by an officer.

Officers' Reserve Corps. An organization under the National Defense Act whose purpose is to provide a reserve of officers available for military service when needed. Reserve officers for the combatant arms are drawn from the R. O. T. C. and C. M. T. C. First organized in 1916, the Reserve Corps furnished many officers to the National Army in the World War.

Official. With or having proper authority. Pertaining to legal or proper authority.

Official Copy. A document exhibiting the official stamp or seal of a bureau or office and usually the signature of an executive officer or adjutant.

Official Correspondence. Correspondence on official matters carried on by offices, officers, or departments of the service.

Official Courtesies. The interchange of official visits and compliments between military and naval officers and civil officials of the same or of different nations.

Offset. A horizontal displacement, angular or linear, as from a gun to an observation post. A distance measured perpendicularly to a traverse or other line. An angle subtended at any point by two other points. A short trench perpendicular to another. A recess. Parallax. A method of lithographic printing. Out of line with adjacent elements. Balanced or compensated. Echeloned.

Ogival. The shape usually given the head of a cylindrical projectile in order to reduce the resistance of the air and keep the projectile head on during flight, and to increase its penetration.

Ogive. The curved and pointed head of the cylindrical projectile used in rifled firearms.

Oil Bomb. A large oil drum containing oil and high explosive which scatters burning oil in all directions when exploded.

Old Glory. An affectionate term applied to the flag of the United States.

On Duty. In the active exercise or performance of military duties.

On Guard. Serving as a guard or member of a guard. On watch. The initial position in fencing or bayonet exercise.

On Post. Engaged in walking a designated sentinel's post. Stationed in observation against the approach of an enemy. On post to maintain internal discipline, or to guard stores or prisoners.

On Right (Left) into Line. A drill movement in which line is formed from column to the right (left) flank by each element marching beyond and successively executing a turn to the right (left) and forming on the left (right) of the preceding element.

On the Alert. In a state of vigilance or activity.

On the Base Line. A battery is on the base line when the plane of fire of the base piece is directed on the base point and the planes of fire of all pieces are parallel.

One Hundred Per Cent Zone. The area beaten by all the shots fired at a single sighting of a machine gun. The area beaten by all the shots fired by a unit or battery with the same elevation or range setting. See Beaten Zone; Dispersion Diagram; Zone of Dispersion.

One-and-a-half-Plane. A biplane in which the span of the lower plane or wing is decidedly shorter than that of the upper.

One-pounder. A small cannon firing a projectile weighing one pound. The term is commonly applied to the 37mm gun, but in the later models its projectile weighs more than one pound.

One-track Road. A road capable of accommodating only one column of transport throughout its length.

One-way Road. A road on which traffic is permitted in one direction only. A road designated as such by proper authority.

Open. Expanded. Uncovered. To open or expand a battery sheaf of fire.

Open Arrest. A state of arrest which implies restricted limits and no detailed performance of duty, but not actual confinement.

Open Circuit. See Circuit.

Open Country. See Terrain, Open.

Open Fire. To commence firing. To begin a bombardment.

Open Market Purchase. See Purchase, Open Market.

Open Ranks. A formation, usually for inspection, in which the distance between ranks is two steps, or 60 inches, greater than the normal distance. At close ranks the normal distance is resumed.

Open Sheaf. A battery sheaf covering a maximum front without sweeping.

Open Sight. A rear sight or sight slide having a curved notch instead of an aperture.

Open Traverse. A traverse, in sketching or surveying, which does not end at its point of beginning or some other point whose location is known.

Open Warfare. Warfare of maneuver, as distinguished from stabilized warfare.

Open Work. A fortification that is open or unprotected on the side away from the enemy, usually called the gorge. A field emplacement for a weapon, having no protective roof or overhead cover.

Opening of the Trenches or **Parallels.** The first breaking of ground by the besiegers in order to carry on their approaches toward the place besieged.

Operate. To act. To carry out one or more acts of a tactical or strategical nature. To conduct operations against an enemy.

Operating Crank, Handle or **Lever, Breechblock.** A crank or lever which unlocks and withdraws a breechblock from the breech and returns it to its firing position, by a continuous motion.

Operation. A group of movements, engagements, marches, etc., closely related to each other, carried out for the attainment of a certain goal during a particular phase of a war. Any tactical activity, as a march, attack, retreat, etc.

Operations. The procedure of an army in campaign and battle, involving movement, attack, defense, etc.

Operations Map. A map used in planning operations, or on which data concerning operations or the progress of operations is plotted.

Operations of War. The strategical and tactical methods in the conduct of war.

Operations Officer. An officer charged with the duty of preparing plans of operations and supervising their execution as the representative of his commander.

Operations Orders. Orders which deal with strategical and tactical operations.

Operations Plan. A plan adopted or proposed for the employment of means for the accomplishment of a military object.

Operations Section. That section of the general staff charged with the planning, preparation and supervision of military operations. In the United States service it is also charged with training. G-3 of the general staff.

Oral Order. An order delivered by word of mouth.

Order in Battery. A formation in which the artillery pieces and their caissons, placed for action, are in line.

Order in Double Line, Double Section Column. Similar to order in line and order in column, except that each section is in double section (piece and caisson abreast).

Order in Flank Column. A formation of a battery in double column, the pieces being in one column and the caissons in another, with sufficient interval to allow a wheel into line.

Order in Line. A formation in which the sections of a battery are in line, carriages limbered, and pieces all in front or in rear of their caissons.

Order in Section Column. A formation in which the vehicles of a battery are in column, each piece followed by its caisson.

Order of Battle. The general or geometrical disposition of troops for battle. At various times the following orders have been recognized: simple parallel; parallel with a crotchet; parallel reinforced on one or both wings; parallel reinforced in the center; simple oblique; oblique reinforced on the attacking wing; perpendicular on one or both wings; concave; convex; echelon on one or both wings; echelon on the center; combined order in columns for simultaneous attack on the center and one wing; converging; cinquain; sixain; salient; boars-head. These formations were formerly of great importance. They are less so today as dense masses can no longer be employed against modern firearms, and troops are widely deployed in battle.

Order of March. The disposition of troops for a march, or their order in column of march, stated in a march order.

Orderly. An enlisted man who attends a superior officer to carry messages and communicate his orders, or to render other service.

Orderly Room. A colloquial term applied to the office room of a company, or other similar organization, where the company records are generally kept and the company business conducted.

Orders. The will of a commander expressed orally or in writing; they include routine orders and combat orders, which see.

Ordinate. The vertical height of any point of a trajectory above the muzzle of the gun. Considering the trajectory as a rigid curve, ordinates may also be measured perpendicular to the plane of site. One of the two rectangular coordinates of a point, the other being the abscissa. See Coordinates.

Ordnance. All property of whatsoever nature supplied by the Ordnance Department, consisting of all kinds of cannon and artillery, together with the equipment and apparatus for their maneuver and use; all small arms, accouterments and horse equipments; and all ammunition, tools, machinery, etc., for the ordnance service. In England the term includes supplies of various other classes.

Ordnance Department. That branch or service which supplies all ordnance material, and maintains arsenals and depots for its manufacture, storage and distribution.

Ordnance Depots. Depots which contain the necessary or prescribed reserves of arms, ammunition and other equipment furnished by the Ordnance Department.

Ordnance Maintenance Company. An organization whose duty is the repair and maintenance of ordnance materiel in the field.

Organ Gun. An early hand gun consisting of a number of barrels placed side by side like the tubes of an organ.

Organic. Permanently assigned to and forming an integral part of.

Organic Antiaircraft Defense. The antiaircraft defense provided by the organic weapons of units, other than antiaircraft artillery.

Organization. The science of building up a symmetrical and serviceable whole from a number of units or elements. The scheme of arrangement of an army in branches having definite functions and units of various sizes, with the various grades of personnel assigned to each. The theory of organization is based on individual responsibility and subordination, each within its proper sphere. Military organization comprises the three elements of command, combat, and service. An organized military unit.

Organization Day. The date on which an organization was first organized, each anniversary of which is commemorated.

Organization for Combat. The measures taken to insure that the troops are so grouped that they can most efficiently carry out their assigned combat missions.

Organization, Military, Elements of. The basic elements of military organization are: (1) Command elements, including commanders and their staffs; (2) Combat elements, including all combat troops; and (3) Service elements, including the administrative, supply and technical troops.

Organization of the Ground. Preparation of the ground for defense by taking advantage of its natural features and strengthening the position by works of field fortification. See Lines of a Defensive Position; Defensive Area, etc.

Organizational Requirements. Supplies necessary for an organization to function as a unit.

Organizational Equipment. See Equipment, Organizational.

Organize. To arrange or distribute into units or elements with the proper officers to carry out a preconceived plan or scheme. To prepare a position or terrain for defense. To create a military organization.

Organized Land Forces. See Army of the United States, Organization of.

Organized Reserves. Partially trained and partially organized reserves maintained in time of peace to facilitate the formation of a citizen army in the event of war; including the Officers' Reserve Corps and the Enlisted Reserve Corps.

Orient. To determine one's position on the ground with respect to a map or to the four cardinal points of the compass. To identify directions on the terrain. To place a map so that its meridian will be parallel to the imaginary meridian on the ground, and all points on the map in the same relative positions as the points on the ground which they represent. To inform or explain, to make another conversant with. See Orientation.

Orientation. The act of adjusting oneself, or of orienting a map or instrument. The process of adjusting the circles of guns or instruments so that they will show correct azimuths. Placing a map in such a position that the meridian and all other lines on the map will be parallel to the corresponding lines on the ground. Orientation is construed to include all of the surveying and astronomical methods employed in connection with artillery firing.

Oriented. A map, plane table or sketching board is oriented when all lines on the map or plane table are parallel to the corresponding directions on the ground. An instrument is oriented when it gives correct positions. A gun sight is oriented if, when the line of sighting is on the aiming point, the reading is the azimuth of the axis of the bore.

Orienting Line. A line whose direction (azimuth) is known and over one point of which it is possible to place an angle measuring instrument.

Orienting Point. A point by which the survey work of a unit is tied in to the map or firing chart.

Origin. The point at which any measurement begins. Beginning or source.

Origin Line. A line selected near the center of the field of observation to which angular measurements are referred.

Origin of Coordinates. For any map, the intersection of the zero X-line and zero Y-line, which is always off the map, to the south and west, to avoid negative coordinates.

Origin of Fire or of the Trajectory. The center of the muzzle of the piece at the instant of firing.

Ornaments. Those parts of the dress or uniform of a soldier which are intended more for appearance or decoration than for practical use, such as breastplates, shoulder straps, etc.

Out of Doors. In the rendering of salutes and courtesies structures such as drill halls, riding halls, gymnasiums, and other large roofed inclosures used for drill or exercise of troops are considered as out of doors.

Outer Flank. The flank opposite the inner or directing flank. An exposed flank, not protected by a natural obstacle or adjacent unit.

Outfit. The articles of equipment. A common reference to one's organization or unit. To equip.

Outflank. To pass around or turn the flank or flanks of an enemy. To increase the front so as to extend beyond the flanks of the enemy's line.

Outflanking Maneuver. A maneuver which passes by or around the enemy's flank.

Outgeneral. To gain an advantage by the display of superior military skill or tactics.

Outguards. Groups posted in the line of observation of an outpost. The stationary elements of an outpost nearest the enemy; they are designated according to strength, from 4 men to a platoon; each group maintains one or more sentries.

Outline. The trace of a fortification or other work. The use of an individual or number of individuals to represent the forces supposed to be participating in a tactical exercise or maneuver. To trace the outline or contour of.

Outlying. At a distance from the main body or works.

Outmaneuver. To gain a tactical or strategical advantage by maneuvering.

Outpost. Stationary bodies of troops placed at some distance from the main body, while at a halt, in camp or bivouac, or in battle position, to protect it from observation and surprise by the enemy, and to keep him under observation. A stationary covering force, security detachment or cordon.

ost Area or Zone. A belt of terrain lying in front of a battle position, pied by the observation or outpost elements, and sometimes partially nized. See Outpost Position.

OUTPOST

Outpost Line of Resistance. A designated line on which the fires of the elements of the outpost and its supporting weapons are to be coordinated.

Outpost Position. An organized or partially organized position in front of a battle position. It is lightly held and its purpose is to oppose raids or local attacks, to observe the enemy, and to give warning of and delay an attack in force.

Outpost Sketch. A sketch showing the location of the subdivisions of the outpost, and the terrain immediately in front of the outpost line, made without going beyond the line of observation.

Outpost Wire. A light, twisted pair of insulated electric conducting wires, used in the field for short telegraph and telephone lines. See Field Wire.

Outrank. To be superior in rank.

Outrigger. An inclined strut used to steady a mobile gun carriage during firing.

Outworks. Defensive works constructed between the scarp of the main work and the crest of the glacis. They include the tenaille, the counter-guard, the ravelin or demi-lune, the ravelin redoubt, the tenaillon, etc.

Over. A shot whose point of impact is over or beyond the target. Above in authority or rank. A surplus of property beyond that for which one is accountable or responsible.

Over, Short, and Damaged Report. A report on property delivered by a common carrier, transport or parcel post, which does not agree with the quantity listed on the shipping ticket, or which is received in damaged condition.

Overhaul Park. A field establishment for the repair and maintenance of motor transportation.

Overhang. Unequal length of span of the wings of a biplane. It is positive when the upper wing is the longer.

Overhead Cover. A cover or shield which extends over the heads of troops on a firing line. The depth of overhead cover afforded by a shelter, bombproof, casemate, etc.

Overlay. A sheet of translucent paper or cloth, for laying over a map, on which various locations, as of artillery, targets, field works, enemy positions, etc., are shown.

Overseas Expedition. A joint army and navy undertaking for the purpose of conducting military operations on shore at the end of a voyage which is under naval control.

Overseas Operations. Operations conducted with a view to the establishment of a base for military operations. Operations conducted on land after the landing of an overseas expedition.

Overshoot. To shoot over or beyond a target.

Over-short Adjustment Chart. A chart which provides means for determining accurately and rapidly the adjustment corrections during fire for effect, when using the bracketing method.

Own Misconduct. See Not in Line of Duty.

Ozalid Process. A method of actinic printing from a tracing as an original, which produces reddish brown lines on a white background.

P

Pace. The length of the full step in quick time. Rate of movement at quick time, double time, trot, etc. Gait.

Pace Setter. The member of a mounted unit appointed by the commander thereof to maintain the proper gait during the march.

Pace Tally. A small hand instrument for counting paces in military sketching.

Pacifism. The spirit and temper which opposes military ideals, emphasizes the defects of military training and the cost of war and the preparation therefor, and advocates disarmament and the settlement of international disputes by arbitration.

Pacifist. An advocate of peaceful relations between nations. One who is imbued with pacifism, or who has impractical ideas concerning the maintenance of peace—often used in an unfavorable or derogatory sense.

Pacifistic. Of or pertaining to or characteristic of pacifists or pacifism.

Pack. The roll and pack carrier of a soldier. The load of a pack animal. To prepare a pack. To load with a pack. To carry. See Infantry Equipment.

Pack, Infantry. See Infantry Equipment.

Pack Animals. Horses or mules which carry loads on their backs. Sumpter mules.

Pack Artillery. See Artillery, Pack.

Pack Carrier. A canvas or webbing holder for the roll, which is attached to the haversack.

Pack Equipment. Equipment made up for pack transportation. The equipment of a pack animal or train.

Pack Reel. A reel for field conductor wire carried on the back of a pack animal.

Pack Saddle. A saddle used on pack animals to support the load carried.

Pack Train. A group of pack animals with the necessary personnel and equipment.

Pack Transportation. The personnel, animals, and equipment used for transporting articles loaded on the backs of pack animals, and the riding animals and equipment used by the personnel.

Packer. A soldier or civilian who loads pack animals, or serves with a pack train.

Packer's Blind. A leather blind-fold used on an animal while it is being saddled or loaded.

Palisades. A barrier made of long wooden posts set firmly in the ground.

Pan. That part of a matchlock or flintlock which holds the priming.

Pan Coupe. A short length of parapet by which the salient angle of a work is sometimes cut off or blunted. A straight line replacing the point of a salient angle in a fortification.

Pancake, To. To level off an airplane at a greater altitude than normal in a landing, thus causing it to stall and to descend on a steeply inclined path with the wings at a very large angle of attack and without appreciable bank. To stall near the ground in landing.

Panel. A device made of canvas, bunting or other similar material of distinctive shape laid on the ground to mark a position, identify a unit, or convey code signals to aircraft. Marking, identification and code panels. A part of a truss between two perpendicular or parallel web members.

Panel, Rip. A strip of fabric, inserted or fitted in the upper part of the envelope of a balloon or semirigid or nonrigid airship which is torn or ripped open when immediate deflation is desired.

a ponton bridge. One of a pair of baskets used to carry loads on pack animals.

Panoramic Sight. A sight for guns and howitzers, including a vertical telescope of periscopic type, which affords the gunner a wide field of view in a plane perpendicular to the axis of the telescope. Some models have an elevation micrometer, which permits view of objects at higher or lower levels, and a mechanism for setting off the angle of site.

Panoramic Sketch. A pictorial representation of the terrain in perspective as seen from one point of observation.

Panzer Division. A German armored division.

Paper Work. The army term for official papers and correspondence. Clerical duty.

Parachute. An umbrella-like device to retard the descent of a falling body, used to descend from or bail out of an aircraft, to support a flare or pyrotechnic signal, etc.

Parachute Grenade. A grenade which carries a parachute to prevent fragments flying to the rear on explosion.

Parachute Harness. A combination of straps, buckles, and fastenings used to attach a parachute to the wearer.

Parachute Light. A bright light or flare supported by a small silk parachute; released from an airplane or from a shell, as a landing light or to illuminate a battlefield.

Parachute Troops. Troops moved by air transport and landed by means of parachutes.

Parade. A formal military ceremony. An assembly or display of troops. Any march or procession. The place where troops form or are drilled. The site in rear of the parapet of a fortification. To assemble in military formation for evolutions and inspection. To form in a regular and prescribed manner and march in review.

Parade Ground. A ground or field on which troops are paraded or formed for ceremonies.

Parade Slope. The rear wall or slope of an emplacement.

Parados. A mound of earth constructed at the rear of a trench for protection from rifle fire or shell fragments from the rear.

Parados Traverse. A traverse projecting forward from a parados to afford protection from fire and shell bursts.

Paragraphs of an Administrative Order. The six paragraphs of an administrative order are as follows:

Par. 1. Supply.	Par. 3. Traffic.	Par. 5. Personnel.
Par. 2. Evacuation.	Par. 4. Trains.	Par. 6. Miscellaneous.

Paragraphs of a Field Order. The five paragraphs in the body of a field order are as follows.

Par. 1. Information of the enemy and of friendly supporting troops.

Par. 2. The plan of the commander issuing the order.

Par. 3. The mission and necessary information for each element of the command.

Par. 4. Administrative instructions, especially supply and evacuation.

Par. 5. Instructions pertaining to signal communications, especially locations of command posts.

Parallax. The difference in direction, or azimuth, to a point as seen from two other points. The angular difference, due to displacement, between the direction from a gun to the target, and the direction from the directing point or directing gun to the target.

Parallel Sheaf. A battery sheaf of fire in which the planes of fire of the guns are parallel.

Parallels. Trenches or lines of a defensive position, whether intrenched or not, whose general direction is parallel to the front. Trenches on a parallel are primarily for fire purposes but may also provide for lateral communication.

Parallels and Approaches. The system employed and works utilized in the reduction of a besieged town or fortress. Successive intrenched lines, generally parallel to the enceinte, are established, and each is used as a base for a farther advance by means of approach trenches or saps, zig-zagging or obliquing to the front to avoid enfilade.

Parapet. A wall or elevation of earth or other material thrown up in front of a trench or emplacement to protect the occupants from fire and observation, and from which fire may be delivered. A breastwork.

Paratrooper. A member of a parachute unit.

Paravane. A hydrovane used to ward off or cut the cables of submarine mines.

Park. The ground occupied by the pieces of a battery or the vehicles of a train in garrison or in camp. A place where supplies of all kinds are stored for distribution in the field. To arrange the pieces of a battery or the vehicles of a train in an orderly manner in garrison or camp.

Parlementaire. One who accompanies a flag of truce. An agent in the nonhostile intercourse of belligerent armies.

Parley. A conference with an enemy, under a flag of truce.

Parole. A word used as a check on the countersign in order to obtain more accurate identification of persons. A pledge made on honor by a prisoner of war to obtain his release or certain privileges. The status of being free on parole. To set at liberty on parole.

Paroled Prisoner. A prisoner or convict who has been paroled.

Parrot Gun or Mortar. A cast-iron rifled cannon, strengthened by shrinking a band or wrought iron sleeve over the rear part of the reinforce. A cast-iron mortar lined with a steel tube.

Parry. A defensive movement in fencing, bayonet and saber exercise. To ward off. To prevent.

Partisan. An adherent usually engaged in harassing the enemy. The commander of a partisan group.

Party. A small detachment of men employed on any kind of duty or special service; as a battery commander's party.

Pass. A written authority given a soldier to absent himself from station or duty for a short period. A narrow passage or gap through hills or mountains. A channel connecting two larger bodies of water.

Pass in Review. To march in review at a ceremony.

Pass Muster. To pass through an inspection or muster without censure.

Passade or Passado. A lunge in fencing. A sudden movement to the front.

Passage of a Defile. The movement of a body of troops through and clearing a defile.

Passage of Lines. A relief of a front line unit in which the rear unit moves forward through the already established line, instead of remaining in place, while the replaced unit may remain in position or move to the rear.

Passive Defense. A class of defense which seeks only to deny the attacker access to the defended area or position, or to expel him by counter attack if he gains access. The purpose of passive defense is to prevent the enemy from gaining decisive results.

Passive Elements. All of the parts that offer resistance to forward motion of an airplane without contributing to its support. Those elements of a defense which are passive in nature.

Passive Operations. Operations whose only object is to prevent the enemy from advancing or gaining any success.

Passport. A safe-conduct. A license granted for the removal of persons and effects from a hostile country.

Paster. A small circular sticker of paper used to paste up a bullet hole in a target, or as a marker or aiming point.

Patch. A jacket of harder metal encasing the lead core of a small-arm bullet, or all of it except the nose. Formerly, a wrapping of leather, cloth or paper placed around a bullet to make it fit the bore. A small piece of cloth used to clean the bore of a small arm. A piece of brightly colored cloth sewed on a uniform as an insignia.

Patched Bullet. A jacketed bullet. Formerly, a rifle bullet which was smooth, having a thin layer of paper wrapped about its cylindrical portion.

Patrol. A moving group or detachment, from two men to a company, sent out from a larger body on an independent or limited mission of reconnaissance or security, or both. The act of patrolling.

Patrols, Classification of. There are two general classes of patrols, based on their mission: (1) Information or reconnaissance patrols, whose mission is to gain information of the enemy, the terrain, etc.; and (2) Security patrols, whose mission is to provide protection in camp, on the march, or in line of battle. There are many subclassifications based on the particular duty to be performed, the organization of the patrol, or its location, etc., e.g., combat patrols, employed during combat; visiting patrols, in an outpost; point of an advance or rear guard; flank patrols; mounted or dismounted patrols; officer's patrols, commanded by officers; etc., etc.

Patrol System of Outposts. A system of outposts in which the groups on the line of observation are placed at relatively wide intervals, the intervals being constantly patrolled. The same method may be employed for an interior guard.

Pattern. The figure formed by a group of shots on any surface, or by a series of air bursts; or the dimensions of such a figure. A model used for castings, sheet metal work, etc. The appearance of an area of ground as seen from above or depicted on a vertical photograph, including the form, arrangement, shadows, texture and color of its various features. See also Fuze Range Pattern.

Paulin. A long sheet of canvas usually waterproofed, for covering purposes. A tarpaulin.

Pay. The prescribed salary for the grade held, including all increases for length or nature of service.

Pay and Allowances. The prescribed salary plus such sums as are allowed for the salary period for payment of rent, light and heat, rations, etc.

Pay Periods. Successive periods of service, usually 3 or 5 years, for which increases of pay are allowed.

Pay Roll. A form showing the names of persons entitled to pay and the amounts due each.

Peace Establishment. The military establishment of a nation in time of peace.

Peace Offensive. Propaganda circulated by a country at war, designed to undermine the will of its opponent to continue the war and thus secure a peace more favorable than might otherwise be possible, or to gain time for a breathing spell in which to recuperate preparatory to undertaking a new offensive.

Peace Strength. The reduced strength of a military establishment or organization in time of peace.

Pedersen Semi-Automatic Rifle. A semi-automatic rifle which has fired 80 aimed shots per minute. It weighs 8 pounds and 14 ounces and uses a clip which contains 10 cartridges.

Pedestal Mount. A fixed mount for a small or medium caliber gun, in the general form of a truncated cone.

Peep Sight. An adjustable rear sight slide pierced with a small hole in its center to peep through in aiming.

Pellet. A bullet or ball for firearms. A small charge of explosive, as used in a cap or fuze. The inflammable chemical (magnesium, etc.) in the base of a tracer projectile.

Peloton. A unit commanded by a lieutenant. A platoon. (Fr.)

Penalty Envelope. An official envelope requiring no postage, which may be used for official purposes only. A franked envelope.

Penetrate. To pierce the hostile front. To effect a gap in the enemy's line. To enter or pass through, as a projectile.

Penetration. A form of attack which consists in driving salients into, or piercing the hostile front, and then enveloping one or both of the flanks created by the penetration. The act of penetrating. The gap created by a penetration. The distance which a projectile travels into the ground, armor or other protection before coming to rest or exploding.

Penetration of Armor. The passage of any part of a projectile through all or any part of armor plate. Complete penetration is called perforation.

Pennon. Any flag or banner. A pennant. A small flag carried by a knight on his lance, bearing his coat of arms. A flag carried by a regiment of lancers.

Pennsylvania Rifle. Rifling for hand arms originated in Pennsylvania about 1740 and is usually credited to Rosser of Lancaster. By the time of the Revolutionary War "Pennsylvania Rifles" of various non-standardized models were in use throughout the American colonies.

Pension. An annuity granted as compensation for wounds or disability, or in consideration of service.

Pension Bureau. An agency of the federal government charged with the administration of all matters pertaining to pensions; now a part of the Veterans' Administration.

Pensioner. A person who receives a regular allowance in reward for long service or injury in the service, or an heir of such a person. One in receipt of a pension.

Penthouse. A small house for the protection of a gun and its carriage from the weather.

Percent Power. Percent of maximum or of rated power of the engines of an aircraft that is employed at any particular time.

Percentage. As used in the measurement of slopes or grades, the number of units of vertical rise corresponding to 100 units of horizontal travel.

Thus a slope of 10% is one that rises 10 feet in each 100 feet of horizontal travel. The natural tangent of an angle of slope, multiplied by 100.

Percentage Corrector. A device for converting range correction, either ballistic or adjustment, into a percentage of the range, and applying to other ranges as a like percentage.

Percussion. A shock or blow. An explosion caused by a shock or blow. Designed to explode by percussion.

Percussion Bullet. A bullet provided with a percussion cap which, on impact, explodes a powder charge within the bullet. An explosive bullet.

Percussion Cap. A cap of fulminate or other high explosive which explodes on percussion, as from the blow of a hammer or firing pin.

Percussion Fuze. A fuze which is ignited by percussion on impact of a projectile.

Percussion Lock. A musket which was discharged by percussion of the hammer on fulminating powder contained in a cap or cartridge, which superseded the flintlock about 1840. Any firing device for a weapon, that operates by percussion.

Performance. The speed, rate of climb and ceiling reached by an aircraft under given conditions of load. Any achievement or accomplishment of a person, machine, unit, etc.

Periscope. An optical instrument which enables one to see over or around a parapet, barrier or obstacle by means of a combination of mirrors, prisms and lenses. A similar instrument is used by tanks and submarines.

Permanent Defenses. Defenses of a permanent nature which are constructed in time of peace in furtherance of a military policy.

Permanent Fortifications. Fortifications of an elaborate and permanent nature intended to insure the possession of points which are of importance to the safety of a country.

Permanent Posts. Military posts designated for permanent occupancy by troops; they are called forts or barracks, in contradistinction to temporary posts which are called camps.

Permanent Property. Permanently installed machinery, appliances, and apparatus.

Permanent Rank. Rank which does not cease with the conclusion of any particular service or condition.

Permanent Shelter. Includes all federal and state owned construction of a permanent character which is located on federal or state reservations.

Permissible Error. In gunnery, an error made in computing or estimating the range which does not render the fire ineffective.

Permit. A formal, written authority. To allow or authorize.

Persistency. The time during which a chemical agent remains effective when released upon the target. The characteristic of persistent agents.

Persistent Chemical Agent. See Chemical Agent, Persistent.

Personal Equipment. The equipment of all kinds necessary for the individual use of a soldier. The equipment used by a soldier in the field, which includes the clothing on his person, his weapons, ammunition and instruments, his haversack and pack and their contents, and any personal equipment that may be carried in the combat train. See Infantry Equipment.

Personal Reconnaissance. Reconnaissance made by a commander in person.

Personal Staff. The aides-de-camp of a general officer.

Personnel. The human element, taken collectively, of an organization. One of the two physical subdivisions of which a military force is composed, the other being materiel.

Personnel Adjutant. An officer charged with the duty of keeping records pertaining to the enlisted personnel of a unit or command.

Personnel Carrier. A motor vehicle, sometimes armored, designed primarily for the transportation of personnel and their weapons to and on the battlefield.

Personnel Specifications. Statements which define briefly the personal characteristics, skill, and knowledge needed to perform efficiently every type of service required in the army.

Petard. An ancient, short gun, of the shape of a mortar, used to blow down gates of fortifications. It was placed against or very close to the gates.

Petrol. Gasoline.

Phalanx. The ancient Macedonian military organization; it was not rigid, but was materially changed from time to time. Under Alexander the Great, the unit was a file of 16 hoplites, commanded by the front rank man. Sixteen of these files, or 256 men, in a solid square formed a syntagma, which was the tactical unit. Four syntagma, or 1024 men, made a chiliarchia, and four chiliarchias, or 4096 men, made a simple phalanx, to which light troops and cavalry were attached. It was commanded by a phalangiarch, or brigadier general. Two phalanges formed a double phalanx, or division, commanded by a diphalangiarch; and two double phalanges formed a quadruple or grand phalanx, commanded by a tetraphalangiarch. The army consisted of a grand phalanx, with such additional troops, horse, and foot, as were available, with a strength of about 29000 men. In battle, in front of the hoplites of the phalanx were groups of slingers and archers, who acted as skirmishers and retired around the flanks when driven back. On the wings were heavy horsemen. The principal arm was the spear, or sarissa, said to have been 18 to 24 feet long. The first five ranks advanced their spears horizontally or with points slightly depressed, the front rank spears extending fifteen feet in advance. The other eleven ranks held their spears vertical or leaned them on the shoulders of those in front.

Phase Line. One of the several lines or terrain features, intermediate between the line of departure and the final objective, which troops in attack are directed to reach by a specified time. Such locations should afford a measure of concealment and cover. Commanders take advantage of halts on phase lines to regain control of their troops, reorganize, replenish supplies, and generally to prepare to resume the attack.

Phases of the Attack. See Attack, Phases of.

Philippine Scouts. Natives of the Philippine Islands enlisted and organized for service in the United States Army.

Phillips Pack Saddle. The pack saddle adopted for use in the United States Army in 1929, replacing the aparejo. It employs the principle of the pack frame with pads, in which the frame is an integral part of the saddle.

Phosgene. A very poisonous, lung-irritant gas made from chlorine and carbon monoxide gases.

Phosphorus Bomb. A bomb filled with phosphorus, whose fragments, after explosion, continue to burn.

Photogrammetry. The making of maps by means of photographs, particularly aerial photographs. Phototopography.

Photograph, Aerial. Any photograph taken from an aircraft. See Vertical and Oblique Photographs.

Photolithography. The production of maps, illustrations or other prints, by a combination of photography and lithography.

Photomap. An aerial photograph upon which information commonly found on maps has been placed, including at least a scale and a directional arrow.

Photostat. An instrument for producing photographically on bromide paper, copies to any scale of maps, drawings, documents, etc. A copy produced by a photostat. To produce such a copy.

Physical Fitness. Freedom from any physical or mental abnormality which disqualifies, or which may ultimately incapacitate the individual for full military duty.

Physical Standards. The physical qualifications required for entrance to the army.

Pick Mattock. A tool having one pick point and one chisel (mattock) blade, used in digging.

Picked-up Message. A written message picked up from near the ground by an airplane in flight.

Picket. A detachment of an outpost sent out to perform the duties of an outguard at a critical point, the detachment being stronger than an ordinary outguard and establishing sentinel posts of its own. A stake of wood or steel used in the construction of revetments and obstacles. To guard, as a camp or road, by an outlying picket. To post as a picket.

Picket Guard. A guard held in readiness in case of alarm. A guard for a picket line.

Picket Line. A line held by pickets. A rope or cable to which animals are tethered. Animals tethered to a picket line.

Pickets. An early form of military punishment in which the culprit was held by a raised arm in such position that his whole weight fell on one foot which was supported on a blunt pointed stake. Sharpened stakes of wood or metal used in the construction of obstacles and revetments.

Picric Acid, Picrates. High explosives, used as bursting charges for shells.

Piece. A general term for a gun or firearm of any nature. A single gun. See Gun.

Piecemeal Action. A kind of action in which the units are engaged one by one as the need arises or as they become available, and not as the result of a well prepared plan.

Piecemeal Attack. An offensive action in which the elements of a command are not committed to action in coordination.

Pike. An ancient weapon consisting of a long wooden shaft with a pointed steel head, sometimes as much as 24 feet in length. The sharp pointed staff on which the color or guidon of a dismounted unit is carried. A heavy lance.

Pill Box. A covered, concrete emplacement for one or more machine guns.

Pilot. One who operates the controls of an aircraft in flight. The operator of an aircraft, irrespective of sex.

Pilot Light. The particular light of a searchlight group which is employed normally in searching for, covering, identifying, and indicating to the remaining lights of the group the aircraft whose presence and approximate position have been determined previously by the listening apparatus.

Pilot Model. An experimental model of a new weapon or other apparatus.

Pilum. The heavy lance carried by the Roman legion, which varied from three to five and a half feet in length. See also Hasta.

Pin Point. An isolated aerial photograph of a particular object or small area.

Pintle. Any pin used as a hinge or axis as for a gun or breechblock. A pin or hook with a latch, on the rear of a towing vehicle, such as a limber, by means of which a gun or other trailer is coupled. The vertical bearing about which a gun carriage revolves in traversing. A center pintle is one which is exactly underneath the center of gravity of the piece; a front pintle is one which is in front of the center of gravity.

Pintle Center. The center line or axis of a pintle.

Pintle Socket. A socket in the bottom carriage of a gun into which a pintle fits.

Pintle Traverse. Traverse about a vertical pivot (pintle) attached to a gun carriage.

Pioneer. A military laborer employed in building roads, digging trenches, etc. Any soldier who regularly performs pioneer work. A soldier of a combat engineer regiment. To act as a pioneer.

Pioneer Infantry. Infantry organized for and performing pioneer labor duties.

Pioneer Platoon. An infantry unit trained and equipped to perform simple engineering duties in campaign and battle.

Pioneer Tools. Engineer tools, especially those used in field fortification, which for convenience are of smaller size than corresponding commercial patterns.

Pioneers. Engineer troops.

Pistol. A small firearm, with a short barrel, designed to be aimed and fired from one hand. The pistol has an obscure history, but is believed to have been invented by Carminelleo Vitelli at Pistoja, Italy, whence it received its name, about 1540-1550 A. D. The first known pistols, fitted with wheellocks, date from 1547 and have barrels 20-25 inches long. Short, heavy pistols, called daggs, were in common use around 1650. Except for occasional use in duelling they dropped out of sight for 200 years. The earliest known specimen of a partly automatic pistol appears to have been invented by Ordea in 1863. In 1893 appeared the Borchardt, the predecessor of the German Luger Parabellum. In quick succession came the Mannlicher, Bergman (1894); the Charola-Anitua, Simplex (1897); and others. But it was not until the Mauser made its appearance in 1898 that the self-loading pistol had a definite place. In 1900 the Webley-Fosberry was introduced. In 1903 Browning's patents were taken up by the Colt Firearms Company.

The pistols at present in use include the Astra; Bayard; Browning; Dreyse; French Ordnance Type; Frommer; Luger; Mauser; Nambu; Savage; Steyr; Walther; Webley. See also Revolver.

Pistol Belt. A belt for the attachment of a pistol holster, with pockets to hold extra magazines.

Pistol Carbine. A pistol with a detachable shoulder stock, so that the weapon may be fired from either the hand or the shoulder.

Pistol Grip. A shape given to the small part of the rifle stock to afford a better hold for the hand. The handle of a pistol.

Pistol Marksman, Sharpshooter, Expert. Grades of classification in pistol marksmanship.

Pit Commander. An enlisted man who commands a pit of a mortar battery.

Pit Salvo. The simultaneous firing of all the mortars in a pit.

Pitch Camp. The act of establishing a camp.

Pitched Battle. A bitterly fought engagement in which both combatants seek decisive results.

Pitching. Angular motion about the lateral axis of an aircraft. The setting up of a tent.

Pits. Individual fire trenches. Small holes to afford shelter from fire or to form an obstacle. Sunken emplacements for mortars or howitzers. A line of holes, or a continuous trench in front of the butts or backstop of a target range, to shelter the men who operate and mark the targets.

Pivot. The soldier or element upon whom a line of troops executes a turn. See Fixed Pivot; Moving Pivot.

Pivot Flank. In a drill, that flank upon which a turn of subdivision is made.

Pivot of Maneuver. That part of a force which first engages the enemy and seeks to immobilize him. A fortification or fortified area which serves as a supporting point for an army in the field.

Place Mark. A point whose coordinates and elevation are known, marked on the ground near an artillery position and used as an origin for surveys.

Place Sketch. A topographical sketch of an area made from a single point of observation.

Places of Arms. The points within a fortification where the troops assemble for duty, usually on the covered way.

Plain. A drill ground. A name formerly applied to a field of battle.

Plain Wire Spiral. A portable obstacle made of barbed wire coiled as a spiral, used to reinforce or thicken wire entanglements.

Plan. A scheme or design, specifically for any military operation. A course of action or method of procedure decided upon and adopted by a commander, and which is the basis for his orders to his command. The outline or trace of a fortification upon the ground or a sheet of paper. A horizontal projection. A drawing showing the details of any structure. To formulate or prepare a plan.

Plan Directeur. Operations maps, generally on a scale of 1:20,000 (Fr.).

Plan of Action. A plan based upon the available information of the enemy, the terrain, the intentions of higher commanders, the supporting troops, and the mission assigned the troops for which the plan is prepared. The culmination of an estimate of the situation.

Plan of Attack. The plan of disposition for an attack together with the tactical mission of each subdivision of the unit for which the plan is prepared.

Plan of Campaign. A comprehensive general plan based upon the available information of the enemy forces, the theater of war, the condition of our own troops, the object of the war, and the end to be attained.

Plan of Defense. A plan for the defense of a position, including the distribution and missions of troops and weapons, organization of the ground, etc.

Plan of Maneuver. See Scheme of Maneuver.

Plan of Operations. The general method of carrying out the plan of war in a particular theater of operations.

Plan of Signal Communication. A plan for the coordination of signal communication throughout the command for which issued.

Plane. An airplane. An airfoil or hydroplane, especially a supporting surface, a wing.

Plane, or Hydroplane. To move through the water at such a speed that the support derived is due to hydrodynamic and aerodynamic rather than to hydrostatic forces.

Plane of Defilade. The plane through the point from which defilade is desired, tangent to the covering mass or mask.

Plane of Departure. The vertical plane containing the line of departure.

Plane of Flight. The vertical plane containing the line of flight of an aircraft.

Plane of Position. A vertical plane containing a line of position.

Plane of Sighting. The vertical plane containing the line of sighting.

Plane of Site. A plane containing the line of site and a horizontal line perpendicular to it. A plane containing the right line from the muzzle of the piece to any assigned point of the trajectory and a horizontal line perpendicular to the axis of the bore.

Plane of Symmetry. The vertical plane containing the longitudinal axis of an aircraft, and which divides it into two equal parts.

Plane Table. A small table usually mounted on a tripod, used in the field in plane table surveying. The military sketching board is a form of plane table.

Planimetric. In one plane; usually a reference to the horizontal detail of a map.

Planned Data. Data used in conduct of fire with air observation, when the target is designated in advance to both observer and battery commander.

Plaster Shot. A demolition charge placed on the outside of an object or structure and tamped with mud, earth, or sacking.

Plat. The flat or broad side of a sword. A map or plan. To plot, as in preparing a map.

Platoon. A number of squads formed for the purpose of instruction, discipline, control, and order. One of the component elements of a company or similar organization. Usually, the smallest military unit commanded by an officer.

Platoon Columns. A formation executed from skirmish line in which the platoon is formed in double column of files.

Platoon Defensive Area. See Defensive Areas.

Platoon Leader. The officer or enlisted man in charge of a platoon.

Platoon Scouts. The soldiers, usually two in each squad of the platoon, trained and designated as scouts; often used together as an advance guard or to cover the platoon during combat.

Platoon Sergeant. The senior sergeant of a platoon.

Plot. The act of plotting. A scheme, secret design, or strategem. A map, chart or drawing. To place data upon a map or chart. To determine data by means of plotting.

Plotted Point. See Tracking.

Plotter. The person who has charge of a plotting room immediately under the range officer.

Plotting. The process of drawing to scale on a piece of paper. Representing graphically, as a piece of terrain, the position of a stationary target or of the course of a moving target, etc. Representing locations or placing data of any kind in correct positions on a plotting board or map.

Plotting and Relocating Board. A board for plotting and relocating. See Plotting Board.

Plotting Board. A board on which is represented to scale the field of fire of a battery, provided with arms, scales and attachments. It is used to plot quickly and accurately the data from the position finders and, in connection with range and deflection boards and other instruments, to determine corrected data for firing.

Ploy. To diminish front. To form column from line. The opposite of deploy.

Pneumatic Buffer. A pneumatic cylinder which eases the shock of recoil or counterrecoil by the resistance of air, etc., to compression. An air cushion.

Pneumatic Gun. A cannon in which the force of compressed air is used to propel a projectile containing a large charge of explosive. An air gun.

Pneumatic Gun Carriage. A carriage in which compressed air in a cylinder is utilized to check the force of recoil and return the gun to its firing position.

Point. The patrol or reconnaissance detachment which forms the leading element of an advance guard, or the rear element of a rear guard. A movement executed in fencing. The sharp end of a lance, spear, sword, bayonet, or other thrusting weapon.

Point d'Appui. A fixed point in the rear of an army or on its flanks, to which it can repair in case of necessity. Any point designated to furnish shelter or facilitate defense or resistance.

Point Fuze. A fuze inserted in the point of a shell.

Point of Aim. The point aimed at, at the instant of discharge of a weapon.

Point of Alignment. The point upon which troops begin a formation and by which they dress. The point of rest.

Point of Attack. The particular point or part of the enemy's lines upon which the attack is being, or is to be made.

Point of Burst. For a time-fuzed projectile, the point in the air at which the burst of the projectile takes place.

Point of Curve, Tangent. The point where a railroad curve or tangent begins.

Point of Fall. The point in the descending branch of the trajectory at the same altitude as the muzzle of the gun.

Point of Graze. The point where a projectile grazes the ground, prior to reaching the target.

Point of Impact. The point at which the projectile first strikes the ground or other material object, or the water (point of splash).

Point of Rest. The point at which a formation begins, or the point toward which the units are aligned in successive formations.

Point of Splash. A point of impact on a water surface.

Point-blank Range. A range within which the trajectory is so nearly a straight line that it is unnecessary to set the sight for distance.

Pointers, Electrical, Mechanical. In antiaircraft fire, the electrical dial pointers are operated from the director. The mechanical pointers are operated by elevating and traversing the gun. When the mechanical pointers are matched with the electrical pointers the gun is laid with the proper data, it being unnecessary to read the scales.

Pointing. The operation of aiming a missile weapon, specifically in cases where a sighting on the actual target is utilized for part or all of the process. There are four technical cases of pointing a cannon. See Case I Pointing, etc.

Pointing Corrections. In antiaircraft fire, that part of either the lateral or vertical deflection which corrects for factors other than the travel of the target, such as wind, drift, density and muzzle velocity as they apply, and also for adjustment corrections.

Police. The cleaning up of a post or camp. The state of cleanliness of a post or camp. To clean up. See also Military Police.

Police Party. A detachment engaged in cleaning up a post or camp.

Polygonal Trace, Front or System. A system of fortification dating from the 18th century, in which the trace or magistral is a polygon without bastions, the ditch being defended by caponiers extending from the scarp in the middle of each face or at salient or re-entrant angles.

Pommel. The knob on the hilt of a sword. The knob or horn at the front and top of a saddlebow.

Pompom. An automatic gun, as the Nordenfelt or Vickers-Maxim, so called from the noise of its firing.

Pompon. An ornamental tuft worn on the front of a shako or other military headpiece.

Poncho. A prepared piece of rubber or painted cloth with a slit in the middle for the head to pass through, used as a protection against the weather.

Poniard. A small dagger. To stab with a poniard.

Ponton, Pontoon. A flat bottomed boat of wood or metal, cylinder or canvas-covered frame used as a support in the construction of floating bridges. An aircraft float.

Ponton Bridge. A floating bridge constructed with pontons for supports.

Ponton Bridge Units. Organizations which transport and maintain bridge equipage. In the United States service they include light ponton companies and heavy ponton battalions. The organizations having charge of the old equipage are known as light and heavy bridge trains.

Ponton Equipage. A designated amount of floating bridge equipment allotted to certain tactical organizations. The standard floating bridge equipment of the army.

Pool. A combination of property or equipment to further a joint undertaking or to permit more economical use. The placing of a certain class of equipment, especially transportation, in one group, from which issues are made as required.

Port of Embarkation or **Debarkation.** Any port established by the War Department for the embarkation or debarkation of troops and supplies, and designated as such in general orders.

Portable Firearms. Rifles, carbines, pistols, revolvers, machine guns of all types, automatic rifles and small cannon which can be carried by soldiers in the field.

Portable Obstacles. Obstacles capable of being moved.

Portable Wire Entanglements. Entanglements which are prepared in rear and then carried forward and placed in desired positions.

Portcullis. A grating of iron or wood pointed with iron, hung over the gateway of a fortress, which can be raised or lowered to open or close a passageway. Designed as a measure of security against a sudden rush of the enemy.

Portee Artillery. See Artillery, Portee.

Porthole. An opening in a wall. An embrasure or loophole.

Position. The area or locality occupied by any one or all of the combat elements of a command, especially in defense. A battle position, occupied or prepared for occupation. The manner in which a weapon is held, as in the manual of arms or manual of the saber. An attitude assumed in firing a rifle, as standing, kneeling, sitting or prone. The firing position of a weapon, as a machine gun or cannon, or of a battery of such weapons. An emplacement for a weapon. The location of a target, either terrestrial or aerial. The angle of site or of position, or the actual elevation of a target or point observed, above the level of a gun, position finder, or observer.

Position, Assembly. See Assembly Area.

Position at Observation. The observed position of a moving target at any instant.

Position Defense. A form of defense in which the tactical dispositions are strengthened by the utilization of one well organized position, for which a minimum of 6 hours is available for preparation, on which the primary defense is, or is to be, conducted.

Position Defilade. A gun is said to have position defilade when located behind a crest so that the gun is invisible from the target, but an observer standing at the gun position can see the target; sometimes called site defilade.

Position Exercises. Part of the preparatory instruction in rifle marksmanship, which teaches a soldier how to assume correctly the various firing positions.

Position Finder. An instrument for determining ranges and directions. Three general types are used, known as vertical base (depression); horizontal base (using two stations); and self-contained base.

Position Finding. The process of determining a range and direction of a target, or of a predicted point, from a battery. See Position Finder.

Position Finding System. The particular system used to determine the range and direction of a target from a battery or directing point.

Position for Immediate Action. A position of a field artillery unit in which it is prepared to open fire immediately.

Position in Observation. A position for a field artillery unit from which it can observe its probable field of fire, and from which it is prepared to open fire when ordered. A position occupied by any force for the purpose of observing the enemy, or a designated terrain.

Position in Readiness. A tactical expedient employed when the situation is uncertain, in which a body of troops is held in a central location to await developments or contingencies. Necessary security measures are taken and the troops are so disposed that they can march or deploy for combat in any probable direction in a minimum of time. A location of a field artillery unit, close to one or more possible firing positions, into any one of which it is prepared to move quickly, when so ordered.

Position Lights. Small cylinders of illuminating composition, in white and colors, used to mark positions or furnish illumination.

Position of Assembly. A place, preferably out of sight and out of range of the enemy, where attacking troops assemble and make preliminary dispositions for an attack.

Position Sketch. A sketch of an area in which a position or part of a position is located. Any area sketch in which the sketcher has access to the entire area.

Position Warfare. A phase of warfare, characterized by strong defensive positions in close contact, which results when neither of the combatants has sufficient strength to maintain the initiative consistently, or to drive the other from his defenses. Trench warfare.

Positions of the Target. In fire on naval targets three positions are considered: the position at instant of observation; the predicted point; and the set-forward point. In fire on aerial targets the three positions are: the position at instant of observation; the position at instant of firing or present position; and the predicted or future position, at end of time of flight.

Positive Information. Information which states definitely that a certain condition or situation exists, or that a certain event occurred.

Post. A military station. A place where a body of troops is stationed, usually in time of peace. The limits of a sentry's beat. To assign to a station, as to post a sentinel. To publish or give notice.

Post Exchange. A shop or general store maintained at a military post or camp, for the sale of necessities and luxuries, the profits of which are divided among the funds of the organizations belonging thereto.

Post Flag. The national flag, with 19 feet fly and 10 feet hoist, which is issued to all occupied posts and displayed in pleasant weather.

Post Records. All official books of record, reports, etc., kept at a post.

Post Schools. Schools maintained and intended for the educational and vocational training of enlisted men.

Post Trader. A civilian sutler whom the Secretary of War was formerly authorized to appoint for each military post.

Post War. A term signifying after or since the war.

Postern. A subterranean passage in a fortification communicating between the main ditch and the interior of the works.

Postern Gate. Any small gate, other than the main entrance to a fort, usually at the rear.

Pot Shots. Shots fired at persons within easy range and when at rest. Shots fired simply to kill without reference to military necessity or sportsmanship.

Powder. An explosive compound or mixture used in gunnery, usually as a propelling charge. See Gunpowder.

Powder Bags. Bags of raw silk which contain the propelling charge in separate loading ammunition.

Powder Blast. The outrush of air, gases, and powder fragments from the muzzle when a piece is fired. The scored area in front of the muzzle of a gun, resulting from its firing.

Powder Chamber. That portion of the bore of a weapon designed for the reception of the propelling charge of powder.

Powder Charge. See Bursting Charge; Propelling Charge.

Powder Fouling. Fouling in the bore of a firearm from the residue of the powder charges.

Powder Marks. See Powder Blast.

Powder Rings. Cloth rings containing propelling explosive added to the cartridge in certain types of trench mortar for the purpose of increasing the range.

Powder Train Fuze. A fuze whose time element depends on the rate of burning of a train of compressed powder.

Powder Tray. A tray on which a powder charge is carried to the breech of a cannon. See also Shot Truck.

Powder Velocity. The muzzle velocity actually developed by a particular lot of powder.

Power, Margin of. The availability of engine power, in an aircraft, above that required to maintain flight at the optimum angle, under any particular conditions.

Power of a Tackle Rigging. See Mechanical Advantage.

Powers. The inherent qualities, strength, or potential capacity, of any person, group, arm, or weapon. Nations of power and influence, which usually implies the maintenance of an adequate army or navy, or both.

Powers and Limitations. Considered collectively, what can and cannot be done by any person, group, arm, or weapon.

Practice. Actual performance. Usage. Habit. To exercise in. To teach. To train.

Practice Dummy. A dummy cartridge used in training; easily distinguished from a live cartridge.

Practice Marches. Marches made with a view to hardening the men and animals and instructing the personnel in the duties incident to campaign.

Precede. To have precedence. To go or come before. To be or move in front of. To guide on or maintain a designated distance from a unit marching in rear.

Precedence. Priority as to rank, dignity, time, position, etc.

Precision Adjustment. An adjustment executed for the purpose of determining a single elevation or range which will place the center of impact of fire at or near the objective.

Predicted Point. A point at which it is expected that a moving target will arrive at the end of the predicting interval.

Predicting. The process of determining a point on the expected course of a target, in advance of its present position, at which the target will probably be at a particular instant in the future.

Predicting Interval. The time which is allowed after an observation on the target in order to work out new firing data, point the guns with that data and fire them.

Prediction. The process of determining the probable (future) position of a moving target.

Prediction Scale. A scale for measuring the rate of linear travel of a target and determining the predicted point. It may also be used, with a set-forward rule or chart, to locate the set-forward point.

Predictor. An instrument used in connection with a plotting board to locate the predicted or set-forward point.

Preferment. Promotion, especially when over the heads of one's former seniors. Selection for a favorable detail or desirable duty.

Prejudice of Good Order and Military Discipline. See Conduct to the Prejudice of Good Order and Military Discipline.

Preliminary Bombardment. The intensive period of the artillery preparation by an attacker immediately preceding an attack.

Preliminary Service Practice. A service practice fired as a preparation for record service practice. Instruction practice on a rifle range.

Preparatory Command. That part of a drill command which is given first and which indicates the movement to be executed. See Command of Execution.

Preparatory Training. In rifle marksmanship training, all instruction given prior to actual firing on the range. It includes five steps, as follows: (1) Sighting and aiming exercises; (2) Position exercises; (3) Trigger squeeze exercises; (4) Rapid fire exercises; (5) Examination. Basic military training.

Preparedness. A state or condition of military and naval preparedness for defense in case of possible hostilities. The state of being prepared. Readiness for war.

Preponderance. Excess of weight or force. Inequality in the weight or moment of a gun in front and in rear of the trunnions. Lack of balance.

Present Position. The position of a moving target at the instant of firing at it.

President of a Court-martial. The senior member present.

Presidio. A place capable of strong defense. A military post or fort. A garrison. A guardhouse. (Sp.)

Pressure Gauge. A device inserted in the mushroom head of the obturator of a cannon which measures the maximum pressure of the powder gases; also used in the cylinder of an engine. A crusher gauge.

Primary Armament. Seacoast guns, howitzers, and mortars 12 inches or greater in caliber.

Primary Items. See Items, Primary.

Primary Mission. The principal or most important mission assigned.

Primary Position. A position selected for the employment of a weapon or fire unit, from which it can carry out its primary mission or the most important part thereof. See also Alternate Position, Supplementary Position.

Primary Port. Any War Department port which is the headquarters of a line of army transports.

Primary Target Area. The target area assigned as the principal fire mission of a weapon or unit.

Prime. To prepare a charge for firing by the insertion of a primer or other means. First in rank, dignity or importance.

Prime Contractor. A contractor who enters directly into a contract with the government; a principal contractor.

Prime Mover. A motor, engine, turbine, etc., which furnishes the power which operates machinery of any kind. The source of power.

Primer. A device for igniting a propelling charge. A cartridge containing a high explosive cap, electric or non-electric, used to detonate a charge of explosive in blasting or demolition.

Primer Fouling. Fouling in a firearm resulting from the discharge of primers.

Priming Charge. See Igniting Charge.

Priming Compound. A compound or mixture of very high explosive, used as the initial percussion element in caps and fuzes. Chlorates and fulminates have been commonly used, but are unsatisfactory, and other compounds having guanyl as a base are now employed.

Principal Lateral Deflection Angle. In antiaircraft fire, that part of the vertical deflection angle due to change in the azimuth of the target during the time of flight.

Principal Vertical Deflection Angle. In antiaircraft fire, that part of the vertical deflection angle due to change in angular height of the target during the time of flight.

Principles of Training. The principles prescribed by the War Department for the development of the individual soldier to play his part as a member of the combat team.

Principles of War. Certain fundamental and immutable principles applicable to the conduct of war which have been deduced from the study of all military history, and whose method of application varies with the situation. As enunciated by the United States War Department, they are nine in number, as follows:

1. *Cooperation.* Teamwork throughout the unit or units concerned is essential to success in military operations.
2. *Economy of Force.* The mass to be employed in the main effort is secured by reducing the employment of units elsewhere to the minimum consistent with safety.
3. *Mass.* Success in war is attained by the employment of mass or combat power in the main effort.
4. *Movement.* Maneuver should be employed to bring mass where it can be used to the best advantage.
5. *Objective.* Every military plan or operation must have a definite objective. This objective is generally the neutralization or destruction of the combat power of the enemy.
6. *Offensive.* Decisive results in war are attained only by offensive action seeking to gain a decision.
7. *Security.* Necessary measures must be taken to guard against observation and surprise, and to retain complete freedom of action.
8. *Simplicity.* Simplicity should be the keynote in all plans and operations. Frequent changes in plans are to be avoided and unity of command should be observed.
9. *Surprise.* Surprise in some form is an important means to success. Its principal elements are secrecy in preparation, and rapidity and vigor in execution.

Priorities. Definite rulings which establish, in order of time, the precedence of shipment, movements of rail, road, water or other transport; or performance of several tasks.

Priority of Traffic. The order of precedence of traffic over any artery of communication.

Prismatic Compass. A type of compass, so named because provided with a prismatic eyepiece for reading graduations.

Prison Guard. Enlisted men detailed to guard prisoners while at work. Specially enlisted men for duty at military prisons.

Prisoner. A prisoner of war. A person in confinement under guard.

Prisoner of War. An individual whom the enemy, upon capture, temporarily deprives of his personal liberty because of his participation, directly or indirectly. in the hostilities, and whom the laws of war prescribe shall be treated with certain consideration.

Private. A soldier below the grade of noncommissioned officer.

Private, First Class. The highest grade of private.

Prize. Property captured from the enemy.

Prize Money. Money which is paid on a proportionate basis to troops who are present at the capture or surrender of a place, etc., which yields booty.

Probability. The law of chance or averages. The likelihood that any particular event will occur. The ratio of the chances of occurrence to the whole number of chances; e.g., that any card drawn from a pack will be a spade, 13:52, or 1 in 4.

Probability Curve. A curve showing graphically the law of probability. See Curve of Accidental Errors.

Probability Factor. A factor used as an argument in entering the probability tables. It is equal to the error not to be exceeded, divided by the probable error.

Probability Table. A table which shows the size of error, in terms of the probable error, which can be expected to cover any percentage of the shots fired.

Probable Error. In gunnery, the amount or magnitude of error that will as often be exceeded as not. Probable errors may be in range or direction; or, on a vertical surface, horizontal or vertical. See Range Probable Error.

Probable Intentions of the Enemy. A deduction arrived at from a consideration of the several courses of action open to the enemy as to his probable line of action.

Problem of Decision. A problem requiring, in concise terms, the decision and general plan of the commander in a particular situation.

Problem of Execution. A problem in which the general plan of the higher commander in the particular situation is given to the leader or student, and he is required to state the steps to be taken, either in summary or in full detail of technique, to carry out the plan of the commander. A troop leading problem.

Proceedings of a Court-Martial. All the happenings, including arraignment, pleas, evidence, incidents, rulings, findings, and sentence of a court-martial in any case. These are made of record, the form of the record being prescribed for courts of all kinds.

Processing. An organized procedure in carrying out the various steps incident to the induction of recruits into the military service, necessary when they are received in large numbers. It includes such matters as classification, preparation of initial records, physical examination, vaccination, initial issue of clothing and equipment, etc., etc.

Procurement. The administrative process of acquiring supplies or services by purchase, requisition, establishment of credits, or a system of automatic supply. The obtaining of personnel.

Procurement Authority. An authorization by an arm, service, or bureau of the War Department to procure supplies, equipment or services.

Procurement Plan. A written plan to insure the procurement of any item, material, or equipment, either for a supply arm or service or for a prime or sub-contractor.

Professional Soldier. An officer or soldier who serves regularly and continuously for pay. A member of the regular service.

Professor of Military Science and Tactics. An officer so detailed at a civilian educational institution that maintains courses of military instruction.

Proficiency. The possession by a military person of a knowledge of the duties and responsibilities of his grade, together with the ability and determination to perform and exercise them under all circumstances.

Proficiency Tests. Tests conducted to determine the absolute and relative efficiency of individuals and units.

Profile. A drawing in outline, as in vertical section or the like. A cross section of a parapet or other work. A line cut from the surface of the ground by any vertical plane. In portraying such a profile it is usual to increase or exaggerate the vertical scale relative to the horizontal scale (e.g., horizontal scale, 1 inch = 1000 feet; vertical scale 1 inch = 50 feet) whereby the slopes and elevations are made more apparent.

Profile or Cross Section of a Fortification or Trench, Nomenclature of. (1) Plane of site; (2) Trench; (3) Parapet; (4) Ditch; (5) Parade; (6) Banquette slope; (7) Banquette tread; (8) Interior slope; (9) Superior slope; (10) Exterior slope; (11) Berms; (12) Scarp; (13) Counterscarp; (14) Glacis; (15) Foreground; (16) Interior slope of glacis; (17) Superior slope of glacis; (18) Depth; (19) Relief; (20) Command; (21) Fire step; (22) Fire crest; (23) Bottom (of trench); (24) Parados.

Profiling. The process of making a profile. The erection at various points of templates to show the cross section of a trench, battlement, etc.

Profitable Target. A favorable target so situated that the probable results of fire upon it will justify such fire. See also Favorable Target.

Progressive Concentration. An accompanying fire which precedes the infantry advance by lifting from target to target, but without the regularity of a rolling barrage.

Progressive Explosives. See Low Explosives.

Projectile. A missile, particularly the elongated type used in all modern firearms. A missile for a firearm. The elongated projectile was invented by the Bishop of Munster or one of his followers in 1662.

Projection. The representation on any surface of objects as they appear to the observer. A system for representing the curved surface of the earth on a sheet of paper (map projection).

Projector. A lens or mirror for projecting beams of light. A seachlight. A device for projecting flame or gas.

Prolonge. A heavy rope used to drag guns by hand.

Promote. To advance to higher rank. To foster or further.

Promotion. Advancement to higher rank than the one previously held.

Promotion List. A list in which officers are arranged in order of their priority for promotion, as distinct from their lineal or relative rank.

Prone. Lying flat on the belly. One of the prescribed positions for firing a rifle.

Prone or Lying Trench. See Skirmisher's Trench.

Propaganda. Effort directed systematically toward the gaining of support for an opinion or course of action. Information or arguments distributed to influence public thought or opinion.

Propellants. Explosives used for propelling charges in firearms.

Propeller. Any device for propelling a craft through a fluid, such as air or water; especially a device having blades which, when mounted on a power-driven shaft, produces a thrust by its action on the fluid.

Propelling Charge. The explosive placed behind a projectile in the bore of a gun and used to impart motion to the projectile. The substance which, when ignited, expels the projectile from the bore of a gun.

Property Return. A periodical accounting for government property by an accountable officer.

Prophylaxis. An inoculation or vaccination which confers individual immunity against some communicable disease, notably smallpox, typhoid and yellow fevers.

Protected Position. A firing position for artillery which affords a degree of protection, natural or artificial, or both, for the guns, ammunition and personnel. The term is entirely relative. See Artillery Position.

Protection. A covering or defense against hostile fire or observation. A safe conduct or passport.

Protective Barrage. A standing barrage laid in front of a position to protect the troops occupying it from attack. A defensive barrage.

Protective Chemicals. See Defensive Chemicals.

Protective Clothing. Clothing especially prepared to protect the body against the effects of vesicant chemicals.

Protective Obstacles. Obstacles whose chief purpose is to prevent a sudden incursion of attacking forces into a defensive work or area. See also Tactical Obstacles.

Protocol. A preliminary memorandum of an agreement arrived at in nego-
tiations between belligerents as a basis for a final convention or treaty.

Proving Ground. Ground used for the testing of cannon, powder, projectiles,
etc.

Provisional Defenses. Defenses constructed during or preparatory to a state
of war to supplement permanent defenses or to strengthen strategic points.

Provisional Force or Unit. A temporary detachment or organization provided to meet an emergency or for a special purpose, whose components
are drawn from different organizations, or from different arms or services.

Provisional Map. A map produced by compiling existing map detail or by
tracing data from aerial photographs. It may contain form lines or contours.

Provost. A temporary prison in which prisoners are confined by military
police. One in charge of such a prison. A provost marshal.

Provost Court. A military tribunal for the trial of all offenders against the
laws of war and under martial law, other than those in the military service.

Provost Guards. Guards used in the absence of military police, usually in
conjunction with the civil authorities at or near large posts or camps, to
preserve order outside the limits of the interior guard.

Provost Marshal. The commander of the military police at army, corps, division, section, district, or other headquarters.

Provost Sergeant. A sergeant who is charged with police duties.

Public Armed Forces. All the forces pertaining to or belonging to the military and naval establishments of a nation.

Public Exigency. A national emergency requiring the employment of armed
forces and an immediate supply of articles for the military service.

Public Property. Property of every nature belonging to a government.

Public Relations. Includes all duties pertaining to publicity, counterpropa-
ganda, relations with the press and with civic and patriotic organizations,
the authorization and control of visitors, censorship, and the dissemination
of military information to the public.

Pull-out. To change from a dive to horizontal flight in an aircraft.

Pull-through. A cord with a cloth at one end used to clean the bore of a firearm.

Pull-up. To force an aircraft into a sharp climb. See also Zoom.

Pulmonary Irritants. Chemical agents which cause irritation and damage
of the respiratory passages, often resulting in death by asphyxia.

Pummel or Pommel. The hilt of a sword. The end of a gun. A saddlebow.

Punctured Wounds. Gunshot wounds involving puncturing or complete perforation.

Pump Gun. Any repeating rifle, some forms of which are reloaded by means
of a lever, suggesting a pump handle.

Purchase, Open Market. A purchase of supplies in which the solicitation of
sealed bids by public advertisement or other formal invitation to prospec-
tive bidders is dispensed with, price quotations being solicited informally
from two or more manufacturers or dealers, orally, by telegraph, or letter.

Purchase Without Competition. A purchase of supplies without having solicited bids from more than one regular dealer in or manufacturer of the
article desired.

Purple Heart. A decoration established by General Washington in 1782, recently revived; awarded to persons who, while serving in the Army of the
United States, perform any singularly meritorious act of fidelity or essential service.

Pursue by Fire. To subject a retreating enemy to long range fire.

Pursuit. The act of following a retiring enemy closely in order to complete
his capture, demoralization, or destruction.

Pursuit Aviation. See Aviation, Classes of.

Pusher Airplane. An airplane in which the propeller or propellers are in
rear of the main supporting surfaces.

Puttee. A gaiter or leggin of leather or cloth, or strips of woolen cloth
wrapped spirally around the leg below the knees.

Pylon. A post, tower, or the like, as on a flying ground, marking a prescribed course of flight. A structure supporting the propeller on each side
of a rigid airship, the propeller being driven by shafting and gearing. A

mast or post. A V-shaped structure from the point of which stays are led to brace other members.

Pyramidal Target. A canvas-covered framework in the form of a pyramid, mounted on a float; used as a stationary or towed target for seacoast artillery.

Pyramidal Tent. A tent, about 16 feet square, with side walls and rising to a point, having a single pole.

Pyrotechnics. Fireworks used in transmitting signals, including colored lights by night and colored smoke by day.

Q

Quadrant. An instrument used for setting off elevation in laying a gun. A quarter of a circumference. The horizon between adjacent cardinal points as the northeast quadrant.

Quadrant Angles. See Angles, Quadrant.

Quadrant Sight. A combination of quadrant and panoramic, or collimator sight, by means of which a gun is pointed in both range and direction.

Quadrilateral. A combination of, or the area inclosed by, four fortresses not necessarily connected, but mutually supporting each other.

Quadruplane. An airplane having four main supporting surfaces, placed one above the other.

Qualification. The rating of individuals according to the showing made in different tests. Capacity or eligibility for certain duties.

Qualification Course. The prescribed course in small arms record practice designed to test the soldier's proficiency in the use of his weapon. Any course for the purpose of establishing qualification.

Qualification in Arms. A demonstration on the target range of a certain degree of proficiency in the use of the weapon or weapons of his branch, which usually entitles a soldier to increased pay.

Qualification Pay. Additional compensation allowed enlisted men for qualification in the use of the weapon or weapons they are required to operate.

Quarantine. Measures taken to prevent the spread of communicable diseases. Specifically, restrictions upon the liberty of persons who may have been or are in danger of being exposed to contagion, for the full period of incubation of a disease. To establish or place in quarantine. See also Isolation.

Quarrels or **Quarries.** Missiles used in crossbows.

Quarter. The sparing of the life of a defeated or captured enemy. To furnish with shelter. To establish in quarters or billets.

Quarter Guard. An old term signifying a small guard which was posted by each battalion about 100 yards in front of the regiment.

Quartered. Furnished with quarters. Billeted.

Quartering. The assignment of soldiers to quarters. Said of a wind making any acute angle with the range on which firing is being conducted.

Quartering Area. The area in which an organization is quartered.

Quartering Parties. Details sent out to reconnoiter for billets or quarters. Billeting parties.

Quartermaster. An officer at each post, camp, or station, and on the staff of a large unit, charged with duties relating to the quartering and supply of troops, and transportation.

Quartermaster Corps. A noncombatant service of the army which provides subsistence, clothing, quarters, transportation, and, in general, all supplies and equipment except those whose procurement is specifically allotted to another supply service.

Quartermaster General. The chief of the quartermaster corps.

Quarters. A place of lodging for officers or enlisted men.

Quick Time. A cadence of 120 steps to the minute, the length of the step being 30 inches. The normal cadence for drills and ceremonies.

Quickstep. A lively, spirited piece of military march music.

Quitting Post. Leaving or abandoning without authority the post to which assigned.

Quiver. A sheath or case for arrows.

Quoin. A wedge; specifically one placed under the breech of a gun to fix its elevation; used with old types of cannon. An external corner or angle in a wall or building, or a stone cut to fit at a corner. The hinge of a vertical lock gate or of any heavy gate or door.

Quota. A proportional part or share required to make up a certain number or quantity, as of money, supplies, or troops.

R

Racer. That part of certain gun or mortar carriages which rests upon and turns on the traversing rollers.

Rack Stick. A short stick used to tighten wires or lashings.

Racking. The impact of heavy projectiles moving at low velocities.

Racolage. The kidnapping of recruits. Enticing men to enlist. (Fr.)

R. A. Corrector. A data computer, a range-azimuth corrector.

Radio. Any form of communication through the medium of electromagnetic (Hertzian) waves. Wireless telegraphy and telephony. To send a message by means of radio.

Like many other notable scientific achievements, radio has been a gradual development, extending over more than a century, and to which many have contributed. In 1887-88, Hertz enunciated and demonstrated the theory of electromagnetic waves. Radio, as a commercial application, owes its birth to Guglielmo Marconi. In 1895 he transmitted radio messages to a distance of one mile. By 1900 he had increased this to 200 miles, and on December 12, 1901, in his first attempt, he bridged the Atlantic Ocean (2000 miles) by radio. Since that date the development of radio has been rapid, until it is now almost a household necessity, and voice communication reaches to all parts of the world.

Radio has been increasingly important as a means of signal communication in the armed forces, who have been among the leaders in its development. It is a valuable means of communication between all headquarters and over all distances. Its principal advantages are that it is independent of roads and traffic, can be quickly set up as it requires no wire installation, and can be used for communication between constantly changing or rapidly moving units. Its chief disadvantage is that it is subject to interception and interference by the enemy, so that important messages must usually be sent in code.

Recent developments have been the improvement of the two-way portable voice-radio or radiotelephone. These sets have various ranges, depending upon the use to be made of them. A small set having a range of 5 miles is easily carried on the person of one man. Because of its limited range it does not interfere with other sets in distant areas. Larger sets operate up to 15, 50 or 100 miles, and there is in fact no limitation on their range except the weight of the equipment, if portable sets must be used, and the equipment is being constantly simplified and improved.

Radio Beacon. A radio transmitting station in a fixed geographical location which emits a distinctive or characteristic signal for enabling mobile stations (as aircraft) to determine bearings or courses.

Radio Channel. A radio transmission path between a transmitter and receiver, characterized by the band of frequencies transmitted; the width depends upon the type of transmission.

Radio Compass or Goniometer. A receiver which indicates the direction to the transmitting station. A direction finder used for navigational purposes.

Radio Control. Control of mechanism, other than signalling apparatus, at a distance by radio waves.

Radio Directional Beam. See Radio Range Beacon.

Radio Goniometry. The use of specially designed sets for the purpose of locating enemy sending radio stations.

Radio Intercept. The use of certain radio sets or stations to intercept enemy radio messages.

Radio Landing Beam. A beam projected from a field to indicate to the pilot his height above the ground and the position of the airplane on the proper path for a gliding landing.

Radio Navigation. The method of conducting an aircraft from one point to another by radio aids, such as the radio beacon, radio direction finder, or radioed bearings.

Radio Net. The radio stations of a superior unit and its next subordinate units, with attached or supporting units, considered as a coordinated whole.

Radio Range Beacon. A radio transmission of directive radio waves that provide a means of enabling an aircraft to keep its proper course toward the beacon.

Radio Receiver. An apparatus for converting radio waves into perceptible signals, usually pertaining to a command post. The receiving apparatus of radio.

Radio Station. A place from which radio messages are sent or at which they are received; usually pertaining to a command post. A radio broadcasting station.

Radio Transmitter. A device for producing and modulating radio-frequency power, for purposes of communication. The sending apparatus of a radio set.

Radiophone. A receiving or transmitting set for radiotelephony.

Radiotelegraphy, Radiotelephony. Wireless telegraphy or telephony.

Radius of Action. The distance an aircraft can fly before having to return to its base, with a designated margin of fuel and oil. Flying range.

Radius of Burst, Effective. The mean radius of an approximately circular area within which serious wounds may be inflicted by the fragments of a bursting shell, bomb, or grenade.

Radius of Rupture. The distance to which the rupturing effect of a subterranean charge of explosive extends, depending on the type of explosive, the weight of the charge, and the nature of the soil.

Rafale. Sudden and repeated bursts of artillery or infantry fire delivered for the purpose of producing a paralyzing effect.

Raffia. A kind of grass cloth used in camouflage.

Raid. A sudden and rapid hostile or predatory incursion. A small movement, offensive in character, directed against an enemy's lines or forces for the purpose of obtaining information, especially by the capture of prisoners, as a measure of training, or to raise the morale of the raiding troops and lower that of the enemy, etc. To make or take part in a raid.

Railhead. A point on a railroad in the theater of operations at which supplies for troops are unloaded, and from which they are distributed or forwarded to refilling or distributing points by other means. The railhead is, whenever practicable, advanced sufficiently close to the troops to permit the division trains to refill at that point.

Railway Artillery. See Artillery, Railway.

Railway Battalion. A special engineer unit, including a maintenance of way company, a maintenance of equipment company, and a railroad operating company. It is assigned to the maintenance of a railway division, the battalion commander being the division superintendent.

Raise a Blockade. To break up or remove a blockade.

Rake. To bring enfilade fire to bear upon. To fire along the length of a position or body of troops.

Rally. To bring disorganized troops back to order. To assemble. To unite.

Rallying Point. A designated point where a unit commander assembles his unit for further operations. An assembly point. A rendezvous.

Rammer. A rod or staff used in forcing home the projectile or charge of a gun.

Rammer and Sponge. A combined rammer and sponge for cleaning the bore of a gun.

Rampart. The embankment surrounding a fortification on which the parapet is raised. The embankment including the parapet. That which fortifies, defends, or secures against attack. A defense or bulwark. To fortify with ramparts.

Ramrod. A rod used in charging a gun, especially a muzzle-loading gun. A rammer. A rod used in cleaning a rifle.

Randing. The process of weaving brush wattling in and out between pickets, as in making a gabion or hurdle; or the weave thus made.

Range. The distance to the enemy objective or target. The horizontal distance from the origin to the target, point of fall or point of impact. The horizontal distance to which a shot may be projected. A place where shooting is practiced. The horizontal travel of an aircraft bomb, from release to point of impact, measured on the level of the point of impact. The radius of action of an aircraft. To obtain an actual range by firing.

Range Card. A card used in attack or defense for recording the ranges and azimuths to various points in the field of fire for ready reference.

Range Determination. The act of finding the range to any object or target by the best available means, including range finding and range estimation.

Range Deviation. The difference between the range to the target and the range to the point of impact, measured along a line parallel to the line of position.

Range Difference Chart. See Difference Chart.

Range Disc. A disc on the gun, graduated in fuze ranges, by means of which super-elevation is given to the gun. A circular range scale.

Range Drum. A drum, graduated in range, forming part of a sight or of the elevating mechanism of a fixed gun or howitzer.

Range Dummy. A dummy cartridge which closely resembles a live cartridge, used for instructional purposes on the rifle range.

Range Error. The divergence of a point of impact from the center of impact, measured along a line parallel to the line of position.

Range Estimation. An estimation of the range without the use of a range finder or other measurement.

Range Finder. An instrument designed to determine the distance to an object or target. The three principal types are the vertical base, the horizontal base, and the self-contained base. See also Position Finder.

Range Officer. An officer charged with the care, upkeep, police, etc., of a target range and its accessories. An officer who is in charge of the position finding equipment and the range section of a seacoast battery command, and whose post is at the battery plotting room.

Range Percentage Corrector. A device for converting range corrections into percentages of the range for application proportionally to the ranges.

Range Probable Error. The probable error in range measured parallel to the line of position or site.

Range Section. A detail of battery personnel charged with the duty of range finding, including observing, spotting, and plotting details.

Range Setter. An enlisted man charged with the duty of laying a gun for range. An elevation setter.

Range Tables. Tables of elevation prepared for each gun or class of guns, of periodical times, showing also corrections to be applied for variations from standard meteorological and other conditions.

Range-correction Board. A devicec for obtaining the resultant of the range corrections for wind, atmosphere, tide, velocity, rotation of the earth, and weight of the projectile; used in connection with a percentage corrector.

Range-deflection Fan. A series of concentric range circles and direction rays diverging from a point, constructed on translucent paper, for the rapid determination or ranges and deflections in terrestial fire. See also T. A. B. Clock.

Ranges, Classification of. The ranges of firearms vary from a few hundred yards to many miles. The following terms are used to describe variations in range: point-blank; close; short; medium; mid; long; distant; and extreme; also battle and effective; these terms are self-explanatory, but sometimes loosely employed. For many weapons arbitrary limits for the different ranges have been assigned; these are naturally subject to variation. Ranges are also classified as distant, serious, and decisive.

Ranging. The placing of troops in proper order for battle or the march. The act of determining the range to a target by actual firing. Firing for the determination of range.

Ranging Shots. Shots fired for the purpose of adjustment or correction of firing data.

Ranging-shot Method. A kind of trial fire in which four shots are fired directly at a naval target, either by single shots from one or more guns or by battery salvo, and full correction is made for the deviation of the center of impact.

Rank. A line of men, horses, teams or vehicles placed abreast of each other. That characteristic or quality bestowed upon military persons which marks their station and confers eligibility to exercise command or authority in the military service within the limits prescribed by law, and which is divided into degrees or grades which mark the relative positions and powers of the different classes of persons possessing it. Relative precedence in the military service. To have precedence, or to exercise such precedence. To possess higher rank than another.

Rank and File. The body of soldiers constituting the mass of an army and including all ranks from corporal downward.

Ranking. Having or taking precedence.

Rapid Fire Exercises. Preparatory exercises intended to teach a man to hold the rifle and manipulate the bolt properly in rapid fire.

Rapid Fire Gun. A single-barrelled, breechloading cannon, using fixed ammunition, and capable of fire at a rapid rate. It is provided with an eccentric breechblock, or one that can be opened or closed by a single motion of an operating lever. In general, any cannon using fixed ammunition is a rapid fire gun.

Rapid Preparation of Fire. The determination of approximate data for opening fire when time is limited or facilities, such as surveys and accurate maps or charts, are not available.

Rapid-firing. Capable of being fired rapidly.

Rapier. A long, narrow sword, used for thrusting only.

Rate of Advance. The rate prescribed for troops to move forward when following a rolling barrage, usually expressed in terms of a hundred yards in a certain number of minutes.

Rate of March. The rapidity of march calculated in terms of miles per hour or per day.

Rated Men. Enlisted men who have qualified for and been rated as gun commanders, gun pointers, observers, etc.

Rating. A statement of the relative degree of military proficiency of an individual or an organization, based on observation, tests, or inspections.

Ration. The prescribed allowance of the different articles of food for the subsistence of one person or one animal for one day; in the United States Army there are five classes: garrison, field, travel, reserve, and emergency.

Ration and Savings Account. A voucher upon which is computed the value of the ration and indicated thereon for each organization the period covered, number of rations due, amount of credit due, value of stores purchased, and the balance due the organization or the United States, as the case may be.

Ration Certificate. A certificate furnished by the quartermaster to the commanding officer of an organization or detachment changing station showing the date to which and by whom its ration and savings account has been settled, to be attached to the first ration return submitted at the new station.

Ration Party. A detail sent to the rear to bring up rations for the front line troops.

Ration Return. A requisition on the quartermaster for rations, signed and submitted by an officer under whom persons entitled thereto are serving.

Ration Saving. The amount of the money savings of an organization to which rations are issued, on the monetary ration allowance for any period settlement being made monthly.

Ration Weights. The weights of the prescribed rations are as follows: garrison ration, 5.46 pounds; field ration, 4.81 pounds; reserve ration, 3.0 pounds

Rations in Kind. Actual issue of rations instead of the money value thereof

Ravelin. A small outwork with two faces forming a salient angle.

Ravitaillement. Supply. (Fr.)

Raw Troops. New and inexperienced soldiers.

Readiness. A state or condition of alertness or preparedness. See Position in Readiness.

Ready. A preparatory word of command. Prepared. Prompt.

Real (actual) Speed of an Aircraft. Speed measured with reference to the ground. See also Ground Speed.

Rear. That part of a force which comes last, or is stationed behind the rest Behind. Back of the front. The direction away from the enemy.

Rear Area. An area within which all localities are beyond the range of hostile medium artillery, and which is not immediately affected by the tactical dispositions of combat troops, nor normally subject to ground attack.

Rear Boundary. A designated line defining the rear boundary of an area or position.

Rear Echelon. Such elements of a unit or headquarters as are not required with the forward echelon for combat purposes. A rear line or element of a force or unit in attack or defense.

Rear Guard. A detachment which follows the main body and affords it protection on the march; used in a retreat or retrograde movement.

Rear Guard Artillery, Cavalry, etc. Artillery, cavalry, etc., attached to a rear guard.

Rear Guns. Machine guns emplaced or held in the rear part of a battle position.

Rear Party. The detachment from the support of a rear guard which follows and protects it on the march.

Rear Rank. The rank of a body of troops that is in rear.

Rear Sight. The sight which is nearest the breech of a weapon. See Sight Leaf.

Rear Slope. The slope in rear of the battery parade in a fortification. The rear wall of a trench.

Rearward. Toward the rear.

Rebel. One who revolts against his recognized government. To revolt. To be disobedient to authority.

Rebellion. An insurrection of considerable magnitude. The act of rebelling.

Rebounding Lock. A gunlock in which the hammer rebounds to half-cock after discharge, for safety.

Recall. A signal by which troops are recalled from drill or other duty. To call back.

Receive. To convert incoming radio waves into perceptible signals.

Receiver. That part of certain firearms in which a cartridge is received from the magazine before being seated in the breech. The receiving instrument of a telephone, telegraph or radio set.

Receiving Set. An apparatus for receiving radio signals.

Receiving Table. The table at the lower end of an ammunition hoist on which the projectiles are placed preparatory to raising.

Reception. The steps incident to the induction of an enlisted man into the service, up to and including his initial assignment. The receiving of messages or communications by any auditory or visual means; the clearness or facility with which they may be seen or heard.

Reception Center. A place to which draftees are sent from their induction station where clothing and personal equipment is issued them, and where they are vaccinated, inoculated and classified. See also Induction Station and Replacement Training Center.

Reciprocal Laying. A method of bringing the planes of fire of two pieces parallel by each piece sighting on the sight of the other.

Reclassification. The determination of the duty or class of duties an individual may be qualified to perform when, for any reason, he is considered not qualified for the duty or class of duties to which he has been assigned.

Reclassify. To classify again, as an individual, when it has been determined that he is not qualified to perform the duties pertaining to his earlier classification.

Recoil. The backward movement of a gun on discharge. To retreat. To draw back. To rebound.

Recoil Brake. That part of the recoil system which limits and controls the movement of the gun and cradle or top carriage, to the rear after discharge.

Recoil Buffers. Devices on gun carriages for the purpose of reducing the length of recoil, and the shock due to abnormally excessive recoil or counterrecoil.

Recoil Cylinders. Hydraulic or hydro-pneumatic cylinders designed to reduce the length and speed and ease the shock of recoil.

Recoil, Length of. The length of recoil in modern cannon is automatically varied according to the elevation of the piece, being greatest at low elevations.

Recoil Mechanism or System. The mechanism which limits and controls the recoil of a cannon, usually including also the counterrecoil mechanism or recuperator, by which the piece is returned to its firing position. A number of methods have been employed to check the recoil of cannon, including ropes, counterweights, sliding friction of the top carriage on the chassis, gravity, springs, and pneumatic and hydraulic cylinders. In most modern cannon the recoil brake consists of one or more hydraulic (oil) cylinders with throttling devices by which the resistance to recoil is gradually increased. The piece is returned to its firing position (counterrecoil) by means of a counterweight, pneumatic cylinder or springs, which are lifted or compressed by the force of recoil. A counterrecoil buffer, usually hydraulic, pneumatic or spring, prevents shock at the end of the counter-recoil. In automatic weapons the force of recoil is partially absorbed in performing the operations of extraction, feeding, etc., and is checked by means of spring or pneumatic buffers.

Recoil Rollers. Rollers on which the top carriage of a cannon moves in recoil and counterrecoil.

Recoil-operated. Operated by the force of recoil, as the mechanism of an automatic weapon.

Reconnaissance. The procurement in the field of information of military value, as of the enemy, terrain, etc., by military personnel in uniform sent out for that purpose. As distinguished from observation, recon-naissance implies movement. Reconnaissance is a continuous process during active operations, being conducted prior to, during, and subsequent to combat, and the information obtained is considered in the preparation of all operation orders.

Reconnaissance Cavalry Patrols. Detachments sent out from reconnai sance groups or troops to gain contact with the enemy's main body and secure information of its strength and movements.

Reconnaissance Group. A strong and mobile force, often mixed, for distant reconnaissance, or for seizing and holding important positions and locali ties in advance of the army, and later to act as a mobile reserve.

Reconnaissance in Force. A demonstration or attack by a considerable force for the purpose of determining the positions and strength of the enemy.

Reconnaissance Officer. An officer on the staff of any unit who is charged with supervision of the duties of reconnaissance and observation, surveys selection of routes and positions, procurement and issue of maps, etc.

Reconnaissance Strip. A series of overlapping vertical photographs made from an airplane flying a selected course.

Reconnaissance Troops. Troops or squadrons pushed forward from the main body of army cavalry for the purpose of receiving and transmitting reports of reconnoitering patrols, to support these patrols and to provide reliefs for them.

Reconnoiter. To make a reconnaissance. To inspect with a view to military operations or to gain military information.

Reconnoitering Cavalry. Cavalry, usually a large body, sent out well in advance of the army to gain and maintain contact with the enemy's main body and to furnish information regarding its movements.

Reconnoitering, Reconnaissance or Information Patrol. A patrol which is primarily charged with obtaining information, generally of enemy troops but sometimes of terrain, means of communication, supply, temper of in habitants, in short, of anything that might be useful to commanders in formulating their plans for future action.

Record. An official written document or account. The report of the pro ceedings of a court-martial or board of officers.

Record Adjustment. An adjustment of fire on a check point recorded for later use in firing on targets in its vicinity.

Record of Fire. An adjustment on a check point, immediately following an adjustment on a target by means which may later not be available. The record of the data thus obtained.

Record Practice. In target practice, a firing to obtain a record of the prog ress of the soldier, which is used in determining his marksmanship rating for the year.

Record Range. The adjusted or corrected range to a check point.

Record Service Practice. A service practice conducted for the purpose of affording experience to and testing the efficiency of a battery in firing.

Record Transfer. A transfer of fire based on a record of fire.

Recording Cone. An apparatus which registers upon the film of a photo graph certain pertinent data, such as serial number, date and hour of ex posure, altitude, etc.

Recording Panel. A landscape target printed in black and white, which is placed above the corresponding target printed in colors, and which receives the shots aimed at the vari-colored target when rifles are harmonized.

Records. Official accounts or papers of various natures.

Recruit. A newly enlisted soldier. An individual whose induction into the service has been completed but who has not been assigned to a unit.   To enlist men for the military service.

Recruit Depots.   Designated military posts to which recruits are sent from general recruiting stations for final examination, enlistment and distribution to the army.

Recruiting Ground.   A territory or area in which recruits are sought.

Recruiting Party or Detail.   A detachment engaged in seeking recruits.

Recruiting Station.   An office where applicants for enlistment are given a preliminary examination and conditionally accepted for service.   A general recruiting station is one that may accept applicants for any branch of the service.

Recruitment.   The operation whereby personnel is obtained by voluntary enlistment or selective service for military duty.

Recruitment Rate.   The estimated rate at which manpower may be procured, either by voluntary enlistment or by operation of the selective service, examined, classified, and assigned to military employment.

Rectangular Coordinates.   See Coordinates, Rectangular.

Rectify the Alignment.   To improve the alignment of troops formed in line.

Recumbent.   Lying down or prone; applied to wounded who must be disposed and transported in a recumbent position.   As distinguished from prone, recumbent means lying on the back.

Recuperator.   A pneumatic counterrecoil mechanism.

Red Cross.   The Geneva cross.   The Red Cross Society.

Red Cross Society.   A charitable organization which takes an active interest in the welfare of the sick and wounded in war.

Red Hot Shot.   Cannon balls heated to redness and used for incendiary purposes.   (Old.)

Redan.   An outwork or detached work of fortification, consisting of two faces forming a salient, and usually open at the gorge.

Redan Batteries.   Batteries, with mutually flanking fire, located at the salient and reentrant angles of a fortification.

Redan Line.   A line of redans usually connected by curtains or rifle pits.

Redoubt, Redout.   A work completely closed by a parapet, thus allowing all around fire.   A small, and usually rough, inclosed work of varying shape, usually temporary in nature and without flanking defense.   A small, detached fieldwork.

Reds.   The term used to designate the hostile forces in the preparation of map exercises, map problems, etc.

Reduce.   To degrade to a lower rank or grade.   To bring to terms.   To conquer.   To capture or subdue.

Reduce to the Ranks.   To degrade a noncommissioned officer to the grade of private.

Reduction Coefficient.   The ratio of the distance observer-target to the distance gun-target, by which observed deviations are multiplied to transform them for use at the guns.

Reduit.   A central work within a fortification intended as a place of last resort.   A keep.

Reel Cart.   A cart carrying a revolving drum for the handling of signal wire in the field.

Reenlist.   To enlist anew.

Reenlistment Allowance.   See Enlistment Allowance.

Refer.   The piece being laid for direction, to refer the piece is to bring the line of sighting on a chosen aiming point, called a referring point, without moving the piece.   To announce an aiming point and measure and record the direction of a piece which has been laid for direction.

Reference Line.   A line to which directions or azimuths are referred.   A line of zero azimuth for a particular system would be a reference line.

Reference Numbers.   Arbitrary numbers used in place of the actual values in the graduation of certain scales of instruments employed in gunnery to avoid the use of plus or minus, right or left, or up or down in making corrections.   The elevations of contours in feet above the datum plane, usually mean sea level.

Reference Piece.   In calibration, the piece to which the other pieces of a battery are compared.

Reference Point. An unmistakable object used to identify less prominent objects by describing their relation to or angular direction from the reference point. A prominent point on the terrain by reference to which objectives may be readily identified.

Refilling Point. A place at which supplies are transferred to the supply columns of divisional and higher units.

Refuse. To dispose certain troops in rear of the general alignment. To draw the flank or center elements of a line or command to the rear in order to protect against a possible enemy flank attack or penetration.

Regiment. An organization composed of a headquarters and two or more battalions or similar units and normally commanded by a colonel. The largest permanent organization in the military service.

Regimental. Pertaining to a regiment.

Regimental Court-martial. An inferior court appointed by a regimental commander from among the officers of a regiment; superseded by the special court-martial.

Regimental Munitions or **Combat Train.** A combined train composed of the combat wagons of the regiment and attached units.

Regimental Reserve Line. One of the lines of a defensive position, from 400 to 900 yards in rear of the battalion reserve line, occupied by the units in regimental reserve. In a prepared position it is marked by more or less continuous trenches.

Regimentals. Uniform dress. The uniform clothing prescribed for and worn by the troops of the regiment, such as blouses, breeches, caps, etc.

Regimentation. The classification into groups and employment of the civilian population under enforced governmental regulation and supervision.

Register. To adjust fire on several selected points in order that these may later serve as auxiliary targets or transfer points.

Registration. Adjustment on a number of points to assist prompt opening of fire when targets appear.

Registration Points. Points in the objective zone on which adjustment of fire has been made.

Regular. A member of the regular army. A professional soldier.

Regular Approaches. Trenches zigzagging toward a fortified position which is too strong to be taken by open assault. See Parallels and Approaches.

Regular Army. The army maintained in time of peace to furnish the necessary protection of our frontiers, to garrison overseas possessions, to assist in the training of the civilian components of the Army of the United States, and to form the nucleus for expansion in time of emergency. The standing army of a nation, composed largely or wholly of professional soldiers.

Regular Practice Season. That part of the year designated by proper authority during which the prescribed course of target practice will be carried out.

Regulating Officer. The officer in charge of a regulating station.

Regulating Station. A place on a line of supply and evacuation, specifically on a railroad, at or near the rear boundary of the combat zone, where the movement of troops and materiel to or from a certain area, is controlled. There is usually one regulating station for each army in the combat zone, which includes a railway yard and storage facilities for at least one day's supply.

Regulations. Administrative rules for the government, conduct, training etc., of the armed forces. Regulations emanate from an executive department, as the War Department, and are not a part of statutory law, but have all the force of law when approved by proper authority.

Regulator. A mechanism used to control counterrecoil.

Reichswehr. The national army of Germany, recruited by voluntary enlistment under the terms of the Treaty of Versailles, up to March, 1935.

Reinforce. To strengthen, as by the addition of fresh troops, materiel, or support. That part of a cannon immediately in front of the breech, which is strengthened to resist the maximum pressure of the powder gases.

Reinforced Brigade. A detached infantry brigade strengthened by attached artillery and auxiliary troops.

Reinforced Unit. A unit having attached to it troops not organically a part thereof.

Reinforcements. Additional troops sent or used to augment the strength of another body of troops, especially for combat purposes.

Reinstate. To place an individual in the same position or rank as he previously held.

Relative Points. Selected points by which the heads of columns of troops marching on different roads are kept abreast.

Relative Rank. Comparative rank. The order of seniority in the same grade. Corresponding rank in the army and navy, viz.: second lieutenant, ensign; first lieutenant, lieutenant, junior grade; captain, lieutenant, senior grade; major, lieutenant commander; lieutenant colonel, commander; colonel, captain; brigadier and major general, rear admiral; general, admiral.

Relay Point. A transfer point in the system of supply established when the distances between railroad and refilling points are too great to be spanned with the transportation normally with the army.

Relief. Fresh troops which replace others on duty or at work. A division of a working party or of a guard which is at work or on duty for a limited, specified time. Personnel or units which relieve others. The act of relief. Variations in the elevations of the earth's surface, such as hills, valleys, etc. The height of an embankment above the immediately adjacent ground or the bottom of a ditch in its front. The height of the parapet or fire crest of a trench above the bottom of the trench.

Relief of a Front Line Unit. The act of moving up a unit from the rear and replacing a unit that is in contact with the enemy, the replaced unit moving to the rear.

Relieve. To release from a given post or duty, by substitution or otherwise. To replace. To change.

Relocation. The process of determining the range and azimuth of a point or target from one station when the range and azimuth from another station are known.

Remaining Velocity. The velocity of a projectile at any point of the trajectory.

Remission. A partial exercise of the pardoning power by relieving a person of the punishment, or a part of it, without pardoning the offense.

Remount. The furnishing of a horse to replace one killed or disabled. The horse so furnished.

Remount Depot. A place where newly purchased animals are sent to be trained and conditioned for issue to the service.

Remounts. New horses obtained for the military service.

Remunerative Target. See Profitable Target.

Rencontre. A meeting engagement. To meet a foe unexpectedly.

Rendezvous. A place of meeting. A place at which the personnel of a unit assembles upon receipt of mobilization orders. An assembly point designated for a particular purpose, as after a withdrawal or retreat, or the scattering of a patrol or other body. An assemblage. To assemble or bring together.

Renegade. A deserter. One who goes over to the enemy. A turncoat.

Renseignements. Information, intelligence (Fr.).

Rental Allowance. An allowance for the rental of quarters.

Reorganization. The act of restoring order among troops which have become mixed and disorganized during combat or retreat. The reestablishment of combat units on a functioning basis following an assault; a phase of the attack. The process of changing the details of organization, or of organizing anew.

Repeating Firearms. Firearms, especially small arms, designed to deliver several shots from a single loading. Magazine weapons.

Replacement. A trained officer, nurse, or soldier available for assignment. The act of replacing. The fact or condition of being replaced.

Replacement Center. An establishment in the zone of the interior where individuals are held as replacements. Normally it will include a number of schools to train officers and enlisted men.

Replacement Depot. An establishment in the theater of operations for the reception and distribution of replacements.

Replacement Pool. A pool comprising all the replacement centers in the zone of the interior.

Replacement Training Center. A place to which draftees are sent from their reception center for basic training and where inoculation and records are completed, before they are sent to organizations to which assigned.

Replacements. Officers and soldiers sent to fill the vacancies in an organization.

Replenish the Attack. To add fresh impetus to the attack by the employment of supports and reserves.

Replenisher. A small re¯ervoir which supplies a deficiency or takes up an excess of oil in the recoil cylinders.

Report. A statement of facts, officially rendered, for any purpose. The noise caused by the discharge of a firearm.

Report of Survey. A report on the condition of property surveyed by an inspector or board, with findings as to the reasons for unserviceability, and recommendations as to disposal.

Represent. To play the role of a given person or unit, as in field exercises. To simulate.

Representative Fraction. The scale of a map. A fraction whose numerator is one and whose denominator shows the distance on the ground represented by unity on the map. The ratio between any distance on the map and the corresponding distance on the ground.

Reprimand. An official rebuke administered as a punishment. To censure formally.

Reprisal. The seizure of property or the punishment of an enemy in retaliation for some breach of international law. An act of retaliation, resorted to by one belligerent against enemy individuals or property for illegal acts of warfare committed by the other belligerent, for the purpose of enforcing future compliance with the recognized rules of civilized warfare.

Requirements. The computed essential needs, embracing all things necessary to meet any emergency that may confront our armed forces.

Requirements, Initial. Items of prescribed individual and organizational equipment and supplies required initially to equip individuals and units.

Requirements, Maintenance. Items of individual and organizational equipment and supplies required to replace similar unserviceable items of initial equipment.

Requisition. A demand for funds, supplies or services made upon the people of an invaded country. A formal request for funds, supplies, or services. To demand or request funds, supplies or services.

Reservation. An area of land set apart for military purposes.

Reserve. That portion of a body of troops which is kept to the rear, or withheld from action at the beginning of an engagement, with a view to its use to influence the later action; all units larger than a company habitually employ a reserve. The largest element of an advance or rear guard. To keep back. To hold for future use.

Reserve Ammunition. The supply of ammunition provided for the purpose of replenishment when the original supply is exhausted or nearly so.

Reserve Battle Position. Any battle position in rear of the main battle position, that has been reconnoitered and staked out, and generally partially organized, for use in case the troops are driven out of the main battle position. In general, it is placed at such distance in rear of the position next in front that the enemy after having successfully attacked the forward position would have to advance the bulk of his artillery before undertaking its attack.

Reserve Echelon. The units constituting the reserve of any force. The general line on which the reserve is disposed.

Reserve of the Advance or **Rear Guard.** The principal maneuvering and fighting element.

Reserve Officers Training Corps. Organizations maintained at civil educational institutions for the military training of students. The R. O. T. C includes two branches; the senior division at colleges and universities, and the junior division at high schools and secondary schools.

The R. O. T. C. is the principal feeder for the Officers' Reserve Corps especially in the combat branches. It had its inception in the Morrill Act of 1862, which provided for grants of public land to state educational institutions, and required, amongst other provisions, that such institutions maintain courses of military instruction. The National Defense Act of 1920 extended and improved the provisions for military training in schools and colleges, which now receive a considerable measure of federal support.

Reserve Ration. The ration prescribed for use in the field when the field ration is not available.

Reserve Trench. A trench dug for the use of reserves.

Reserved Road. A road reserved for the use of certain traffic only.

Reserves (Supply). Supplies accumulated in excess of immediate needs to insure continuity of an adequate supply. Battle reserves are supplies accumulated near the front, in addition to individual and unit reserves. Individual reserves are those carried on the soldier, animal, or vehicle for his or its individual use in emergency. Unit reserves are prescribed quantities carried as a reserve by the unit.

Reservicing Point. A place where tanks coming out of action are supplied, reconditioned, and units reorganized.

Resignation. The act of resigning one's commission as an officer.

Respirator. A simple form of gas mask. A gauze worn over the mouth and nose as a protection from infection.

Responsibility. That for which one is answerable. A duty, trust, or obligation. The state of being responsible for the discharge of a duty, or the care of property, etc.

Responsible Officer. An officer to whom military supplies are issued, either for his personal use or for the use of an organization, and who is responsible for their proper care and use in the military service.

Rest. A support for a gun in aiming and firing. An order permitting men in ranks to relax, talk, etc. To cease from action or motion.

Rest Areas. Areas established for the purpose of providing facilities where divisions, either wholly or in part, can be brought from front line duty, rested, reequipped, filled with replacements and given the necessary training in the time available.

Restitution. The process of adding data to a map from aerial photographs or other maps.

Restricted. A document will be marked "Restricted" when the information it contains is for official use only or of such nature that its disclosure should be limited for reasons of administrative policy, or should be denied the general public.

Retaliation. Reprisal. Revenge. A reprisal justified by the laws of war.

Retardation. In ballistics, negative acceleration or loss of velocity of a projectile.

Retire. To move, voluntarily and in good order, to the rear or away from the enemy, with a view to avoiding or postponing action, or for the purpose of regaining freedom of action or otherwise improving the tactical situation. To fall back from a position. To withdraw. To retire from active duty.

Retired List. A list of officers who, by reason of advanced age, any disability, or the completion of a prescribed number of years of service, are relieved from active service, but receive a certain percentage of their active service pay until death.

Retirement. A voluntary retrograde movement of the main forces which, while contact with the enemy is not an essential condition, is generally made for the purpose of regaining initiative and freedom of action by a complete disengagement. A movement made to forestall a decisive engagement, to attract the enemy in a desired direction, or to gain time for the reorganization of the forces preparatory to renewed efforts against the enemy. The act of retiring from duty on the active list of the army, or the status of a retired officer.

Retiring Board. A board of officers convened to consider and report upon the incapacity of an officer for active duty.

Retraction Gear. The gear used in retracting a gun from its firing position to its loading position when the force of recoil is not utilized.

Retreat. An involuntary retrograde movement forced on a command as a result of an unsuccessful operation or combat. The act of retreating. A beat of drums, or sounding of bugles or trumpets, which takes place at sunset. To retire from any position or place. To withdraw.

Retrenchment. An additional line or lines of defense provided within or in rear of the principal line, for use in case of the capture of the latter.

Retrograde Movement. A movement to the rear.

Return. An official report rendered to higher authority. An accounting for troops, property, or supplies, for which an officer is accountable. A short branch leading off from a trench.

Returns of Troops. Monthly reports of the strength of troops serving under their command, required of post and higher commanders.

Reveille. The first bugle call, sounded at about the break of day as a signal that it is time for the soldiers to arise. The first morning assembly of troops in camp or garrison.

Reveille Gun. The morning gun fired at the first note of reveille, or at the commencement of the first march, if marches are played.

Reverse. A partial defeat. A change for the worse.

Reverse Arms. A position in which the rifle is held or carried between the right elbow and the body at an angle of 45 degrees, the butt upward.

Reverse Slope. A slope which descends away from the enemy and forms the masked or sheltered side of a covering ridge. The rear slope of a position on elevated terrain.

Reverse Slope Position. A position on a reverse slope concealed from the position of the enemy.

Reverse Turn. A rapid maneuver to reverse the direction of flight of an airplane, made by a half loop and half roll.

Reversed. Upside down, said of arms when carried with the butts upward.

Revet. To place a revetment, or to support by means of a revetment.

Revetment. Any artificial means provided to cause earth to stand at a steeper slope than it would naturally assume. The usual forms of military revetments are retaining walls, which stand by their own weight, and thin superficial coverings anchored or secured to the revetted surface, and protecting it from abrasion and the effects of the weather.

Review. A ceremony, including a formal examination or inspection of troops. To go over or examine critically. To make a formal inspection of the state of, as troops, and the like.

Reviewing Authority. The officer whose duty it is to review and take action upon the proceedings of a court-martial or board, and to approve or disapprove the sentence or recommendations.

Revolt. A rebellion. To rebel against the constituted authority or government.

Revolution. An extensive change in the constitution or government of a country, suddenly accomplished by force. A successful rebellion.

Revolver. A single-barreled small weapon with a revolving cylinder containing several cartridges, thus enabling several shots to be rapidly fired without reloading. Firearms having multiple revolving barrels or several chambers and a fixed barrel date back to the 16th century. In nearly all of the early weapons only one wheel or flint is employed, the barrels rotating. The invention of a percussion priming lock by Forsythe in 1807 simplified firing mechanisms. The first percussion revolvers were of the type known as "pepperbox," in which several barrels rotated about a central spindle. Many variations of this type were produced between 1825-1835. In 1836 Colonel Samuel Colt patented his well-known revolver, which differed from the "pepperbox" in having a revolving cylinder and a single barrel. Many types have been produced since that time, but, in general, the service revolvers of this and later periods have been very similar. Since about 1890 the principal revolvers in military use have been the Colt, Smith and Wesson, and the Webley. See also Pistol.

Revolving Target. Two targets connected by a crosspiece on which they revolve, so that they may be alternately exposed.

Ribard Cylinder or Wire. A portable wire entanglement in the form of a cylinder, consisting of circular frames of heavy wire with barbed wire strung upon them. It can be collapsed for carrying and extended for placing.

Ricochet. A glancing rebound of a projectile after impact. A projectile which bounds along the ground in approximately its original direction. To rebound.

Ricochet Battery. A battery which fires with very small charges of powder and with just enough elevation to clear a parapet.

Rifle. Any rifled firearm, but ordinarily applied to a rifled small arm fired from the shoulder and having an attachment for a bayonet. In artillery, the term is sometimes used to distinguish a gun with a relatively long barrel from a howitzer or mortar. It is generally accepted that the rifle was invented by Gaspard Zollner of Vienna, and it made its first appearance at a target practice at Leipsic in 1498. The first rifle grooves were made parallel to the axis of the bore to diminish the friction of loading forced or tightly-fitting bullets. It was accidentally discovered that greater accuracy could be secured by spiral grooves. About 1600, the rifle began to be used as a military weapon employing special bullets. In 1729, it was found that good results could be obtained by the use of oblong bullets eelipsoidal form. The great difficulty of loading the rifle, generally by blows of a mallet or a stout iron ramrod, prevented its general use. The major steps in the development of the modern magazine rifle from the old musket (flintlock) rifle are as follows: (1) The percussion cap. (2) Rifling and the elongated bullet. (3) Breechloading. (4) Use of a magazine (repeater). (5) Smokeless powder and smaller calibers.

Rifle Bullet. A bullet designed to be used in a rifle.

Rifle Grenade. A bomb or grenade projected by means of a rifle.

Rifle Grenade Discharger. A special device which fits on the muzzle of a rifle to permit the projection of grenades. A tromblon.

Rifle Grenadier. See Grenadier.

Rifle, Magazine. See Rifle, Repeating.

Rifle Marksmanship. The art of shooting a rifle. Training in the use of the rifle, which includes sighting and aiming, positions, trigger squeeze, rapid fire, examination prior to range practice, and range practice.

Rifle, Military. A hand firearm, the weapon of the individual infantry and cavalry soldier, used in generally similar forms in all armies. Rifled small arms had made their appearance in the 16th century, and certain military units and individuals had, from time to time, been armed with them. They were used in battle as early as the middle of the 17th century. But the general use of the rifle as a military weapon may be said to date from 1800, when the first British regiment was equipped with the Baker rifle, caliber .615. Similar weapons were later adopted by Austria, France, Prussia, Russia, and other European states.

The first successful breechloading rifle was produced by Hall, an American, for the United States Army, but it was not generally adopted. About 1841 the Prussian Army was equipped with the celebrated Dreyse needle gun, caliber .61, a bolt-action breechloading rifle. In 1849 Captain Minie brought out the Minie expanding bullet and the Minie muzzle-loading rifle, which were at once adopted by the French Army. Great Britain adopted the Enfield rifle in 1855, and converted it to a breechloader in 1865 by the use of the Snider breech mechanism. The use of breechloaders now became general. France adopted the Chassepot in 1869, followed by the Gras in 1874. In 1871 Germany discarded the needle gun, which had proven decidedly inferior to the Chassepot, in favor of the Mauser. During the American Civil War many of the troops, including the bulk of the cavalry on both sides, were armed with breechloaders, some of them repeaters, yet it was not until 1873 that the old Springfield camlock, caliber .45, was adopted as the weapon of the United States Army.

The next advance was the introduction of magazine or repeating rifles. Germany was the first nation to arm with repeating rifles, which she did in 1884 by converting the Mauser of 1871. In 1885 France adopted the Lebel, caliber .315, the first weapon in which smokeless powder was used. Austria adopted the Mannlicher in 1886; Great Britain the Lee-Metford in 1887, followed by the Lee-Enfield in 1902; Germany replaced the converted Mauser with a new Mauser in 1889; Italy adopted the Mannlicher-Carcano in 1891; and in this same year Russia adopted the Berdan or Nagant Three Line rifle.

Colt, an American, had produced a repeating rifle as early as 1840. The Henry and Spencer rifles, which were improvements on the Colt, were actually used, especially by cavalry, in the Civil War. The famous Winchester had appeared in 1867. Nevertheless, the United States Army was slow to adopt the magazine rifle, but in 1892 adopted the Krag-Jorgensen, caliber .30. This was succeeded by the new Springfield, Model 1903, a clip-

loading, bolt-action magazine gun, with a capacity of 5 cartridges. With
modifications and improvements these are the rifles now in use by the
above named nations. The present trend of development is toward a semi
automatic rifle. See also Crossbow; Harquebus; Long Bow; Firearms
Flintlock; Hand-arms; Matchlock; Musket; Wheellock.

Rifle Pit. A hasty excavation to shelter one or more riflemen, and from which
they can fire. See Fox Holes.

Rifle Practice. Practice with the rifle as prescribed in Training Regulations

Rifle Range. An area of ground prepared for rifle practice; usually includ
ing targets permanently installed.

Rifle, Repeating. A rifle capable of delivering several shots from a single
loading of a magazine contained in or attached to the weapon, but in which
extraction and reloading are performed by hand and not by automatic ac
tion of the recoil or powder gases. A magazine rifle. See also Rifle, Mili
tary and Semi-automatic.

Rifle, Semi-automatic. A repeating or magazine rifle in which the actions o
extraction and reloading are performed automatically by action of the re
coil or powder gases. A self-loading rifle. See also Rifle, Military.

Rifle Shot. One who shoots with a rifle. A shot fired by a rifle.

Rifle Unit. A combatant unit whose principal weapon is the rifle, e.g., a rifl
squad, platoon, company, or battalion.

Rifled Cannon. A cannon whose bore is rifled.

Rifleman. A soldier armed with a rifle and instructed in its use.

Rifleman's Insignia. Insignia indicating skill in marksmanship, issued t
those who qualify in firing the record practice.

Rifling. The spiral grooves cut in the bore of a gun for the purpose of im
parting a rotary motion to the projectile on discharge.

Rigging. The science of handling cordage and wire rope and chains, in vari
ous block and tackle and lever combinations, to raise and move heavy loads
An arrangement of blocks and tackle for such a purpose.

Right. The right flank, extremity or element of a body of troops.

Right About. A turning directly to the rear by the right, as of an individua
or squad.

Right Bank. The bank of a stream which is on the right of the observe
when facing downstream.

Right (Left) Front into Line. A movement in which line is formed from
column to the right (left) front by the units in rear obliquing to the righ
and forming on the right of the preceding unit.

Right (Left) Turn. A movement in which a line effects a change of directio
to the right (left).

Right Wing. The portion or element to the right of the center or middl
element of a force deployed for battle.

Rights of War. See Usages of War.

Rigidity of the Trajectory. The assumption that the relation between th
trajectory and a chord thereof of any length, joining the muzzle of the gu
and a target or point of impact, is the same whether the chord be horizonta
or inclined.

Rim, Rimless and Semi-rim Cartridges. A rim is provided at the base o
certain cartridge cases to prevent the cartridge from passing too far int
the bore, to sustain the blow of the hammer or firing pin, and to provide
grip for the extractor. As the rim interferes with automatic feeding, rim
less cartridges or cartridges having a very small rim (semi-rim) are use
in automatic weapons, a grip for the extractor being provided by a groov
just above the base of the case.

Rim Fire Cartridges. Rim cartridges in which the priming composition i
contained in the hollow of the rim, which is struck by the hammer o
firing pin.

Rimbase. The shoulder of the stock on which the breech of the rifle rests
A shoulder of metal forming the junction between a trunnion and the trur
nion band or body of a cannon.

Riot. A breach of the peace committed by an assembly of 12 or more persons

Riot Duty. Military duty in the suppression of rioting.

Rip Cord. The rope running from the rip panel to the car or basket, the pull
ing of which tears off or rips the panel and causes immediate deflation of

balloon or nonrigid airship. The cord, together with the handle and fastening pins, which, when pulled, releases a parachute from its container.

Rise and Run. The vertical rise and the corresponding horizontal distance
of a slope or grade, often expressed as a fraction with the rise as numerator
and the run as denominator.

Rise from the Ranks. To obtain a commission after serving as an enlisted man.

Road Blocks. Barriers in the form of logs, wire, excavations, or specially
constructed traps placed in strategic points on a road to prohibit, limit, or
control traffic thereon.

Road Capacity. The number of vehicles or passengers that can be moved
over a road in a given time, and which depends on the nature of the
road surface, the width of the road determining the number of tracks available, and the class of vehicles employed. A road which is of but sufficient
width, throughout its considered length, to accommodate but one line of the
usual wheeled traffic is called a one-track road; if it is of sufficient width
to accommodate two lines of traffic, it is termed a two-track road; and so on.
The terms one-way road and two-way road are also used, but their correct
use is in designating the direction of traffic rather than the number of lines
of traffic the road can accommodate.

Road Net. The roads, considered together, available for use within a certain
area and placed under traffic control.

Road Reconnaissance. An inspection of a road with a view to its military
use. An examination of a proposed route, preliminary to a location survey.

Road Screen. See Screen.

Road Sketch. A hasty sketch, made on a road, showing the line of the road
and the topography of the adjacent country.

Road Space. The length of the space occupied by an element or command in
column of march on a road. The distance from head to rear occupied by an
organization on the road.

Rocket. A projectile set in motion in open air by the force of explosives
contained within it.

Rocket Board. A board with graduated arc and pointing arm, used for the
purpose of identifying signals from rocket posts.

Rocket Posts. Posts established for the purpose of sending rocket signals.

Rodman Guns. Cast-iron cannon up to 20 inches caliber, mounted on barbette
carriages; formerly extensively used in the United States seacoast fortifications.

Rogues' March. Derisive music played when a soldier is drummed out of a
post.

Roll. A list of the names of all the members of an organization. A long
beat of a drum. An aerial maneuver in which a complete revolution about
the longitudinal axis is made, the horizontal direction of flight being approximately maintained. An angular displacement about an axis parallel
to the longitudinal axis of an aircraft.

Roll Call. The act of calling over a list of names of members of a military
unit in order to ascertain who are present.

Roll of a Drum. The continuous and uniform beating of a drum for a limited
time.

Roller Type Range Finder. A self-contained horizontal base range finder
operating on the coincidence principle.

Rolling Barrage. An artillery barrage that precedes infantry troops at a
predetermined rate in their advance during the attack to protect them and
facilitate their advance. A creeping or jumping barrage.

Rolling Ground. Terrain which has a succession of sloping elevations and
depressions. Undulating ground.

Rolling Kitchen. A kitchen mounted on wheels to accompany troops on the
march or in the field, in which a fire can be maintained while in motion in
order that hot food can be served to troops while on the march.

Rolling Reserves. Reserve supplies carried in the corps and division trains.
In general, any reserve supplies carried in wheeled vehicles.

Rolling Shields. Shields about 3 feet high and 3 feet wide so arranged that
they can be rolled to the front by, and afford protection to one man.

Rolling Stock. The locomotives and cars of a railroad.

Rolling Up the Hostile Flanks. An attack against the hostile flanks which drives them toward the center. Widening a penetration of the enemy's front by attacking to right or left, or both.

Rope Ferry. A ferry in which a raft or boat is drawn along a rope or chain crossing a stream at right angles. If the current is swift it may be utilized to propel the boat by setting the latter at an angle to the current, such an arrangement being called a trail ferry.

Ross-Enfield Rifle. The rifle used in the armies of Estonia and Lithuania. It has a caliber of .305 inches and weighs 9.13 pounds.

Roster. A list of officers and men for duty, together with a record of the duty performed by each. A roll or list. See Guard Roster.

Rotating Band. A copper band encircling a projectile near its base for the purpose of engaging the lands of the rifling and giving the projectile an angular rotation in passing through the bore.

Rotation Coefficients. Quantities entering into such firing tables as involve comparatively long ranges. They are used in computing the effects on range and deflection which are caused by the rotation of the earth.

Rotating Crank. Specifically, the crank by which a slotted-thread breech-block is rotated to lock it or release it from the breech.

Round. One shot discharged by each soldier, gun or cannon of a command. A general discharge of firearms by each member of a body of troops. The ammunition for a single discharge of a piece. A personal inspection of an officer through a particular circuit of ground, to see that all is well.

Round Shot. Spheres of iron or steel fired from smooth-bore guns. Cannon balls.

Rounds. An officer or noncommissioned officer, accompanied by a small party, who visits the sentinels on post in order to ascertain whether they are vigilant.

Rout. An overwhelming defeat. A disorderly flight. To break the enemy's ranks and throw him into confusion. To defeat disastrously.

Route. A course, road, or way. The course followed. The order to march, indicating the way to be taken, and the location of headquarters for each evening.

Route Column. A body of troops which, in close order formation, follows a prescribed route. A close order formation suitable for marching when not in the immediate presence of the enemy; usually column of squads for dismounted troops.

Route March. A march used to conduct troops from one place to another, or for the purpose of assembling the elements of a command in a manner to conserve the comfort of the men, who are not required to keep step, maintain silence, or the like. A march at route step.

Route Order. Route step. The order in which a march is conducted.

Route Sketch. A sketch of a road or other route of travel showing the military features of the terrain for some distance on either side.

Route Step. A step used on a march in which the men are not required to maintain cadence nor silence, and are permitted to carry their pieces at will, provided the muzzles are elevated.

Routine Orders. Military orders covering matters not directly connected with any phase of operations in the field. They include general and special orders, bulletins, circulars, memorandums, etc.

Roving Guns. Detached pieces of artillery assigned to special missions, generally involving fire from a number of positions. They are usually employed to deceive the enemy as to the strength and positions of the artillery.

Roving Light. A searchlight used to search extended areas beyond the reach of fixed lights.

RSOP (Arsop). Reconnaissance, selection and occupation of positions for artillery.

Rucksack. A kind of flexible knapsack or bag carried on the back.

Ruffle. A low, measured beat of a drum, not as loud as the roll.

Ruffles and Flourishes. Ruffles on drums and flourishes on bugles or trumpets used at ceremonies.

Rules and Articles of War. The laws which govern the conduct of all persons in the military service, at all times and in all places.

Rules of Land Warfare. Certain well established rules, both written and unwritten, known also as the laws of war, which are generally applicable to the conduct of war amongst civilized nations. They apply particularly to the treatment of personnel, and their object is to ameliorate the sufferings and hardships of war.

Run. The swiftest gait at which an individual on foot can move, used in combat but not in close-order drill. An airplane flight undertaken in connection with the firing of a portion of an outlined event. A brook or small stream. A runway. The horizontal component or projection of a slope or grade. To move at a run.

Run a Blockade. To evade a blockade and enter or clear from a blockaded port.

Run the Guard. To pass a sentinel or line of sentinels without authority.

Runaway Gun. A malfunctioning of an automatic weapon which causes it to continue firing after the trigger is released.

Runner. A foot messenger. A skid or slide, as on a sled or airplane fitted for moving or landing on ice or snow. One who operates a machine, as an engine-runner.

Runner Tackle. A tackle rigging consisting of a single running block attached to the load, which has a mechanical advantage of 2.

Running Block. A block in a tackle rigging that is not attached to a point of support, and moves when power is applied.

Running End. The free end of a rope, wire or cable.

Running Fight. A battle in which one contestant retreats and the other pursues, the fight continuing meanwhile. A fight in which the enemy is continuously pursued and continuously resists.

Running Part of a Rope. See Tackle Rigging.

Running the Gauntlet. A punishment consisting of running between two ranks of men and receiving a blow from each in passing.

Runway. An artificial landing strip which permits airplanes to land or take-off under all weather conditions. An airfield or airport is usually provided with several runways in various directions to allow for changes in the direction of the wind. See Landing Strip.

Ruptured Cartridge Case. A cartridge case which is torn apart during discharge so that the case or a part of it is not extracted by the mechanism of the weapon.

Ruptured Cartridge Extractor. A tool for extracting ruptured cartridge cases which stick in the bore of a small arm.

Rush. To attack impetuously. To capture swiftly.

Russian Ponton. A very light type of ponton consisting of framework covered with canvas.

Russian Sap. A sap carried forward below the surface of the ground.

S

Sabarcand. A blow gun used by the Malays and other tribes.

Saber, or Sabre. A weapon, similar to the sword, generally having a slightly curved blade, used for cutting and thrusting. The traditional weapon of cavalry in mounted combat, not now used in the U. S. service. To strike, cut or kill with a saber.

Saber Exercise. Exercise with the saber to give suppleness to the wrist and increase the dexterity and confidence of the men in the use of the weapon.

Saber Knot. An ornamental knot attached to the hilt of a saber. A leather thong on the hilt of a saber for attaching it to the wrist.

Saber-tache. A leather pouch hung from a sword belt.

Sack. The pillage or plunder of a captured town or city. To pillage a place. To storm and plunder a town. To ravage.

Saddle. A device placed on the back of a horse to support the rider. A support for a hinge placed in a ponton bridge to allow for variations in the water level. A depression or low ground between two adjacent hills.

Saddlebags. Bags, usually of leather, designed to be attached to the saddle for the purpose of carrying small articles.

Saddlebow. The bow or arch in the front part of the saddle, or the bow formed at the front by the union of the side pieces; the pommel.

Saddlecloth. A cloth which covers the saddle blanket under the saddle. It is usually decorative and bears in its rear corners the insignia of the arm, service, or regiment of the rider.

Saddletree. The frame of a saddle.

Safe Conduct. A written authority by a military commander permitting an individual or number of persons to pass into or out of certain localities, to move goods or supplies, or to carry on trade within prescribed limits.

Safeguard. A written instrument given by the commander of belligerent forces for the protection of an enemy subject or enemy property. A detachment of soldiers posted or detailed by a commander of troops for the purpose of protecting some person or persons, or a particular village, building, or other property.

Safety Angle. The angle of elevation which must be applied to a gun in order to insure that the fire will pass over the heads of friendly troops.

Safety Belt. A belt or harness worn by an occupant of an aircraft and attached to the craft to prevent the wearer from being thrown from his seat under any circumstances.

Safety Factor. An increase in range or elevation to insure that fire will not endanger friendly troops in front. See Factor of Safety.

Safety Fuze. See Time Fuze.

Safety Lock. A device which protects against accidental discharges of a weapon.

Safety Officer. An officer who is responsible that practice firing is conducted within the limits of safety.

Safety Stop. A safety lock.

Sag Paste. An ointment for protection of the skin against vesicant gas.

Sales Commissary Unit. A mobile quartermaster corps organization whose function is to provide limited sales store facilities for troops in the field operating under conditions that make it impracticable for the personnel thereof to patronize a regularly installed sales store.

Salient. A portion of a battle line or fortification which extends sharply to the front of the general line.

Salient Angle. See Angle, Salient.

Salient Places of Arms. The parts of the covered way in front of the salients of the bastions and demi-lunes.

Sally. A sudden offensive movement by the garrison of a besieged place. To issue suddenly, as troops from a besieged place, to attack the besiegers.

Sallyport. A large gate or passage in the wall of a fortified place.

Salute. A mark of respect. A salutation. A sign, token, or ceremony, expressing good will, compliment, or respect. A formal mark of deference paid to superiors in rank by their subordinates, and acknowledged and returned by the superior, made with the hand, rifle, sword, etc. To exhibit a mark of deference to a superior. Discharge of cannon as a mark of deference to a flag, persons of high rank, etc.

Salute to the Union. A salute of one gun or one shot for each state in the Union, fired on July 4 of each year.

Saluting Distance. The distance at which an individual may be recognized and at which personal salutes are supposed to be rendered, usually about 30 paces.

Salvage. The collection of abandoned, captured or unserviceable property in the field or at a post or station, with a view to its utilization or repair. Property so collected. To recover or save.

Salvo. The simultaneous or successive firing of a single shot from each of a number of guns. One round per gun fired in a certain order and with a certain time interval between rounds. Two or more airplane bombs released together.

Salvo Interval. The time interval between the successive shots of a salvo.

Salvo Point. A point of known range and azimuth at which fire from one or more batteries may be directed on short notice.

Sam Browne Belt. A uniform belt with shoulder straps worn by officers, designed by an officer of that name in the East Indian Army.

Sanction. Official approval, permission or authority. A penalty imposed upon a nation for violation of its international obligations.

Sand Shot. Small cast-iron balls, which are molded in sand.

Sandbag. A bag used for revetment purposes, which when filled to about ¾ its capacity with earth, measures about 4¾ x 10 x 19 inches.

Sandbag Revetment. A revetment made with filled sandbags.

Sandtable. A relief model of an actual or imaginary terrain, made in sand in a box or curbed table; used for instruction in topography, musketry, tactics, etc.

Sanitary Discipline. The observance of and enforcement of sanitary regulations.

Sanitary Squad. A detachment of medical corps men charged with the inspection of sanitary conditions in an area.

Sanitation. The use or application of sanitary measures. See Military Sanitation.

Sap. A narrow trench which is extended in the desired direction by digging away the earth at its head, from within the trench itself, and throwing it to the front and exposed flank as a cover for the working party. To execute a sap. See Single Sap; Double Sap.

Sap-fagot. A fascine, about 3 feet long, used to close the crevices between gabions.

Saphead. The end of a sap at which work is progressing.

Sapper. One who is employed in sapping. A soldier in the Royal Corps of Engineers of the British Army. An engineer soldier.

Sapping. A method of digging trenches by continually advancing the head of the trench under cover of filled gabions, sandbags, etc.

Sap-roller. A gabion, about 7½ feet long, filled with fascines, which is rolled before a sapper for protection against enemy fire.

Sap-shield. A steel shield mounted on wheels, used for the protection of sappers.

Sarisa, or Sarissa. A Macedonian pike which was about 24 feet long.

Scabbard. A sheath in which a rifle or the blade of a sword, or other cutting or thrusting weapon is inclosed when not in use. See also Holster.

Scabbard and Blade. An honorary military society, established at civil educational institutions maintaining a senior unit of the R. O. T. C. Its members are selected from the senior class.

Scale. A line or rule of definite length divided into a given number of equal parts, and used for the purpose of measuring other linear magnitudes. To climb by means of a ladder. To clamber up. See Graphical Scale.

Scale Armor. Armor consisting of small steel plates riveted together, or fastened upon leather or cloth, resembling the scales of a fish.

Scale Extension. A division of a graphical map scale, to the left of the zero point, which is subdivided into fractional parts.

Scale Line. A line on a vertical air photograph that is used in determining its scale.

Scale of a Map. The ratio of any distance on a map to the corresponding distance on the ground. The scale of a map is expressed in three ways: as a representative fraction; in words and figures (e.g., 3″ equals 1 mile); and as a graphical scale.

Scaling a Map. Determining distances on a map.

Scaling Ladders. Ladders designed to be used in scaling walls.

Scarp. The front slope of a rampart, or rear slope of the ditch in front of it. Any steep slope. To cut a steep slope.

Scarp Gallery. A gallery or casemate installed in the scarp for the defense of the ditch.

Scarp Wall. A retaining wall at the foot of a scarp, or a detached wall in front of a scarp.

Schedule. A detailed plan showing the date, hour, duration of periods, and the character of instruction to be imparted. A plan for coordinated fires in battle.

Schedule Fires. Planned fires, generally executed according to a time schedule or upon a signal or call from the supported troops.

Scheme of Maneuver. A plan for the employment of the available forces in an offensive action, contemplating a decisive blow to be made effective by teamwork, proper combination of fire and movement, and utilization of the terrain to the best advantage. The detailed plan of an attack.

Schneider Gun, Howitzer. Ordnance originally manufactured at the Schneider works in France; used in the French, Russian, American, Japanese, and other armies.

School of Fire. A term sometimes used to describe a school where marksmanship or gunnery is taught.

Science of War. The accumulation of classified knowledge pertaining to war.

Scimitar. A large curved sword used among Eastern nations.

Score. A prescribed number of consecutive shots fired in target practice. To record or register the number of points made in firing.

Score Book. A book in which the target values of successive shots are recorded, together with other pertinent data relating thereto.

Scout. A man specially trained in shooting, in using ground and cover, in observing and in reporting the results of observation. A man who gathers information in the field. To reconnoiter a region or country to obtain information of the enemy or for any other military purpose. To act as a scout.

Scout, Air. A soldier posted to give timely warning of the approach of hostile aircraft.

Scout Car. An armed and armored motor vehicle used primarily for reconnaissance.

Scout Pair. Two men in each rifle squad specially trained and designated as scouts.

Scout Planes. Airplanes used to reconnoiter enemy positions and territory by direct observation and by photography. Observation planes.

Scouting. The operations of a scout or scouts.

Screen. A line of troops or other means intended to prevent the enemy from obtaining information of the troops or terrain in rear of the screen. Smoke released on the ground or in the air to obscure from hostile view. Panels, not camouflaged, placed in front, at the side of, or above a road or any military establishment, to prevent enemy observation. Any device which prevents enemy observation. A sieve or perforated cylinder used to separate sizes of shot, broken stone, gravel, etc. To cover or protect a command usually by a detachment thereof, in such a manner as to conceal its maneuvers. To place a screen or conceal by means of screens.

Screening Agent. A chemical agent which, when released in the field, produces an obscuring or screening smoke cloud. Screening agents are released by means of candles, bombs, grenades, shells, or the exhaust of an airplane.

Screw, Interrupted. See Interrupted Screw.

Screw Picket. An iron picket having a spiral point by which it is screwed into the ground, used as a support for wire entanglements.

Seacoast Artillery. All artillery, fixed, tractor-drawn, and railway, employed either within or without a harbor defense, organized primarily for defense against hostile naval vessels. It is classified, according to caliber, as primary and secondary armament.

Seacoast Carriages. Carriages for seacoast artillery, usually immobile.

Seacoast Mortar. A heavy mortar used in seacoast defenses.

Seal. A closure, as one to prevent the escape of gas or liquid. An authentication. An insignia or coat of arms. The great seal of the United States consists of the eagle, shield, stars and motto, and is used in cap insignia and certain other ornaments.

Seaplane. Any airplane designed to rise from and alight on the water. This general term applies to both boat and float types, though the boat type is usually designated as a flying boat.

Sear. A pivoted pawl or latch in the gunlock of a small arm or automatic weapon, operated by the trigger, which holds the hammer or firing pin at full or half cock until released.

Search. To distribute fire in depth.

Searching a Bracket. Covering a determined bracket with fire for effect.

Searching Light. See Pilot Light.

Searchlight. An electric arc light with a parabolic reflector (mirror) which produces a concentrated beam, together with the mounting by which the beam can be pointed in any direction; used to search for and illuminate targets for night firing. Searchlight equipment for antiaircraft fire includes the searchlight, power unit, sound locator, comparator, and distant electric controller.

Seat of War. The country or countries in which a war is being waged.

Seated. A piece is said to be seated when the trail spade is embedded in the ground.

Seating Distance. The distance from the face of the breech to the base of the projectile when the latter is in position for firing.

Second Classman. The lowest grade of qualification in rifle firing. A cadet in his third year at the U. S. Military Academy, corresponding to a junior in college.

Second Class Pistol Shot. The lowest grade of qualification in pistol firing.

Second Lieutenant. The lowest grade of commissioned officer in the United States Army.

Second Line. A second defensive position, usually at a sufficient distance from the principal position so that the attacker must displace the bulk of his artillery forward in order to attack the second position. A reserve position.

Secondary Armament. All seacoast guns of less than 12 inches caliber.

Secondary Attack. See Holding Attack.

Secondary Target Area. Any target area assigned as a secondary fire mission to a weapon or fire unit to be engaged when it is not required to fire on its primary target area.

Second-in-command. The second in rank in certain organizations, who usually acts as executive officer, and who succeeds to the command in case of the death, removal, or incapacity of the commander.

Secret. A document will be marked "Secret" when the information it contains is of such nature that its disclosure might endanger the national security, or cause serious injury to the interests or prestige of the nation, an individual, or any governmental activity, or be of great advantage to a foreign nation.

Section. A subdivision of a platoon, especially in combat organization, there being usually two sections to a platoon. A subdivision, as of an office or other activity. A cross section or profile. (The rifle platoon is not now divided into sections.)

Section Column. An infantry section in column of twos; an artillery section, vehicles in single file, one section in rear of the other, each caisson in rear of the piece of its section.

Section Journals. Permanent records kept by the various staff sections, which contain briefs of all important messages sent and received, and notations of periodic reports, orders, and similar matter or data pertaining directly to the duties of the section.

Section of Field Artillery. A gun section consists of a gun and its caisson, manned, horsed (or provided with motor transport), and equipped. A caisson section consists of two caissons similarly equipped.

Section Reports. Periodic reports based on available information, rendered by each section of a staff.

Sector. A term of wide application. The area within which the combat functions or responsibilities of a command are exercised, when acting defensively. One of the areas into which a defensive position or system is divided. An area occupied by a unit in defense, or the frontage of such an area. A portion of the front assigned to a support or other subdivision of an outpost for observation, reconnaissance or resistance; or to an observation or snipers' post for observation or sniping. The area to be covered by the fire of a gun or fire unit. One of the subdivisions of a coastal frontier.

Sector Boundary. A line limiting a defensive sector, especially on its right or left, and defined by reference to prominent features of the terrain, e.g., roads, streams, towns, buildings, etc.

Sector, Defensive. A lane or belt of terrain, normal to the front, assigned to a unit in defense. See Sector.

Secure. To gain possession of a position or terrain feature and make such dispositions as preclude any danger of its destruction, obstruction, or loss. To provide security for. To protect.

Securing Latch. A latch on the face of the breech of a cannon which secures the breechblock tray in the loading position.

Security. All measures taken by a command to protect itself from observation, annoyance, or surprise attack by the enemy, and to obtain for itself the necessary freedom of action. The protection resulting from such measures.

Security Detachments. Fractions of a command that are assigned the primary mission of protecting the main body; they include outposts, advance, rear and flank guards, etc.

Security Measures. Measures of any kind taken by a command to protect itself from observation or attack.

Security on the March. The measures taken by a marching column to protect itself from observation or surprise attack; include advance, rear and flank guards.

Security Patrol. A combat patrol, or a patrol sent out from the main body, advance or rear guard, flank guard, or outpost, to add to the security of the main body or of other security groups.

Seize. To occupy a position before the arrival of the enemy. To lay hold of, or take possession of, suddenly or forcibly. To lash the free end of a rope back on the standing part.

Seizing. Lashing the free end of a rope back on the standing part, as in an anchor knot or hawser bend. A lashing used for this purpose.

Selection. A method of promotion by which officers or noncommissioned officers are selected for advancement according to their merits or supposed merits, instead of by seniority in grade. In the United States Army promotions below the grade of second lieutenant and above that of colonel are normally made by selection.

Selective Service. The process whereby, under the law, the personnel required for the military forces of the United States in time of war is selected from the population (unorganized militia) in accordance with a prescribed plan and inducted into the service.

Selectivity. The degree of ability of a radio receiving set or station to receive the impulses to which it is tuned and no other.

Selector. A device placed in a distribution box which enables the operator on shore to fire, at will, any one or all of the submarine mines of a group. A similar device mounted on the control panel in the casemate.

Self-contained Horizontal Base Range Finder. A range-finding instrument in which a short base is contained within the instrument itself. See Horizontal Base System, Coincidence Type; Stereoscopic Range Finder.

Self-Loader. See Semi-automatic Gun.

Self-propelled Mount. A gun and carriage mounted on a self-propelled chassis, usually of the track-laying (caterpillar) type.

Semaphore. A method of visual signalling by the use of two small flags, one in each hand. An instrument with arms for signalling. To signal by semaphore.

Semi-automatic Guns. Guns in which the force of the powder gases or of recoil is ultilized to extract the empty case, reload, and cock the piece. Self-loading small arms.

Semi-automatic Rifle. A rifle which operates semi-automatically. Many automatic rifles can be operated semi-automatically. See also Change Lever.

Semi-fixed Ammunition. A round of ammunition in which the propelling charge and primer are assembled in a single metal container, in which the projectile is loosely fitted and from which it may readily be removed.

Semi-permanent Fortifications. Fortifications combining certain features of both field and permanent works.

Semi-steel Shell. A shell made of semi-steel (cast iron) and used both for high explosive and gas.

Senior. One who is above or ranks another. More advanced in rank or office. Superior in grade or rank.

Seniority. Priority in rank based on date of commission or appointment in grade in the case of an individual, and on date of organization in the case of an organization or branch.

Sense. The location of a point of impact, burst or splash of a projectile with reference to the target, as over or short, right or left, lost or doubtful, without regard to the magnitude or actual amount of the deviation. To determine by observation or spotting the sense of an impact or splash.

Sensing. The process of determining the sense of a shot.

Sensing by Rule. In lateral conduct of fire, the determination of the range sensing from the deviation, when the deflection is correct or nearly so.

Sensing, Forced. Under certain conditions of lateral conduct of fire, a sensing for range or deflection which indicates the sense of the other element.

Sensitive Point. A point which is of particular tactical importance and hence, apt to be subjected to fire; as a road junction, observation post, etc.

Sensitized Paper. Paper treated with actinic chemicals, used in making photographic prints, blue prints, photostats, etc.

Sentinel, Sentry. A soldier posted and performing active guard duty.

Sentry Box. A place of shelter for a sentry in inclement or hot weather.

Sentry Go. Formerly, the call for changing the guard. Duty as a sentry. Guard duty. A slang term signifying the time one is on duty in the front line trenches.

Sentry Squad. A squad posted for security and information with a single or double sentinel in observation, the remaining men resting nearby and furnishing the relief for the sentinels. An outguard of one squad. (Old.)

Separate Arms. The various branches of combat troops, as, the infantry, cavalry, artillery, signal corps, engineers, and air service.

Separate Battalion. A battalion which is not organically a part of a regiment, and designated as such by the War Department.

Separate Brigade. A brigade operating independently, and designated as such by the War Department.

Separate Company. A company which is not organically a part of a regiment, separate battalion, or similar unit, as an ordnance company, ponton company, etc.

Separate Regiment. A regiment not organically a part of a division or separate brigade.

Separate-loading Ammunition. A round of ammunition in which the primer, propelling charge, and projectile are separate units, and are separately loaded into the piece.

Serbian Barrel. A cask used for steam disinfestation of clothing.

Sergeant. A noncommissioned officer next in grade above a corporal. In the United States Army the grades are sergeant, staff sergeant, technical and first sergeant, and master sergeant.

Sergeant Major. A noncommissioned officer who performs important duties as assistant to an adjutant.

Sergeant of the Guard. The senior noncommissioned officer of the guard.

Serial. One or more march units, preferably with the same march characteristics, placed under a single commander for march purposes.

Serial Number. A number assigned to an officer or soldier when commissioned or enlisted, which never changes and is never assigned to another. The identification number of any order, bulletin, message, etc., in a series.

Serried. Formed in successive lines or ranks at close distance.

Serum. A prophylactic or antitoxin that prevents disease; e.g., antitetanic serum. See Prophylaxis.

Serve. To be in the service or do duty. To wrap the cut end of a rope with wire or cordage to prevent unraveling.

Service. Every kind of military duty performed by an inferior under the influence or command of a superior. The period during which a man has done duty, or followed the military profession in an active way. The act of serving the state in peace and war. The duty which a military person may be called upon to perform. The army or navy. Presentation to an accused of a copy of charges and specifications alleged against him. See also Services.

Service Band. A band of frequencies allocated to a given class of radio communication service.

Service Calls. Prescribed calls sounded on the bugle, drum or other instrument. See Bugle Calls.

Service Charge. The full charge of a gun, as used in battle.

Service Chevron. A gold chevron authorized for wear, in the prescribed place on the left sleeve of the blouse or other military coat, one for each six months military service in a theater of operations during the World War.

Service Colors and Standards. The service national color or standard prescribed to be carried at drills, on marches, and on all other occasions of service other than that for which the silk color or standard is prescribed.

Service Command. The military personnel designated for a corps area in the execution of the general mobilization plan.

Service Company, Troop, or Battery. An organization composed of a headquarters, a headquarters platoon, and a transportation platoon. It comprises the personnel charged with regimental supply and transportation.

Service Elements. One of the essential elements of military organization, the others being command and combat. The technical, supply, and administrative services necessary to maintain the fighting efficiency of the combat elements, as well as to carry out any assigned service mission.

Service Flag. In the United States during the World War, a kind of flag displayed from houses, places of business, clubs, etc., to indicate the number of members who were serving in the armed forces, or who had died in the service.

Service Kit. The field kit and the surplus kit.

Service of Supply. The service of supply, replacement, hospitalization, etc., in the theater of operations. The service of the Communications Zone, one of the main subdivisions of the United States Army under the reorganization of 1942; it includes all branches of the army classed as the services.

Service of the Interior. The service which supplies the commander of the troops in the theater of operations, with the personnel and material means necessary for the accomplishment of his mission.

Service of the Piece. The operation of a team-served weapon, including supply of ammunition, loading, laying and firing, conducted in routine fashion, each man having certain definite duties.

Service Practice. Includes all practices fired with the service or target practice ammunition, and all searchlight practice. Firing artillery ammunition for training purposes.

Service Record. The complete military record of an individual enlisted man, kept on a special form provided for the purpose.

Service Ribbon. A small, narrow ribbon worn on the uniform in place of a service medal by persons authorized to wear the latter.

Service Stripe. A stripe worn on the left sleeve of the blouse or other military coat for each full enlistment completed by the soldier.

Service Targets. The various targets used in training.

Service Trains. Trains which transport and distribute supplies to within reach of field and combat trains. They include the division and higher trains, the artillery brigade ammunition train, the ordnance and service companies, certain vehicles of the medical regiment, and vehicles of the air service.

Service Uniform. The uniform prescribed for ordinary wear in the military service.

Service Units. Units, usually organic parts of larger units, whose functions are to provide for the supply, transportation, communication, evacuation, maintenance, construction, and police of the larger unit as a whole.

Serviceable Property. Property which is of the model in current use, or reserved for future use, and is in proper condition for its purposes.

Services. The noncombatant branches of the army. Those components other than the arms. In the United States Army, they include: Adjutant General's Department; Quartermaster Corps; Inspector General's Department, Judge Advocate General's Department; Finance Department; Medical Department; Ordnance Department; Chaplains.   The Engineers, Coast Artillery and Chemical Warfare Service are both combat and supply services.   In the latter capacity they are also under the commanding general of the Service of Supply.   The armed services, i. e., the Army, Navy, and Marine Corps.

Serving Table.   A table for keeping a supply of projectiles convenient to the breech during firing.

Sesquiplane.   A form of biplane in which the area of one wing is less than half the area of the other.

Session.   The actual sitting or assembly of a court-martial, or other tribunal or board.

Set-back.   The relative movement to the rear of a free plunger or striker, or the compression of a spring inside of a fuze when a projectile is discharged; a means by which a fuze is armed and functions.

Set-forward Point.   A point in the course of a target at which it is calculated the target will arrive at the end of the prediction interval plus the time of flight of the projectile.

Set-forward Ruler, Chart, Scales.   A slide rule, a chart, a set of scales, used to locate the set-forward point.   See Prediction Scales.

Setter.   A member of a detail who sets data on a gun or instrument; e.g., altitude setter, rate setter, fuze-range setter.

Setting-up Exercises.   Gymnastic exercises used in the drilling of soldiers to give them an erect carriage and coordination of muscles.

Settling Shot.   A shot fired for the purpose of settling a mobile gun mount firmly in position, driving the trail spade into the ground.

Set-up.   Military bearing.   The military air and carriage of the body which an individual acquires as the result of military training.   Erect and soldierly bearing.

Seventy-five.   A popular name for the 75 millimeter field gun.

Seventy-five Percent Zone.   See Beaten Zone; Dispersion Diagram; Zone of Dispersion.

Shaft.   A vertical passage or portion of a mine.   A passage for light or ventilation.   An elongated missile weapon, as an arrow.   The stock or stem of a spear, javelin, or the like.   A spire, steeple, or column.   A cylindrical bar of steel which transmits power by torsional action.

Shako, Shacko.   A high military headdress of Hungarian origin, sometimes made of fur, worn by certain military organizations on occasions of ceremony.   A drum major's headdress.

Sham Battle.   A pretended battle for exhibition rather than for training purposes.

Sharpshooter.   One who is skilled in the use of the rifle.   A qualification in pistol or rifle shooting below that of expert.

Sheaf.   A cone containing all the trajectories of a series of shots.   The planes of fire of two or more pieces of a battery considered as a group. Sheafs are designated as parallel, narrow, wide, open, converged, or crossed.

Sheaf Ranging.   The simultaneous firing of two or more guns with their range settings differing by equal increments and increased or decreased from the right by the specified increment in yards, observing the relative positions of target and splashes, and making corrections from these observations.

Shears.   A device consisting of two pieces of timber fastened together near the top, and secured by guy lines, used in lifting guns and other heavy weights.

Sheath.   A case for a sword, saber, bayonet, etc.   A scabbard.

Sheathe.   To put in a sheathe or scabbard.

Shell.   A projectile having a cavity filled with high explosive or explosive and chemical (special shell), with a fuze to produce detonation.   Upon bursting it scatters its fragments and the contained chemical.   The container or case of a cartridge, especially an empty case.   To direct artillery shell-fire upon.   To bombard.

Shell, Armor Piercing.   See Armor Piercing Projectile.

Shell Bags.   Bags attached to machine guns in tanks for reception of the empty cartridge cases.

Shell Bullet.   An explosive bullet.

Shell Bursters. Stones or concrete tiles placed at or near the outer surface of a shelter or covered emplacement to cause shells to burst before they can penetrate deeply.

Shell Filler. Any explosive or chemical used as a bursting charge for a shell.

Shell Fougasse. A fougasse loaded with a shell.

Shell Hole. The crater formed by the explosion of an explosive shell.

Shell Jacket. A common term for an undress, tight-fitting military jacket, short in the back.

Shell Shock. A mental or nervous depression caused by the noise and violence of battle.

Shell, Tracer. A projectile containing in its base a chemical which, on ignition, burns with a bright light or emits smoke, so that the flight of the projectile may be observed.

Shellfire. The firing or shooting of shells.

Shellhole Emplacement. An emplacement for a weapon, adapted to, or made to resemble a shell crater.

Shelling. The act of firing shells upon a position or place. Bombardment.

Shellproof. Capable of resisting bombs or shells.

Shelter. Any form of concealment from view or protection against the elements, or the fire of weapons. That which covers or defends. A screen. Protection. To afford or provide shelter. To screen or cover from notice.

Shelter Pits. Pits for the protection of riflemen.

Shelter, Protected. A shelter which affords a degree of actual protection against fire.

Shelter Tent. A small tent made of two pieces of canvas buttoned together, one piece being carried by or for each man.

Shelter Trenches. Hasty trenches usually constructed in the presence of the enemy to provide shelter from fire and to permit riflemen to fire in the prone position.

Shelters, Protected, Classification of. Protected shelters are classified as follows: (1) According to degree of protection, as splinterproof, light, light shellproof, heavy shellproof; (2) According to method of construction, as surface, cut-and-cover, concrete, cave or dug-out.

Shield. A kind of defensive screen, of various shapes and sizes, borne upon the arm to ward off missiles and the strokes of cutting and thrusting weapons. Anything that protects something. Defense. Shelter. A protection. A screen of armor plate, usually attached to the carriage, protecting an otherwise exposed gun and its crew against small arms fire, shrapnel, or shell fragments. A buckler. To cover with a shield. To cover from danger. To defend. To protect.

Shift. The difference in direction between two targets, or a target and a registration point, etc., as measured at the gun. A change of fire from one target to another. To execute a shift. To change position. See Angle of Shift; Fire, Shift of; Fire, Transfer of.

Shipping Ticket. A form of receipt accompanying a shipment to a supply officer, to be signed and returned to the shipper. A combined invoice and receipt.

Shock Action. Actual hand-to-hand combat between opposing troops.

Shock Tactics. Tactics in which shock action is employed.

Shock Troops. Troops especially selected and trained for attack and assault. First class troops.

Shoe-fitting Machine. A machine for determining or testing the proper size of shoes.

Shoot. The discharge of a missile. The reach of a shot. Range. A competition in firing. To discharge a missile from a weapon. To kill or wound with a firearm.

Shop Companies. Organizations which operate machine shops or manufactories.

Shore Cable. An armored electric cable connecting a mine distribution box with the control station on shore.

Shore Commander. An officer who takes charge of all troops after the immediate vicinity of the landing point has been cleared of the enemy, and of all utilities in connection with the landing.

Short. A projectile which strikes short of the target. A short circuit. Having less property or funds than that for which one is responsible or accountable. Brittle.

Short Fuze. A delay or non-delay percussion point fuze, which is shorter than a super-quick fuze.

Short Maneuver. A maneuver which lasts from a few hours to two days and is designed to illustrate only one phase of an operation, or the application of certain tactical principles.

Short Range. See Ranges, Classification of.

Shortage. A deficiency in funds, property or supplies.

Shot. The discharge of a weapon. A projectile designed to be discharged by a firearm by the force of an explosive. A non-explosive projectile. A missile weapon. The flight of a missile. The distance a missile can be thrown. Range. A marksman. Act of shooting. The firing of a single load from a single gun or mortar. An ancient name applied to a man armed with a missile weapon. To load with shot, cartridge or shell.

Shot Cartridge. A cartridge containing powder and small shot. A shotgun cartridge.

Shot Group. The pattern made on any target or area, either horizontal, vertical or inclined, by all the shots fired at it.

Shot Hoist. An elevator for raising projectiles from the magazine level to the loading or truck platform. An ammunition hoist.

Shot In. A term which signifies that a battery has adjusted its fire upon such a variety of objectives within its sector as to be able to switch its fire to a new objective appearing anywhere in its sector without the necessity of further adjustment. A battalion or regiment is said to be shot in when its batteries are shot in.

Shot Tongs. Tongs for handling projectiles.

Shot Tower. A lofty tower for making shot, by dropping from its summit melted shot metal in slender streams; the shot are then sorted by screening.

Shot Tray. See Loading Tray.

Shot Truck. A truck, usually three-wheeled, on which a projectile is brought from the ammunition hoist to the breech of a cannon. It has a tongue which is inserted in the breech, over which the projectile slides, and is sometimes provided with trays to hold the powder charge.

Shotted. Loaded with a shot or projectile.

Shoulder. The upper part of a sword blade. An epaulement. The unpaved portion of a road between the pavement and the ditch.

Shoulder Angle. The angle formed by the meeting of the face and flank of a bastion.

Shoulder Belt. A belt passing over the shoulder.

Shoulder Knots. Ornamental loops of gold cord worn on the shoulders of full dress coats.

Shoulder Strap. A narrow cloth strap on the shoulder of a blouse or other military coat, to which is attached insignia indicating commissioned rank.

Shrapnel. A projectile equipped with a time fuze and containing a base charge and a number of metal balls set in a smoke-producing matrix. Upon functioning of the time fuze the balls are expelled forward without rupturing the case. It is especially used to attack personnel. Sometimes called a flying shotgun.

Shrapnel Cone. The cone formed by the bullets and fragments of a shrapnel on explosion of the bursting charge.

Shuttle, Shuttling. When the organic and assigned transportation of a unit is insufficient to move at once all the personnel and materiel of the unit, the movement may be accomplished by successive trips of the transport. Sometimes the foot troops may march part of the distance.

Sibley Stove. A small, conical stove, without grate bars, used in heating tents or huts.

Sibley Tent. A light, conical tent, about 16 feet in diameter, easily pitched, and erected on a tripod holding a single pole.

Sick Call. A daily call for the formation of the sick squads who are then reported to the medical officer for examination and treatment.

Sick Leave. A leave granted to an officer, under certain conditions, to promote convalescence, which does not count against annual leave.

Sick Rate. The percentage of men in an organization incapacitated for duty at any particular time, because of sickness.

Sick Report. A book in which the names of the sick belonging to a company, or similar organization, are entered, with the dispositions of cases.

Side Arms. Weapons, which, when not in use, are worn at the side, as swords, bayonets, daggers, pistols, etc.

Side Band. A band of radio frequencies on either side of the carrier frequency.

Side Car. An attachment to a motorcycle, usually accommodating one passenger.

Side Step. A step to the right or left.

Side Wind. A wind blowing at, or nearly at, right angles across a front. A cross wind.

Side Slip. As applied to an airplane, to slide sideways toward the center of curvature in making a turn. See also Skid.

Side-slip Landing. A type of landing in which an airplane is allowed to slip sideways, during part of the glide to the surface. The angle of glide is thus increased without increasing the speed, enabling the landing to be made in a shorter distance than would otherwise be necessary.

Siege. A systematic and more or less deliberate attack upon a fortified place, in which the besieger aims to invest the place and capture its fortifications by regular approaches. The encampment of a besieging army. A continued attempt to gain possession.

Siege Artillery. See Artillery, Siege.

Siege Gun, Howitzer, Mortar. Pieces of ordnance originally designed for the attack and defense of fortified places, superseding the ancient ballistic machines, and extensively used in early European wars. The earliest mortars were designed especially to attack those portions of fortifications that could not be reached by weapons of flatter trajectory. The terms are now relatively little used, as sieges of the kind that were common in earlier European history are now of infrequent occurence.

Siege Train. Artillery, and its ammunition and transport, adapted for use in attacking fortified places. See Artillery, Siege.

Siege Works. All the engineering devices resorted to by besieger and besieged in the attack and defense of a fortified place.

Sight. A device to aid in the aiming of a firearm. To give proper elevation and direction to a firearm. To point at a target. See Sight Leaf; Panoramic Sight; Telescopic Sight.

Sight Bracket, Seat, Socket, Stand, Standard. The support by which a movable sight is attached to a gun or carriage.

Sight Defilade. Defilade from view of the enemy. See Defilade.

Sight Extension Bar. An extension rod placed between a sight and its seat or bracket to raise the sight to the desired level.

Sight, Forward or **Front Area, Antiaircraft.** A device for attachment to firearms, used in front areas to aid in firing at hostile aircraft.

Sight Leaf. A movable portion of the rear sight of a rifle or machine gun, which can be raised to a perpendicular position, and which is provided with a slide which can be set to the desired range. The angular width of the sight leaf as viewed with the eye at the comb of the stock, taken as 50 mils, is used in measuring lateral distances in target designation; commonly called sight or sight width.

Sighting and Aiming Exercises. Preparatory exercises intended to teach a soldier how to aim a rifle correctly.

Sighting Bar. A bar on which an eyepiece, rear sight, front sight and target are represented by metal or cardboard plates; used for instruction in sighting and aiming.

Sighting Plane. A vertical plane containing a line of sighting.

Sighting Platform. A platform on a very large gun carriage on which the gunner or gun pointer stands.

Sighting Shot. A trial shot. A shot fired to determine whether the sights are properly adjusted or attached.

Sighting Triangle. Three successive points of aim on a sighting target, indicated by a man looking through the sights of a gun fixed in position. The three points should be coincident, and the size of the triangle is an inverse measure of the consistency of sighting. Sighting triangles are made at both short and long ranges.

Sight-leaf or **Sight-width Measurements.** See Sight Leaf.

Signal. Any prearranged means of communication or of transmitting information, other than by ordinary conversation in close contact; signals are transmitted through the senses of vision, hearing, or touch. The means whereby a message, intelligence, or order is conveyed in communication.

Signal Code. A list of signal symbols, each having an arbitrary conventional meaning assigned to it.

Signal Communication Agencies. Message centers, messengers, radio, visual, and wire agencies.

Signal Communications. All means and methods employed to transmit information, orders, and dispatches, except the use of mail or direct personal agency not a part of signal communication. Communication by means of various technical devices.

Signal Corps. That arm of the service which has general charge of all military signalling and communications, including the procurement, storage, and distribution of the equipment therefor. Methods for conveying orders or information to a distance were used by the ancients; Polybius (204-122 B.C.) describes two methods used by the Greeks. Marine signalling developed at a much earlier date than land signalling; communication between elements of an army by visual signals being of comparatively recent date. Actual signals were of course used throughout the ages in land warfare by means of flags, banners, rockets, lanterns, etc., or transmitted to long distances by systems of wooden windmills, lights, etc. But surprisingly little progress was made in developing methods and instruments for the systematic exchange of military information by means of signals until about the middle of the 19th century. The construction of a practical telegraph instrument by Morse in 1832-1835 served as an incentive and in the years immediately preceding the American Civil War efforts were made to introduce signalling into the United States Army, which was among the pioneers in this matter. As a consequence, Assistant Surgeon Albert J. Myer was appointed first signal officer of the army on July 2nd, 1860. Although starting from nothing the services of the signal corps in the Civil War were of great value. To the telegraph, flags, lights and rockets then in use have since been added the heliograph for long distance work, the semaphore system, various kinds of lights for night signalling, the telephone, and the radio. With the development and improvement of these means the signal corps has become as we know it today.

Signal Equipment. Equipment of any and every nature designed for and used in signal communication.

Signal Flags. Flags designed for use in signalling, by wig-wag, semaphore, and display.

Signal Kit. A set of signal equipment, usually flags, and carried by one man. A flag kit.

Signal Lamps. Lamps designed to transmit signals or messages by projecting flashes of light by means of mirrors and occulting shutters.

Signal Projector. See Projector.

Signal Rocket. A rocket used for transmitting a prearranged signal.

Signal Tank. A tank used primarily for signal purposes, often in the axis of signal communication.

Silencer. A device which is attached to the muzzle of a firearm to reduce the noise of discharge.

Silhouette. A form of field target showing, in outline, the figure of a man, prone, kneeling or standing, and colored black or olive drab.

Sill. A horizontal member of wood, stone, etc., forming part of the foundation of a structure. The lower, horizontal member of a framed trestle.

Silver Star. The silver star is awarded to persons who, while in the military service of the United States are cited for gallantry in action, which citation does not warrant the award of the Medal of Honor or Distinguished Service Cross.

Simple Standing Trench. A hasty trench or individual rifle pit which affords cover for men firing in a standing position.

Simulated Fire. The procedure of going through all the motions of firing but without the use of ball cartridges.

Simultaneous Movement. A drill movement executed by all or corresponding elements simultaneously. See also Successive Formation.

Single File. A column of men marching one behind the other, or in file.

Single Rank. A line of soldiers side by side.

Single Sap. A sap having a parapet or protection at its head and along one side only.

Site. The position occupied by or selected for a gun, work of fortification, building or any other establishment. The position of a target. The height of a target above a gun or directing point. An angle of site or position.

Sitogoniometer. A very small instrument for measuring horizontal and vertical angles.

Sitting Position. A prescribed firing position in which the rifleman sits down.

Sitting or Squatting Trench. A hasty trench or individual rifle pit which affords cover for men firing in a squatting position.

Situation. All the conditions and circumstances, taken as whole, which affect a command at any given time, and on which its plans must be based. They include such items as the positions, strength, armament, etc., of the opposing forces and any supporting troops, considerations of time and space, the weather, terrain, etc., and the mission to be accomplished. A consideration of these conditions and the possible courses of action open constitute the estimate of the situation.

Situation Map. A map on which is shown graphically the situation at any time, or some element thereof. Each command, staff section, etc., maintains one or more situation maps setting forth the particular information in which they are interested.

Situation Reports. Reports intended to keep higher headquarters and neighboring units informed as to the progress of or changes in the situation.

Six-shooter. A revolver having six chambers. A pistol having six barrels.

Size. To arrange men in ranks in order of height, from the right or left.

Skeleton. The framework of anything. A reduced strength of an organization due to sickness, casualties in the field, or other reason.

Skeleton Drill. A drill with a skeleton organization for the instruction of officers only.

Skeleton of the Terrain. See Critical Point; Master Lines.

Skeleton Organization. An organization which has its complement of officers, but no enlisted men, or a few in key positions only. It is maintained in this status with a view to expansion to full strength when conditions require or permit.

Skeletonize. To reduce the number of officers and men in an organization far below the prescribed strength.

Sketch. A hastily constructed map of a limited area or route of travel. To draw a hasty map of a route or area. See Outpost, Panoramic, Place, Position, Route Sketches.

Sketching Board. The small plane table, mounted on a tripod or held in the hands, used in making military sketches.

Sketching Case. A small and compact case containing a sketching board and tripod, together with the instruments used in sketching.

Sketching Pad. A small pad sometimes used for mounted sketching.

Skid. A runner used as a member of a landing gear and designed to aid aircraft in landing or taxying. A slide sidewise from the center of curvature when turning, as of an airplane or vehicle. A timber laid horizontally or on an incline to support a gun or other heavy object which is being moved by sliding or rolling. To move by the use of skids. The act of skidding.

Skirmish. An unimportant combat between small forces. A combat between detachments of hostile armies. To engage in a skirmish. To fight as skirmishers.

Skirmish Drill. Evolutions in extended order, as employed in battle. Extended order drill.

Skirmish Line. A line of skirmishers. A line of soldiers in widely extended order in advance of the main line of battle.

Skirmishers. Soldiers, either mounted or dismounted, deployed in line and in extended order in drill or attack.

Skirmisher's Trench. A hasty trench or individual rifle pit affording a degree of shelter and permitting the delivery of rifle fire in a prone position. A prone or lying trench.

Skoda Gun or Howitzer. A gun made at the Skoda Works in Czechoslovakia, in various calibers.

Slack. Play or backlash in gears or other machinery. The easy movement of a trigger when pressure is first applied. Looseness, as of a guy rope, tie rod, etc. Not tight or taut.

Slacker. A person who slacks a duty or obligation to his country, as by attempting to evade military service.

Slant Range. In antiaircraft fire, the hypothenuse of the right triangle in space, of which the legs are the horizontal range and the altitude. The actual linear distance from the gun to the target.

Slashing. An obstacle formed by cutting down a belt of trees so that their branches are interlocked.

Sleeve. An old term signifying a wing, or the extreme right or left element of an organization in line of battle. One of the many mechanical devices having the form of a sleeve or cylinder, used as a coupling or covering.

Sleeve Target. A target, consisting of a long cloth cylinder which spreads when towed by an aircraft, used in antiaircraft gunnery.

Sleigh. A steel forging forming the immediate support for certain types of gun, and recoiling with the gun.

Sliding-wedge Breech Mechanism. The Krupp type of breech mechanism, in which the breechblock is a tapered rectangular block which slides sidewise, or up and down, in a slot in the breech. The latter type is known as a drop-block.

Sling. A leather strap attached to a rifle for carrying it and as an aid in firing. An ancient weapon formed of a piece of leather, with two cords of about a yard in length attached, used for throwing missiles. The sling as a weapon is probably the earliest kind of device known to mankind by which an increase of force and range was gained. Sling stones from the Stone Age have been frequently found; there are many references in the Bible to slings and slingers, and ancient historians frequently mention them. In medieval times the sling was much used by the Franks and was last used in battle in 1527. A looped rope, chain, or cordage mat used to suspend or lift anything, as a barrel sling. A looped bandage used to support an injured limb. One of the vertical members by which the deck of a suspension bridge is hung from the main cables. To place in a sling, or support by means of a sling. To hurl by means of a sling. See Loop Sling; Hasty Sling.

Sling Adjustments. Adjustments of the rifle sling. See Hasty and Loop Sling.

Slingers. Soldiers armed with slings.

Slit Trenches. Very narrow trenches, parallel to the front, for protection against shell fire, especially in massing troops close to the front. Shell trenches.

Slogan. A rallying cry. The battle cry of the Scottish Highlanders.

Slope. The inclination of the ground to the horizontal, measured usually as a percentage or gradient. Any ground whose surface forms an angle with the horizontal. An up slope is positive; a down slope is negative. The interior inclined surface of a trench (front and rear slopes). See Superior Slope; Exterior Slope; Interior Slope.

Slope Board. A board, usually a sketching board, provided with a plumb line, used to measure slopes by sighting along the edge of the board and reading the slope on a scale where cut by the plumb line.

Slot. A nozzle-shaped passage through a wing whose primary purpose is to improve the flow conditions about an airplane. It is usually near the leading edge and formed by a main and an auxiliary airfoil.

Slotted Screw. See Interrupted Screw.

Slow Match. A slow fuze used to ignite a charge of powder, as in a matchlock musket. It consisted of tightly twisted strands of cotton, sometimes soaked in a solution of saltpeter; or of hempen twine, saturated in a solution of lead acetate and lye (wood ash) to retard the rate of burning. Slow match burned at a rate of 4 to 5 inches per hour.

Slow Tank. A tank having a sustained cross-country speed of less than 10 miles per hour.

Slugs. Pieces of metal discharged from a gun or sling.

Slush Fund. A fund derived from various sources, which is not authorized and not accounted for, and which is used to procure supplies or luxuries for an organization or its members which cannot be procured otherwise.

Small Arms. All firearms smaller than cannon. They include machine guns, guns fired from the shoulder, and hand firearms (pistols and revolvers).

Small Scale Maps. See Maps, Classification of.

Small-arms Ammunition. Ammunition of all kinds for use in small arms.

Small Pickets. An obstacle consisting of straight branches of tough wood about 3 feet long, driven into the ground in a quincunx order, about 12 inches apart and projecting irregularly. Anchor pickets for revetments or obstacles.

Small Units. Refers to units smaller than a division.

Small-bore Target Practice. Target practice conducted at short ranges with a weapon of small caliber, as a .22 rifle.

Smoke Agent. A chemical agent capable of producing smoke in sufficient volume and density to obscure vision and afford concealment. It may incidentally burn or corrode objects with which it comes in contact.

Smoke, Blanketing. Smoke placed directly on the target instead of between the target and the observer, as in the case of a screening smoke.

Smoke Bomb. A bomb which on explosion emits a thick cloud of whitish smoke, used as a screen. A bomb designed to be dropped from an aircraft, that emits smoke; used as a signal.

Smoke Candle. A small smoke generator designed for stationary use, although it can be thrown or carried; employed, usually, in numbers to establish a small smoke screen.

Smoke Defilade. Defilade which conceals the smoke of firing from the enemy.

Smoke Helmet. A headgear for protection against smoke or gas.

Smoke, Irritant. A smoke which has an irritating effect on the eyes and respiratory passages.

Smoke, Obscuring. Screening smoke placed either on friendly troops or between friendly troops and the enemy, and which has no other purpose than to interfere with visibility, as distinguished from irritant smokes.

Smoke Screens. Curtains of smoke employed to mask known or probable enemy observation posts and machine gun nests, and to cover the advance of friendly troops, laid either by airplane, tank, shellfire, or smoke candle.

Smoke Shells. See Smoke Bomb.

Smokeless Powder. An explosive, used as a propelling or bursting charge, which produces no smoke. Nitro-cellulose powder.

Smooth Bore. A gun or the bore of a gun which has no rifling.

Snaffle Bit. A simple horse's bit, consisting of a bar, jointed in the middle. A snaffle bit and bridle. A bridoon.

Snaphance or **Snaphaunce.** An early form of flintlock for a musket. A type of flintlock musket. A spring lock for a musket.

Sneeze Gas. A chemical agent which attacks the nasal passages and throat, causing sneezing, nausea and headache of a few hours duration. Commonly known as irritant gas. A sternutator.

Sniper. A soldier, usually an expert shot, detailed to fire at and pick off individuals of the enemy.

Sniper-scope. A device, utilizing the principle of the periscope, attached to a rifle near the rear sight and projecting downward into a trench, by means of which a soldier can aim and fire the rifle without exposing himself.

Sniper's Post or **Nest.** A concealed or sheltered place from which a sniper fires.

Sniper's Suit. A camouflage suit, usually of grass, worn as a disguise by a sniper.

Sniping. Shooting at individuals of the enemy's force, especially during a period of stabilization.

Soakage Pit. A pit filled with gravel, broken stone or the like, used to dispose of kitchen sullage water by seepage into the earth.

Soar. To fly without engine power and without loss of altitude, as does a glider in ascending air currents. To perform sustained free flight without self-propulsion.

Sod Revetment. A revetment made of pieces of sod placed grass-side down and secured with pegs.

Soldado. A Spanish word meaning a private soldier.

Soldat. A soldier. (Fr.) (Ger.)

Soldier. One who is regularly enlisted and engaged in the military service. A member of the military profession. One who serves in an army. To serve as a soldier. A slang term meaning to make a pretense of working, while doing only enough to escape punishment.

Soldier of Fortune. One who follows a military career in different parts of the world, for profit or adventure.

Soldierly. Like a well trained soldier. Brave. Efficient. Uncomplaining. Honorable. Possessing the soldierly virtues.

Soldiers' Homes. Institutions of the federal and state governments for the domiciliary care of disabled, retired or aged soldiers.

Soldier's Medal. A medal issued to all persons in the military service who, subsequent to July 2, 1926, distinguish themselves by heroism not involving actual conflict with an enemy.

Solid Shot. Solid or cored projectiles containing no bursting charge.

Solid Square. A body of troops formed in a square, having the ranks and files equal in length.

Solo. A lone flight by a single person. A motorcycle without a sidecar.

Sortie. The sudden issue of a body of troops from a besieged place for the purpose of annoying or attacking the besiegers. A sally.

Sortie in Force. A sortie delivered with all the available troops of a garrison.

Sortie Steps, Ladders. Steps or ladders in the front interior slope of a trench to facilitate prompt egress to the front.

Sound Lag. The condition caused in gunnery by the time necessary for a sound wave to travel from its source to the point of reception. The time required for sound to travel a given distance. Its normal rate is about 1100 feet per second.

Sound Lag Corrector. See Acoustic Corrector.

Sound Locator. An instrument for locating the position of an aerial target by its sound.

Sound Ranging. A method of determining the positions of hostile guns, the explosion of projectiles, or other definite sounds, by observations and calculations based on the propagation of sound waves (aerial, ground, or subaqueous).

Sound Ranging Adjustment. An adjustment of the fire of a friendly gun or battery by sound ranging methods.

Sounding Balloon. A unmanned balloon sent up for meteorological or aeronautical purposes.

Soup Kitchen. A portable apparatus for making and distributing soup or other articles of food out of doors. A rolling kitchen.

Sources of Military Information. The sources of information include: reconnaissance and observation, both terrestrial and aerial; air photographs; actual combat; spies; prisoners; local inhabitants; identifications of units and individuals; captured documents and material; newspapers; telegraph files; letters and diaries; etc.

Space Allowances. The areas, or volumes, allowed in military buildings for various purposes; e.g., sleeping space per man in barracks, rooms for officers, kitchens and mess halls for organizations, baths, latrines, etc.

Spade. The end of the trail of a field piece, which is forced into the ground by or previous to the first discharge, and thereafter prevents movement of the chassis on recoil.

Spade Bayonet. A very broad bayonet which may be used also for digging.

Span. The maximum distance measured parallel to the lateral axis, from tip to tip of an airfoil, or of an airplane wing inclusive of ailerons, or of a stabilizer inclusive of elevator. The lateral dimension of an airfoil. The horizontal distance between adjacent supports of a bridge. The superstructure between supports. A number of draft animals harnessed abreast.

Spanish Windlass. An improvised windlass consisting of a vertical stick about which a rope is twisted by another stick used as a lever.

Spanner. The lock of a fusil or carbine. The weapon itself . A wrench.

Spar. A round timber. One of the main lateral members of the wing of an airplane.

Spar Bridge. A bridge constructed of round timbers lashed or bolted together.

Spar Lock Bridge. A military spar bridge in which the supporting trestles are inclined toward each other and locked together.

Spare Parts. Extra parts of machines, weapons, etc., carried for use in an emergency.

Spare Parts Case. A small case containing spare parts for a certain machine, weapon, etc.

Spear. A weapon having a long wooden shaft with a sharp pointed head of steel, flint or other hard substance. To pierce or strike with a spear. See also Lance; Pike.

Spearhead of the Attack. The troops or units which make the principal attack and upon whose success victory is usually dependent.

Special Army Signal Service. Comprises the following: (1) Radio Intelligence Service; (2) Photographic Service; (3) Meteorological Service; and (4) Signal Supply Service.

Special Court-martial. An inferior court-martial, consisting of three or more officers, appointed by a post or regimental commander. It is limited as to its jurisdiction and the sentences it may impose.

Special Duty. Duty, not essentially military in character, to which soldiers may be detailed when the exigencies of the service require it.

Special Evening Dress or **Social Full Dress.** A uniform resembling civilian evening dress, which may be worn on formal social occasions.

Special Operations. Operations undertaken for the accomplishment of a particular mission incident to the operations as a whole; in general, they are of a temporary nature and their tactical execution may require special organization, special equipment, or special training; night operations, trench raids, river crossings, and forced landings on hostile shores are examples.

Special Orders. Orders that include matter concerning individuals or that relate to personnel, and are not of general application or widespread interest. Orders that apply to a single post in an interior guard.

Special Service Schools. Schools for officers and enlisted men, conducted by the various arms and services, at which are taught the tactics and technique of the particular branch in question; **e.g.,** the Infantry School, the School for Bakers and Cooks.

Special Shell. All shell other than high explosive and shrapnel.

Special Situation. A statement, in a tactical exercise, of the conditions confronting and known to one antagonist.

Special Staff. A staff group, subordinate to the general staff of a unit, whose duty it is to assist the commander in the exercise of his administrative, technical and supply functions. It includes the heads of the administrative, technical and supply services, and certain technical specialists. In divisions and higher units the general and special staffs are separate, but in brigades and lower units they merge into each other. A special staff officer may also exercise command in his own branch, as the chief of artillery of a corps, who commands the corps artillery and is also, as a member of the special staff, advisor to the corps commander on matters pertaining to artillery. See also General Staff.

Special Supplies. Those supplies placed by the Assistant Secretary of War on procurement lists of supply branches other than the Quartermaster Corps.

Special Troops. Certain troops attached to the headquarters of the larger units (divisions, corps, army and G. H. Q.) that perform certain special service, sometimes including combat service, for the command as a whole.

Specialist. An officer or enlisted man who possesses knowledge or training of a special nature not included in the basic military training peculiar to the branch of the service to which he belongs.

Specialist Ratings. See Enlisted Specialists.

Specifications. That part of a military charge which alleges in detail the exact offense committed. A clear and accurate description of a material, an article, or a service which is to be procured.

Spent Ball. A projectile which has lost practically all of its velocity. A ball or shell without sufficient force to penetrate or seriously injure.

Sphere of Activity. An area or district within which the activities of an organization, command, or nation are exercised.

Spherical Case-shot. A projectile consisting of a thin shell of cast-iron, containing a number of spherical bullets and a bursting charge, which is exploded by a fuze at any particular moment.

Spherical Projectile. A solid shot or a hollow shell of spherical shape, as used in a smooth bore cannon. A cannon ball.

Spider Wire. A form of wire entanglement consisting of a number of intersecting fences, 3 or more feet in height, each fence consisting of 4 or more strands of barbed wire.

Spike. To disable a gun by driving a spike, or the like, into some part of it, usually the vent.

Spin. A maneuver in which an airplane descends along a helical path of large pitch and small radius while flying at a mean angle of attack greater than the angle of attack at maximum lift.

Spiral. A maneuver in which an airplane descends in a helix of small pitch and large radius, the angle of attack being within the normal range of flight angles. An obstacle consisting of barbed wire coiled or wound into a spiral or cylinder.

Spirit of the Bayonet. The will to close with the enemy and use the bayonet in combat.

Splash. The point of impact of a shell that strikes the water.

Splay. Divergence or spread, as of the cheeks of an embrasure.

Splinter-proof Emplacements. Emplacements constructed so as to afford protection from the enemy's fire, particularly shrapnel and shell splinters.

Splinter-proof Shelter. A shelter which affords protection against rifle and machine gun bullets, splinters of high-explosive shells, and grenades, but not against direct hits by shells.

Splinter-proof Traverse. A traverse intended only as a protection against splinters and fragments.

Split Ring. A ring used in an obturator to insure leak-proof contact with the wall of the breech. A piston ring.

Split Trail. A gun trail in two longitudinal sections to afford greater steadiness and permit wider traverse.

Spoils. Plunder taken from the enemy.

Sponging Solution. A solution of soapy water, used in sponging out the bore of a cannon.

Sponson. A projection on the side of a tank in which one or more guns may be mounted.

Spotter. A small marker used to indicate the position of hits on a small arms target, especially in a group of shots. One who, in small arms firing, announces the value of shots and places or spots their location on a target for the benefit of the firer. An observer of artillery fire, especially seacoast artillery.

Spotting. The process of determining by observation the position of a point of impact, or point of burst, with respect to a target, objective, or adjusting point. Spotting may be axial, unilateral, bilateral, flank, or a combination of these. Spotting usually determines only the sense of a deviation, but its magnitude may also be determined. The term spotting is used by the Coast Artillery. See also Sense; Sensing.

Spotting Board. A device for determining the sense and magnitude of deviations, from spotting observations.

Spotting Observation. See Observation; Spotting.

Spotting Station. A position from which the fall of shots with reference to the target or objective can be seen. A fire observation post.

Spray. The flying fragments of a burst shell; according to the direction in which projected they are classified as nose, side, and base spray.

Sprengel Mixtures. High explosives whose components are separately non-explosive, and are mixed when they are to be used.

Spring Engine. A medieval war engine in which an arm was bent back by a windlass, and when released struck the butts of arrows or javelins placed in an upright holder, projecting them to a distance of about 150 yards.

Spring Gun. A concealed firearm discharged by a spring, when trod upon. A form of booby trap.

Spring Holster. A revolver holster with a latch to prevent the weapon falling out, which is released by pressure on a spring when drawing.

Springfield Rifle, Model 1903. See United States Rifle.

Spur. A projection from a hill or ridge running out some distance from the main feature. A wall that crosses a part of a rampart and joins an inner wall. A tower or blockhouse forming a salient in the outer works before a fort. A buttress of masonry projecting at intervals from a wall which requires reinforcement to resist the pressure of the earth. A short trench projecting from another trench. An auxiliary track connected to a main track at one end only. A goad for a horse worn on the heel of a horseman. To apply the spur.

Spy. A person who secretly enters the enemy's lines in disguise, or under false pretenses, to seek information. One who, acting under cover or secretly, obtains or endeavors to obtain, military or political information in the theater of operations or home country of an enemy belligerent, with the object of communicating this information to the other belligerent. Soldiers in their own uniform are not spies. To watch secretly. To play the spy.

Squad. The elementary unit of organization in the United States Army. A number of soldiers, normally a sergeant, corporal, and ten privates, organized for drill, combat and other purposes. A group or team which operates a weapon. A general term applied to various small groups. A group of men forming a team or detail for the performance of some particular duty; e.g., sanitary squad, ammunition squad, etc.

Squad Column. A combat formation of the squad in single file with extended distances.

Squad Leader. A corporal or selected private who is in charge of a squad.

Squad Right (Left). A turn to the right (left) executed on a fixed pivot.

Squad Right (Left) About. A turn by squad of 180°.

Squad Room. A barrack room in which soldiers sleep.

Squad Trench. A trench of sufficient length to accomodate a squad.

Squadron. A number of troops of cavalry grouped together for tactical purposes; the cavalry combat unit. In some armies the squadron corresponds to our troop. An engineer, medical, etc., unit with a cavalry division.

Squadron, Airplane. Two or three flights, all of the same type of aviation.

Square. A body of troops forming a square or rectangle, with several ranks facing on each side, formerly often employed in battle, especially by infantry for defense against cavalry. See Hollow Square; Solid Square.

Square System of Organization. An organization in which a unit includes two or four subordinate units, as a four-company battalion, two-regiment brigade, etc.

Stability. That property of a body which causes it, when its equilibrium is disturbed, to develop forces or moments tending to restore the original condition. Smoothness and steadiness in flight, with preservation of equilibrium.

Stabilization. See Stabilized Warfare.

Stabilize. To establish a period of comparatively inactive warfare for any purpose, or as a matter of necessity, along a front or any considerable portion thereof.

Stabilized Front. A strongly organized, continuous defensive system, echeloned in great depth and supported on its flanks by other troops or by impassable obstacles.

Stabilized Warfare. A phase of warfare usually characterized by absence of maneuver and an intensive use of field fortifications. A deadlock which results when the opposing forces are nearly equal in fighting strength, or have lost or deliberately relinquished the power of maneuver, due to exhaustion, or to a desire to economize strength for use in another locality.

Stabilizer. Any airfoil whose primary function is to increase the stability of an aircraft by preventing pitching. It usually refers to the fixed horizontal tail surface, as distinguished from the fixed vertical surface.

Stabilizer, Mechanical. A mechanical device designed to prevent an aircraft from departing from a condition of steady motion, or, in case such a motion is disturbed, to restore it. Includes gyroscopic stabilizers, pendulum stabilizers, inertia stabilizers, etc.

Stable Guards. Guards mounted over various stables or parks, for the protection of the animals, stables, forage, equipment, and other public property, and the enforcement of regulations.

Stable Management. The measures and methods necessary for the proper performance of stable duty, and the care of animals in stables.

Stable Police. Enlisted men detailed to clean stables or picket lines, and to help in the care of animals.

Stable Sergeant. The sergeant in charge of the animals of an organization.

Stack Arms. To arrange the rifles of a unit in a line of stacks, each formed by engaging the stacking swivels of three rifles and spreading the butts to form a tripod. Additional or loose pieces may be leaned against a stack.

Stacking Swivel. The swivel attached to the upper band of a rifle to be used in attaching rifles to form a stack.

Staff. A group of officers specifically provided for the purpose of assisting a commander in exercising his command functions. Certain of these officers have no duties other than staff duties, while others have staff duties in addition to their functions as commanders of combat or service troops. The assistants to these officers, although on staff duty, are not included when referring to the commander's staff or the unit staff. The staff of a large unit includes three groups: the general staff; the special staff; and the personal staff (aides). The pole from which a flag is flown. To provide with a staff.

Staff Authority. The authority exercised by a staff officer. A staff officer, as such, has no authority to command. All responsibility rests with the commanding officer, in whose name all orders are issued.

Staff College. A higher college for the training of officers in staff duties and higher command.

Staff Department. Any of the departments or bureaus into which the War Department is divided for administrative purposes.

Staff Duty. Duty in one of the staff departments or on the staff of a military commander.

Staff Officers. Officers charged with the performance of staff duty of any nature.

Staff Organization. The organization of a staff is based upon the duties of the commander whom it serves, and is prescribed in Tables of Organization. As the duties of all commanders can be divided into four principal functional groups, namely: (1) Personnel; (2) Military intelligence; (3) Operations and training; and (4) Supply, these four subdivisions, under a coordinating head, exist in the staffs of all units from the battalion to general headquarters of the field forces. See also General Staff; Special Staff.

Staff Ride or **Walk.** A tactical exercise in which practical staff operations under assumed tactical, strategical, or logistical situations, (the troops being imaginary), are stated and solved on the ground, the solutions generally being expressed in the form required under actual conditions of war.

Stages of Attack. The preparatory stage, the decisive action and the completion. For a more detailed classification see Steps or Phases of the Offensive.

Stages of Defense. The preparatory stage, the decisive action and the completion. The completion consists of the general counter attack and expulsion of the attacker, or the withdrawal and retreat of the defender.

Stagger. Said of the wings of an airplane when one is in advance of the other. Stagger is positive when the upper wing is in advance of the lower.

Staggered. A formation in which the intervals and distances are irregular, taken up to minimize the effects of hostile fire.

Staggered Positions. A formation in which the pieces of a battery are irregularly spaced or echeloned.

Staging Area. One of a series of areas on a route of march, occupied by troops for and during a long halt.

Stall. As applied to a motor, especially a motor of a vehicle or aircraft, to stop as a result of overload, lack of fuel or spark, or any other cause.

Stalling. That condition of an airplane in which, from any cause, it has lost the relative speed necessary for control.

Stand. A halt or stop for the purpose of offering resistance. To refuse to give way or to yield.

Stand at Ease. A command allowing men in ranks to stand in an unconstrained position, while maintaining silence.

Stand Down. An order given in the trenches at early morn to let the men know that their night vigil is ended.

Stand Fast. A caution to a particular part of a unit to remain immovable while the remainder is executing some movement.

Stand of Colors. A single color.

Stand To. To be prepared to take the firing position in a trench.

Stand to Arms. To be on the alert with arms ready for use.

Stand to Horse. A position standing at the head of a horse and holding him by the bridle.

Standard. A flag carried by mounted or motorized units. An upright post. A prescribed construction; e.g., standard trench, standard warehouse, etc. A required degree of excellence as to condition, or proficiency in performance, and the models and tests therefor.

Standard Bearer. An officer or enlisted man who carries the standard.

Standard Conditions. Certain conditions on which standard tables of ballistic data are based. They include an assumed muzzle velocity, ballistic coefficient, normal atmospheric conditions, etc.

Standard Gauge Railway. The standard distance between insides of rails for railways is 4 feet 8½ inches.

Standard Target Areas. Circles or ellipses of definite sizes, reasonably presumed to include certain targets; used in assigning fire missions.

Standard Train. A railroad train containing the numbers and types of cars required for certain purposes, made up and held in readiness for use.

Standardized Signs. Conventional signs used on all maps.

Standards of Proficiency. Instructions published from time to time by the War Department or subordinate agencies, giving approximate time allotment for qualification in each subject of training and the method of grading or testing to determine satisfactory qualification.

Standing Army. The regular or permanent army of a country, which is liable for every species of duty, without limitation being fixed to its service, other than the period of enlistment. The body of soldiers maintained by a nation in time of peace.

Standing Barrage. A stationary artillery or machine gun barrage laid for defensive purposes in front of an occupied line or position. Fire on a line, usually placed across a probable avenue of enemy approach, or an exposed sector of the front, in order to prevent passage of enemy troops. A standing barrage is coordinated with other defensive fires.

Standing Gun Drill. Gun drill by sections, intended to coordinate the different individual movements so as to insure the rapid service of the piece.

Standing Orders. Orders which are of general and permanent application throughout a command and are issued as such to avoid repetition.

Standing Position. A prescribed firing position in which the firer stands erect.

Standing Trench. A trench whose profile affords cover for a rifleman in a standing position.

Star Shell. A thin iron shell filled with pyrotechnic stars and intended to light up the terrain or a hostile position at night.

Stars and Stripes. A popular name applied to the flag of the United States.

Statement of Charges. A list of charges to be deducted from the pay of an enlisted man, which must be authorized by a report of survey, to accompany an organization pay roll.

Station. The place where an organization, unit, or detachment is located. A locality or place for the rendezvous, stay, or location of troops. A designated military post. A point at which a surveying instrument is set up. To place. To set. To assign to a station.

Station Complement. Troops permanently located at any station where other troops arrive and depart. Its function is to operate the utilities and conduct routine administration incident to the maintenance of the station.

Station List. A directory of the locations of all the units of a command.

Stationary Warfare. Stabilized warfare. Warfare which is without the element of maneuver.

Status Quo. The existing state of affairs.

Statute of Limitations. A law setting a limit to the time in which a person may be brought to trial, especially for desertion.

Steam Condensing Device. A device for use with water-cooled guns which condenses the steam generated by firing and thus effects a saving in the amount of water required for cooling purposes, and avoids betraying the location of the gun.

Steam-escape Tube. The tube by which steam passes from the water jacket of a machine gun to the water chest.

Stellung. A battle position. A fortified zone. (Ger.)

Step. A pace. The distance measured from heel to heel between the feet of a man walking or running. The half step and back step are 15 inches long; the right and left step are 10 inches long; the steps in quick and double time are 30 and 36 inches long, respectively.

Step Off. To move from the position of halt.

Step Out. To increase the cadence or length of pace or both.

Steps or Phases of the Offensive. In a planned offensive combat, infantry units normally pass through certain successive steps or phases. See Attack, Phases of.

Step-thread Breechblock. A breechblock having its threaded surface arranged in steps to afford more bearing surface than in the slotted-screw type.

Stereoscopic Fire Director. An observing and data-transmission instrument, with a self-contained horizontal base, operating on the stereoscopic principle; used in connection with the .50 caliber machine gun for antiaircraft fire.

Stereoscopic Height Finder. A self-contained horizontal base range finder operating on the stereoscopic principle, with an altitude element which converts the ranges found into altitudes; used in antiaircraft firing.

Stereoscopic Range Finder. A range finder with a self-contained horizontal base, which operates on the stereoscopic principle.

Stereoscopic Spotting. Spotting with the aid of a stereoscopic height finder.

Sternutators. Chemical agents which cause sneezing, vomiting, general disability and irritation of the nose and throat, but produce no material effect on the lungs.

Stevedore Troops. Labor troops.

Stick. A colloquial term for the control column of an airplane.

Stirrup. One of the attachments to a saddle to support the feet of the rider; it may be open or hooded. Stirrups were invented about 550 A. D.

Stock. The whole of the wooden part of a rifle, pistol, or other small arm. A supply of articles on hand or in storage.

Stock Record Account. A record of supplies in storage.

Stockade. A tight fence of poles or posts set in the ground, inclined to the front, used as a rampart. An inclosure for the safekeeping of prisoners, or animals.

Stocks. Equipment and supplies stored in depots, in the hands of troops, and in transit.

Stokes Trench Mortar. A light, muzzle-loading, three-inch mortar, much used in the World War.

Stop Butt. A parapet, either artificial or natural, designed to stop bullets back of the targets on a rifle range. A back-stop for bullets.

Stoppage. Failure of any mechanism, as a machine gun or other automatic weapon, to function, due to any cause. A jam. Stoppages are classified as temporary and prolonged; a temporary stoppage is one that can be remedied in the field, under fire, by a simple adjustment or replacement. Temporary stoppages in automatic weapons are also classified according to the phase of immediate action required to remedy them, or according to the position in which the bolt stops when drawn to the rear. A prolonged stoppage is one that cannot be remedied within a reasonable time, or with the means available in the field.

Stoppage of Pay. A retention of pay, either wholly or in part, to satisfy an indebtedness to the government. In general, any fine, forfeiture or detention of pay.

Storm Flag. The national flag flown in stormy or inclement weather, having a 9 feet fly and a 5 feet hoist.

Storm Troops. Specially selected and trained shock troops.

Storming Party. A party assigned the duty of making the first assault in a storming attack.

Straddle. To place shots short of and beyond a target in order to facilitate calculation of the correct range. A salvo which has impacts of opposite sense; also called a mixed salvo. To bracket.

Straddle Trench. A form of latrine used in camps of short duration.

Strafe. To inflict damage upon, as by heavy shell fire. To shell or bombard fiercely. To punish severely. To castigate. (Ger.)

Straggle. To wander from the line of march or from the assigned place in a battle formation.

Straggler Line. A line established in rear of troops in combat for the collection of stragglers and their return to their proper units.

Stragglers. Individuals who wander from the column of march or from their organizations in battle.

Straight Duty. A slang term meaning regular duty with one's organization.

Straps. Decorations made of gold, silver, silk, etc., indicating rank, worn upon the shoulders, without epaulets.

Stratagem. Artifice in war. A scheme for deceiving and surprising the enemy. Permissible if it does not involve acts of perfidy or treachery.

Strategic, Strategical. Pertaining to strategy. Characterized by, or used in strategy. The words strategic and strategical are practically synonymous.

Strategic Interception. The interposition of an army between the two parts of a divided hostile force so as to be numerically stronger than either and have sufficient space to maneuver against both.

Strategic Lines. The important lines of operation, defense, supply, retreat, etc., of an army in campaign. Any lines that are of strategic importance, as for defense, communication, etc.

Strategic Marches. Marches conducted for strategic purposes, as in concentrations or movements from one part of a theater of war to another.

Strategic Position. A position taken up or occupied for the purpose of checking or observing a hostile force. A position whose occupation gives strategic advantages.

Strategic Raw Material. Materials essential to the conduct of war which cannot be obtained in sufficient quantities within the confines of the home country, and for which no substitute is produced or can be produced in sufficient quantities.

Strategical Advance Guard. A strong and mobile force, usually of all arms, sent out a considerable distance in advance of the main forces, to gain contact with the enemy, seize important localities, etc. It is not directly responsible for the immediate security of any troops in rear.

Strategical Concentration. The process by which an army is moved into and concentrated in the theater of operations.

Strategical Decisive Point. A point whose possession will be decisive in insuring the success of strategical operations.

Strategical Deployment. The concentration and development of troops in the positions from which they are to commence active operations.

Strategical Flank. The more important or vulnerable flank of the strategical front; usually that flank which is nearer to the line of communication.

Strategical Fortifications. Defensive works executed largely in time of peace for the protection of points of strategic importance, or in conformity with a general strategic plan.

Strategical Front. The military frontier of a belligerent state. The forward part of the theater of operations.

Strategical Maneuvers. The movements of troops, on a relatively large scale, at a distance from the enemy in preparation for battle, or to gain a strategical advantage.

Strategical Mobility. The ability of troops or weapons to travel relatively long distances in relatively short periods in strategical movements or concentrations.

Strategical Operation. A march, movement, or maneuver, which improves or is designed to improve the strategical situation.

Strategical Plan. A plan which outlines the strategical conception of the war with the objectives and plan of operations, and specifies the units and other means to be employed.

Strategical Point. Any point located in the theater of operations whose possession is of great strategical importance to an army.

Strategical Rear Guard. A large body of troops covering the retirement of an army or group of armies.

Strategical Reconnaissance. A reconnaissance made before the two hostile armies are in contact, to locate definitely the hostile forces, ascertain their strength, number of columns, and direction and probable objective of march.

Strategical Reconnoitering Patrols. Patrols directed towards the probable routes of the enemy's approach, areas where he is likely to be found, and to the strategical points he will probably endeavor to occupy.

Strategical Reserves. Forces whose members in time of peace follow their usual occupations, but are called out in time of war. Strong forces held out of action by the supreme commander for probable use at any point where conditions may dictate their employment. G. H. Q reserves.

Strategical Retreat. A retreat made for causes other than local, such as withdrawing troops for use in other localities. A retreat designed to improve the strategical situation.

Strategical Unit. A unit of such size as to enter into strategical considerations, generally considered to be the division and higher units.

Strategist. One skilled or gifted in strategy.

Strategy. One of the two great divisions of the art of war, the other being tactics. The conception of the general plan of a campaign and of the larger movements of forces, which concern the theater of war rather than the field of battle. The art of moving and disposing troops in a theater of operations so as to occupy the critical points and create a favorable situation before resort to battle, in order to increase the probabilities and consequences of victory and lessen those of defeat. See also Art of War.

Streamlined Division. A compact mechanized division possessing great fire power and mobility. A term applied to the new triangular division.

Streamline Form. A body, such as a projectile, vehicle, vessel, or aircraft, of such shape as to cause a streamline flow when it is propelled through air or water, thus reducing the resistance encountered.

Street Fighting. Fighting incident to the capture or defense of a town or during a riot.

Strength Return. A monthly report of the present strength of an organization, showing all changes since the previous return. A consolidated return is made by combining returns of all the component units of a regiment, post, division, etc.

Stretcher. A portable litter. A sandbag, brick, timber, etc., laid with its long dimension parallel to the face of a wall.

Stretcher Bearers. Men detailed or whose duty it is to carry the wounded from the battlefield. Litter bearers.

Stride. Two natural steps or paces. The usual unit of measure in military sketching on foot.

Strike a Camp. To break camp.

Strike a Flag. To haul down a flag in surrender. To haul down a flag as a mark of respect or submission.

Strike a Tent. To lower a tent in breaking camp.

Striker. A hammer or firing pin of a gun. A slang name applied to a soldier who looks after an officer's equipment and renders other personal services.

Striking Distance. That distance within which, under the existing conditions, an attack is practicable or to be anticipated.

Striking Velocity. The speed of a projectile at the instant of impact.

String. A number of shots fired in succession, with the same sighting.

Strip. To disassemble a weapon or machine as to strip a machine gun.

Strip Mosaic. A mosaic compiled by assembling one strip of aerial photographs taken on a single flight of an aircraft.

Stripe. A common name applied to the chevron worn on the coat or shirt of a noncommissioned officer. A stripe of cloth worn on the sleeve to indicate rank or length of service. A blow from a whip or knout, administered as punishment.

Stripped Deviation. In artillery firing, the deviation that would have resulted if there had been no human errors and no adjustment correction.

Stripping. The act of taking a weapon or machine apart. Disassembling.

Stroboscope. A device for protected vision in tanks, by means of slits in a rapidly rotating turret.

Strong Point. A compact area, strongly organized for defense. Formerly, the term was applied to the defensive area garrisoned by a rifle company. See Defensive Area.

Stuka. A dive bomber. (Ger.)

Subaltern. A commissioned officer below the grade of captain.

Subaqueous Sound Ranging. See Sound Ranging.

Sub-caliber Ammunition. Ammunition designed for use in a sub-caliber gun or barrel.

Sub-caliber Barrel or Tube. A barrel or tube of less caliber inserted in the bore of a gun, to permit the use of sub-caliber ammunition for short range target practice.

Sub-caliber Gun. A gun fitted with a sub-caliber tube.

Sub-caliber Practice. Short range target practice with a sub-caliber gun, used as an economical substitute for service practice.

Sub-caliber Projectile. See Sub-caliber Ammunition.

Subdivision. One of the parts or elements of which a larger organization is composed.

Subgrade. The upper finished surface of the foundation of a road, firing platform, etc., on which the paving or platform is placed.

Submachine Gun. A weapon, lighter than the automatic rifle, which fires pistol ammunition on the automatic principle.

Submarine. A boat capable of operating at various depths and of traveling considerable distances under water. Below the surface of the water.

Submarine Mine. See Mine, Submarine.

Submine. A small explosive charge in a bottle or other container, which is attached to a submarine mine for mine target practice, analogous to subcaliber practice.

Subordinate. One who stands in order of rank below another. To place in or relegate to a lower order or class. To make subordinate.

Subordination. A state of perfect submission to the orders and will of superiors. The state of discipline.

Sub-sector. A subdivision of a sector.

Subsistence Allowance. A monetary allowance paid to individuals in the military service for subsistence.

Subsistence Stores. Stores purchased and issued by the Quartermaster Department, which consist principally of the articles composing the prescribed ration.

Substitute. A person who, for a consideration privately arranged, enters the military service in the place of a drafted or conscripted man.

Subterranean Warfare. Attack and defense by means of subterranean passages driven towards the enemy's works, and the use of explosives. Mine warfare.

Successive Approximations Method. In this method, one shot is fired at the target and a full correction based on its deviation is applied to the pointing of the next shot, which is fired from the same gun. The pointings of the third and fourth shots are similarly corrected by amounts equal to one-half and one-third of the deviations of the second and third shots, respectively. The firing of the fourth shot completes the trial fire and a correction equal to one-fourth its deviation gives the data with which to fire for effect.

Successive Attack. A form of attack in which the leading elements engage the enemy, and rear elements, as they are brought forward, are launched into the attack usually by the extension of the front. A form of attack in which the various elements are committed to action successively.

Successive Concentrations. Accompanying artillery fire in attack, which takes the form of concentrations on selected objectives at successively increasing ranges, preceding the advance of the attacking troops.

Successive Formation or Movement. A drill formation in which the different units arrive in their proper places in the formation successively, as distinguished from a simultaneous movement.

Successive Pontons. A method of constructing a ponton bridge, one bay at a time; the usual method.

Successive Positions. Positions selected and occupied successively in order to delay or interrupt the advance of the enemy.

Summary Court-martial. An inferior court-martial, consisting of one officer, which may be appointed by a post, regimental, or detachment commander. It is limited in jurisdiction to minor offenses and as to the punishment it may adjudge, and may not try an officer, warrant officer, etc., nor a noncommissioned officer except by his own consent or by order of higher authority.

Summit of a Trajectory. The highest point of a trajectory. See Maximum Ordinate.

Sumpitan. A blow gun.

Supercharger. An air compressor used with an airplane engine, to force into the cylinders a heavier charge of fuel than can be drawn in under pressure of a rarefied atmosphere when flying at high altitudes.

Superelevation. In antiaircraft fire, that part of the quadrant elevation which allows for curvature of the trajectory; the elevation above the chord of the trajectory or line of position. Firing table superelevation, when corrected for non-standard conditions. The additional elevation given to the outer rail of a track or the outer surface of a road on a curve, to compensate for centrifugal force.

Superficial Revetment. A revetment in the form of a protective facing or covering for the surface revetted, and which is usually held in place by pickets driven into the ground and anchored by guy wires at their tops. A skin or surface revetment.

Superior Officer. An officer of higher rank, or one who has priority in the same rank.

Superior Slope. The upper surface of a parapet. The slope which extends toward the enemy, between the inner fire crest and the exterior slope.

Supernumerary. An individual detailed as such, in excess of those normally required for the performance of a certain duty, and held subject to call in case of need. A person or thing beyond the number prescribed.

Super-quick Fuze. A point percussion fuze with a long neck, which explodes a projectile before actual impact.

Supersede. To deprive of rank or command, or both. To place one officer over the head of another. To displace. To supplant.

Supplementary Fire Position. A firing position assigned a unit or weapon to accomplish secondary fire missions, other than those to be accomplished from primary or alternate positions. See Fire Positions.

Supplementary Season. A period of the target year designated for the instruction of recruits who joined too late to participate in the regular practice season.

Supplies. A general term covering all things necessary for the maintenance and operation of a military command, including especially food, clothing, equipment, ammunition, fuel, forage, and other materials.

Supply. Food, and other daily necessities of troops. The furnishing of all necessary articles to the military forces, including procurement, storage, distribution, and issue. The amount of a commodity on hand or available. To furnish with what is needed. To provide.

Supply Arms and Services. The branches of the service responsible for the procurement of supplies for use in the military service, which include the Quartermaster Corps, Medical Department, Corps of Engineers, Ordnance Department, Signal Corps, Chemical Warfare Service, Air Force, and Coast Artillery. They all have other functions in addition to supply.

Supply Columns. Trains for the supply of troops in the field.

Supply Credits. Lists of specific quantities of supplies in designated places allotted to a commander and available to him on demand either as a whole or in part within any limiting dates which may be specified.

Supply Establishments. Those establishments by means of which the supply functions of chiefs of arms or services are accomplished, such as depots, arsenals and manufacturing plants.

Supply Officer. An officer on the staff of a small unit charged with the procurement and issue of supplies, especially rations, clothing and general equipment, and transportation.

Supply Point. A general term used to include depots, railheads, dumps and distributing points.

Supply Rate. The rate at which troops can be sheltered, fed, clothed, equipped and maintained on mobilization.

Supply Requirements. The needs of troops, camps, supply points, depots, arsenals, etc., arranged according to the program of mobilization.

Supply Train. A train, usually an organic part of a unit, including all personnel, vehicles and animals, employed in the transportation of supplies.

Support. The second echelon of a platoon at the beginning of an attack, usually a squad. A part of a company, usually a platoon, withheld from an attack at the opening therof, with a view to participation as needed. That part of a company in the front lines in defense, usually one or two platoons, which is posted on the support line, and assists the troops in front by fire, reinforcement, or counterattack, or any combination of these. A detachment of infantry or cavalry supporting artillery to defend it against direct attacks by the enemy. One of the groups on the line of resistance of an outpost. To aid or assist, as by fire, reinforcement, etc.

Support Echelon. The line on which the support elements are disposed. The platoon or platoons which support the front line platoons.

Support Line. One of the lines of a defensive position, from 100 to 300 yards in rear of the line of resistance, occupied by the support platoons of the front line companies. In a prepared position it may be marked by more or less continuous trenches. (Old.)

Support of the Advance Guard. That part of the advance guard which marches in front of the reserve and protects it by observation and resistance.

Support of the Rear Guard. That part of the rear guard which marches in rear of the reserve and protects it by observation and resistance.

Support Trench. A fire trench constructed a short distance in rear of the front line trenches to shelter the supports. A trench on the support line of a battle position.

Support Wave. The second line in an attack, which supports the assault wave. It includes the local supports of the troops in the leading wave.

Supported Attacks. Attacks made against a point on a coastal frontier in which the forces that carry out the principal attack are escorted, assisted or supported by considerable naval forces.

Supporting Artillery. Artillery that has been assigned the primary mission of supporting a unit of another branch, while remaining under the administrative control of its usual artillery command. See also Direct Support; General Support.

Supporting Distance. Generally, that distance between two units that can be covered in the time available by one to come to the aid of the other. For small infantry units it is also that distance between two adjacent elements which can be effectively covered by their fire.

Supporting Point. A locality well organized and strongly held, whose possession facilitates or protects tactical operations.

Supporting Troops. Troops which support, or are to support, a tactical front line unit, generally in compliance with the orders of a higher commander.

Supporting Unit. A unit which affords assistance to another given unit in combat, or is designated to do so, by fire, movement, shock action, or otherwise.

Supreme Commander. A commander-in-chief. A generalissimo. In the United States Army the following are considered to be supreme commanders and entitled to a color: (1) President; (2) Secretary of War; (3) Assistant Secretaries of War; (4) General of the Armies; (5) Chief of Staff.

Surface, Control. A movable airfoil designed to be rotated or otherwise moved by the pilot in order to change the attitude of the aircraft.

Surface, Main Supporting. A set of wings, extending on the same general level from tip to tip of an airplane; e. g., a triplane has three main supporting surfaces. The main supporting surfaces include the ailerons, but no other surfaces intended primarily for control or stabilizing purposes.

Surface or Skin Revetment. See Superficial Revetment.

Surface Shelter. A protected shelter built at or partially above the surface of the ground. Any shelter that is not entirely subterranean. See also Cut-and-cover Shelter; Cave Shelter.

Surgical Hospital. An army hospital having facilities for surgical treatment of serious cases, and which relieves division hospitals of such cases.

Surplus Kit. Clothing additional to that worn on the person of the soldier pertaining to Equipment B, and which is not intended to be carried by the soldier.

Surprise. One of the principles of the art of war. A sudden or unexpected attack which takes the enemy in a state of moral or material unpreparedness. The essential elements of surprise are secrecy in preparation and speed and vigor in execution.

Surrender. The act of yielding or giving up to another. To give up. To yield.

Survey. Systematic gathering of data for a map, or the making of a map, in the field. The field work of topography. The process of determining the relative positions of points on the earth's surface. An examination, as of the condition of property. To make a survey. To view anything in its entirety.

Surveying Officer. A disinterested officer designated to survey public property which has been damaged, is worn and unsuitable for use, or has been lost or stolen; and who makes recommendations regarding the responsibility for the condition of the property surveyed, and for its disposal.

Suspend. To cease from action or operation. To deprive an officer of rank and pay in consequence of some offense.

Suspend Firing. The order for a temporary cessation of fire. See also Cease Firing.

Suspension. The connection between the body and running gear of a tank, or other vehicle, whereby the shock of movement is absorbed.

Suspension from Rank. A punishment which includes temporary deprivation of the rights and privileges of rank.

Suspension of Arms. A short truce between hostile forces for the purpose of burying the dead, etc.

Sustentation. The phenomenon of the support of a heavier-than-air machine by the reaction of a deflected air stream; and the flotation of a lighter-than-air machine by displacement of an equal volume of air.

Sutler. A person who follows troops and sells small articles and supplies. A post trader. A camp follower.

Sweep. To distribute fire laterally over a target or area. To remove hostile submarine mines from a water area.

Sweep-back. The acute angle between the plan projection of the leading edge of a wing and a line normal to the plane of symmetry of an airplane.

Sweeping a Mine Field. Dragging a long cable over a submarine mine field to remove or destroy submarine mines.

Swept Space. The danger zone as influenced by the slope of the ground.

Swing Team. The intermediate pair of a 6-line team of draft animals.

Swinging Traverse. The continuous movement of a piece, back and forth, through an arc in order to sweep with fire a certain frontage.

Swishtail. To kill the forward speed before landing by action of the rudder, the rudder action being so controlled that the plane swings right and left over the ground about the vertical axis without appreciable change in the direction of forward motion.

Swiss Hut. A small, portable building used in the World War.

Swiss Rifle. The rifle adopted by the Swiss Army; it has a caliber of .297 inches and weighs 9.9 pounds.

Switch Positions. Oblique positions which connect successive defensive positions. They are employed to maintain the continuity of the defense in case of hostile penetration; and to serve as bases of departure for counterattacks.

Switch Trenches. Trenches whose general direction is inclined to the front, which connect two parallels for the purpose of preserving continuity of the front in case of capture by the enemy of a portion of the line; they are primarily fire trenches, but they may also serve as communication trenches.

Swivel. A linking or coupling device which permits either of the attached parts to rotate independently of the other. A pintle. A mounting for a gun, consisting of a yoke supporting the trunnions and a pintle.

Swivel Gun. A gun fired from a swivel.

Sword. A very general term for an individual hand weapon, usually of metal, consisting of a relatively long blade fixed in a hilt or handle, and used for cutting or thrusting, or both. The origin of the sword is unknown, but it has been found amongst the prehistoric relics of our race, and is mentioned in the earliest records. In the past it was the most universally employed, and with the exception of the club and spear is probably the most ancient of man's weapons. Probably the earliest swords or daggers were made from the horns of animals and of flint. These were followed by weapons of copper and bronze. Even hard wood has been utilized. The first iron swords probably appeared between 1000 and 700 B. C. The advent of firearms presaged a constantly decreasing use of the sword as a military weapon. In our own service it is now used chiefly as a badge of office. The sword is usually carried inclosed in a scabbard and belted or hung at the side. It is peculiarly a personal weapon; its surrender has always been a token of submission, and its breaking a ceremony of humiliation. Its long continued and universal use has made the sword synonymous with armed conflict or war. The sword occurs in a vast variety of forms. Thus compare Acinace; Barong; Bolo; Braquemard; Broadsword; Campilan; Cimeter; Claymore; Cutlass; Falchion, Gladius; Hanger; Kris; Latte; Machete; Parazonium; Rapier, Saber; Scimitar; Scramasaxe; Selictar; Spadroon; etc.

Sword Bayonet. A bayonet shaped somewhat like a sword.

Sword Belt. A belt by which the sword is hung and carried at the side.

Swordsman. One skilled in the use of the sword.

Swordsmanship. The art of handling the sword.

Symbol. A letter, emblem, or sign, used to represent something. A conventional sign used on a map to represent features of the terrain or military units or establishment.

Symbol Numbers. A series of numbers used to designate the accounts of each accountable disbursing officer.

Sympathetic Detonation. Induced detonation.

Synchronization. A process in which the values indicated by all receiver pointers of an electrical data transmission system are made to agree with the values set on the corresponding transmitters. See Pointers, Electrical, Mechanical.

Synchronized Gun Control. The controlling of an aircraft machine gun from some moving part of an engine so that the gun cannot fire whenever a propeller blade is in the line of fire.

Synchronizing Gear. Gear which synchronizes the fire of a machine gun with the revolution of the propeller of an airplane to permit fire of the weapon through the revolving propeller.

Systematic Errors. Errors in artillery fire which follow some law causing them to be either of a constant or of a progressively varying magnitude. The source may be determinate or indeterminate.

T

Tabard. A small banner, bearing a coat of arms, suspended from a bugle or trumpet. A short, heavy outer garment, emblazoned with the arms of a sovereign or a knight, worn by a herald or by a knight outside his armor.

Tables of Allowances. Tables which prescribe the allowances of equipment normally required and issued for use at posts, camps, and stations. They include items of equipment which are not normally taken into the field or away from the station and which are additional to the allowances prescribed for troops in Tables of Organization and Tables of Basic Allowances.

Tables of Basic Allowances. Tables which prescribe authorized basic allowances of organizational and individual equipment with the exception of that prescribed in Tables of Organization; equipment necessary for temporary use for special purposes; that listed in supply catalogues; and recruit clothing and equipment and Alaskan clothing.

Tables of Fire. Tabulated data for each gun for each kind of ammunition. Range tables.

Tables of Organization. Tables which prescribe the personnel and the allowance of weapons, animals, and animal and motor transportation for the organization to which each pertains.

Tachometer. An instrument which measures and indicates velocities.

Tackle Rigging. A combination of blocks and rope used to lift or move heavy objects. That part of the rope between an anchorage and a block is a standing part; the parts between blocks are the running parts; and the part to which power is applied is the fall. Simple riggings commonly used for military purposes are the whip, runner, gun, luff, and burton, See also Mechanical Advantage.

Tactical Control. A control which is exercised to insure tactical cooperation and coordination with higher and adjacent units.

Tactical Employment of Fire. The application of the proper kind of fire in various situations. Fire direction as distinguished from the technical conduct of fire. See Fire, Tactical.

Tactical Exercise. A term which embraces all forms of training exercises, such as map problems, map maneuvers, map exercises, terrain exercises, tactical rides or walks, staff rides, historical rides, field exercises, and field maneuvers.

Tactical Fortifications. Fortifications constructed to provide for the immediate tactical needs of the troops.

Tactical Front. The line connecting the most advanced combatant units when contact with the enemy is established or imminent. The front line.

Tactical Groupings. The balanced grouping of combat units and means within a command to accomplish a tactical mission. It may be organizational or provisional.

Tactical Locality. An area of terrain which, because of its location or features, possessses tactical significance in the particular circumstances existing at a particular time. A defensive area.

Tactical Maneuvers. Maneuvers executed on the battlefield or in the near presence of the enemy.

Tactical Marches. Marches made in the immediate vicinity of the enemy with a view to improving the tactical situation.

Tactical Mechanization. The application of mechanization to combat on the battlefield. In includes the use of mechanized weapons and mechanically propelled vehicles.

Tactical Mobility. The ability of troops or weapons to move easily and rapidly to any desired position on the field of battle, or in immediate proximity thereto.

Tactical Obstacles. Obstacles whose chief purpose is to hold the attacking forces under the effective fire of the defense, especially those used in connection with the flanking fire of machine guns.

Tactical Points. Points on a battlefield whose possession facilitates tactical operations.

Tactical Principles. The principles of war.

Tactical Protection. Active measures of security to assist troops in carrying out their missions without excessive losses in gas casualties.

Tactical Reconnaissance. Reconnaissance for tactical purposes, as distinguished from strategical purposes; it is usually conducted by the opposing forces when in contact, or when contact is imminent.

Tactical Reserves. Reserves retained under control of a higher commander to support the attack, or to cover the retreat in case of a defeat.

Tactical Retreat. A retreat necessitated by local conditions; it may be either voluntary or involuntary.

Tactical Ride, or Walk. A tactical exercise in which a series of military situations, the troops being imaginary, are laid out and solved on the ground, the solution generally being oral.

Tactical Security. Security provided by tactical measures.

Tactical Unit. A unit organized primarily for tactical purposes; it may or may not be also an administrative unit. The organization which is the basis for tactical instruction in each branch of the service, as the battalion, squadron, etc.

Tactical Vehicles. Motorized vehicles used for tactical purposes; they include: (1) Combat vehicles; (a) armored cars, (b) tanks, (c) supporting combat vehicles; and (2) Combat carriers; (a) cross-country or other motor vehicles to carry personnel or weapons, (b) communications vehicles.

Tactician. One skilled or gifted in tactics.

Tactics. The art of maneuvering and disposing troops in immediate preparation for and during actual combat. The leading of troops in battle. The application of the principles of war to the operations of military units. See Grand Tactics; Minor Tactics; Maneuver Tactics.

Tail. The last element of a column on the march. In a retrograde movement, the tail is that part nearest the enemy. The rear portion of an aircraft.

Tail Group. The stabilizing and control surfaces at the rear end of an aircraft, including stabilizer, fin, rudder, and elevator. Also called empennage.

Tail Skid. A skid or short runner at the rear of an airplane, forming its third point of support, and designed to assist in landing and taxying.

Tail Slide. The backward and downward motion, tail first, which certain airplanes may be made to take momentarily after having been brought into a stalling position by a steep climb.

Tail Spin. A maneuver in which an airplane is made to descend with a spinning motion.

Tail Unit. The entire group of tail surfaces of an aircraft. The empennage.

Tail Wind. A wind blowing from the rear, with reference to the direction of flight of an aircraft. A following wind.

Take Arms. To enter upon a state of war. To break a stack of arms.

Take Interval to the Left (Right). A movement which places the individuals of a unit at extended intervals, as for physical drill, etc.

Take Rank With. To be of the same grade.

Take the Field. To enter upon the operations of a campaign.

Take Up. To catch or arrest. To assume responsibility for property or funds found.

Take Up Quarters. To go into barracks, cantonments, etc.

Take-off. The act of beginning flight in which an airplane is accelerated from a state of rest to that of normal flight. In a more restricted sense, the final breaking of contact with the land or water. To rise from the ground or water surface for aerial flight.

Taking Over. The act of relieving a person or force on any duty. Assuming command. Assuming responsibility for supplies.

Tang. The projecting portion of the breech of a musket, by which the barrel is secured to the stock. The part of a sword blade to which the hilt is riveted.

Tangent Sight. The graduated rear sight of the rifle, machine gun, or other weapon, so called because the intercept on the rear sight equals the tangent of the angle of elevation multiplied by the distance between the front and rear sights.

Tank. A self-propelled vehicle of the track-laying (caterpillar tractor) type, combining fire power, mobility, protection and shock action. Since the earliest times man, appreciating his own severe limitations, has desired to effect a combination of mobility, protection, fire power and shock action, in some machine which he could use to attack his enemies on the battlefield. The horse was the first available weapon which to some extent fulfilled these desires. War chariots, sometimes armed with scythes, appeared as early as 3500 B.C. Elephants, both with and without drivers, have been employed in battle. It is reported that about 1600 A. D. the Prince of Orange devised a "sailing ship on wheels." Various protected devices of very low mobility, were used in the attack of fortified places, but were of little use in open battle. But the lack of suitable motive power retarded for thousands of years the development of a really effective vehicle. The power of the defense in the World War, due chiefly to the development of automatic weapons and fortification, called attention anew to the need of a vehicle capable of crossing rough country and obstacles, to bring protected fire to bear on the enemy. The two latest inventions, which made the tank possible, were the internal combustion motor and the caterpillar type of traction. The English secretly built armed and armored vehicles with caterpillar traction and gas engine power. To avoid betraying their plans they gave out that these vehicles were in fact intended as water carriers, hence the name tank. The tank made its first appearance in battle on the Somme, September 15, 1916, when 49 vehicles were employed. Their performance did not arouse any great enthusiasm. Tanks were employed with increasing success throughout the World War, but at its close they were still very unsatisfactory machines. The maximum speeds of the different types varied from 3.5 to 8.5 miles per hour. They were unwieldy and mechanically unreliable. Tank tactics were not sufficiently developed. Since World War I there have been great advances in the mechanical construction and equipment of tanks. Weights have varied from 1¼ to 165 tons. Reliable vehicles having speeds up to 60 m.p.h. on tracks have been developed. The present tendency seems to be toward a few types of moderate weights, discarding the very heavy, and probably the very light. Armored forces, of which tanks are the chief combat element, now form the spearheads of attacks whenever it is possible to use them.

Tank Carriers. Automotive vehicles or carriers designed or used for transporting tanks.

Tank Machine Gun. A machine gun specially mounted for use in a tank.

Tank Obstacles or Traps. Obstacles intended to hinder the advance of tanks, such as mines, ditches, barriers of concrete, etc.

Tankette. A very light tank or armored tractor.

Tanks, Classification by Weight. Tanks have varied in weight from 1¼ tons to 165 tons. There is no rigid classification and the following is approximate only. Very light, (tankette, whippet) 2 to 5 tons. Light, 5 to 16 tons. Medium, 16 to 35 tons. Heavy, above 35 tons.

Taps. The last prescribed military signal at night; at its sounding all lights must be extinguished and all noises cease.

Targ. A marker of metal or other substance used to indicate the intersection of the arms of a plotting board. A pointed marker placed over a gun sight to indicate its exact position.

Target. The objective against which fire is directed. An object presenting a mark to be sighted at or fired at. A marker for sighting in survey operations. A small shield or buckler. To determine by actual firing the proper setting of the sights of a gun in order that the target may be hit.

Target Angle. The angle, 90 degrees or less, between a vertical plane through the keel-line of a ship used as a target and the plane of sight.

Target Area. An area assigned a weapon or unit to cover by fire.

Target Cloth. A coarse cloth used in mounting paper targets.

Target Designation. The procedure of pointing out or describing the location of a target. The usual methods are by tracer bullets, by pointing, by oral description, or by combinations of these. See also Tracer Bullets.

Target Discs. See Marking Discs.

Target of Opportunity. A target not included in the initial fire schedule, which appears during the course of an engagement. A fleeting or transitory target.

Target Offset. The horizontal angle whose vertex is at the target, measured clockwise from the piece to the observation post.

Target Practice. Measures taken to perfect the soldier in the use of a missile weapon, including firing on a range or in the field.

Target-airplane-battery (T.A.B.) Clock. Concentric circles and radial lines on which splashes are located to the nearest intersections of a circle and ray, used in aerial spotting. See also Range-deflection Fan.

Targeted. Said of a gun whose proper sight-setting has been determined by actual firing.

Tarpaulin. A large sheet of heavy canvas, usually water-proofed, used as a protective covering for supplies.

Task. A definite amount of work assigned to an individual or unit, or a relief of workers, specifically in field fortification. In trench work a task is usually a 5 foot length of trench, to be executed by one man, or by two or three successive reliefs. To assign a task. To charge or impose.

Task Force. A provisional tactical unit, composed of elements of one or more arms and services, for the execution of a specified mission.

Tattoo. A drum or bugle signal, usually sounded about 9 o'clock, P. M., as a warning to soldiers that they may repair to quarters at the proper time.

Taxi. To operate an airplane under its own power, either on land or water, without taking off, or after landing.

T-Base. A T-shaped device of wood, on which the tripod of a machine gun is mounted to insure greater stability.

Team. A group of personnel that operates a weapon. Any number of animals attached in draft to a vehicle.

Teamwork. The accomplishment of a number of persons or organizations associated or grouped together and working in coordination for the accomplishment of a common end.

Tear Gas. A chemical agent which causes tears. A lacrimator.

Tear Shell. A shell charged with tear gas.

Technical Control. A control which is exercised for the purpose of standardizing technical instruction, operation, maintenance and inspection.

Technical Reconnaissance. A reconnaissance conducted, usually by technical personnel, for a technical purpose; e.g., an examination of the condition of a road or railroad, bridge, source of water supply, etc.

Technical Regulations. A series of pamphlets supplementary to Training Regulations, and covering subjects of a technical nature.

Technical, Supply and Administrative Staff. See Special Staff.

Technical Troops. Troops organized and trained for duties of a technical nature, as engineers, signal troops, etc.

Telemeter. Any instrument for measuring distance. A range finder.

Telescopic Sight. A sight including a telescope, designed for more accurate shooting at longer ranges. See also Panoramic Sight.

Templet, Template. A pattern or guide, as a frame to show the shape of the profile of a trench or embankment, a gabion form, etc. A stencil or perforated card for laying out standard size artillery target areas on a map.

Temporary Shelter. Tentage and buildings of a temporary character or portable type.

Tenaille. A low outwork in the main ditch of a fortification between two bastions, masking the scarp wall of the curtain and adjacent flanks.

Tenaille Line. A number of redans, of the same or different sizes, forming alternate salient and reentrant angles.

Tenaillon. An outwork consisting of one long and one short face, placed on either side of a demi-lune, or other work, to strengthen it.

Tent. A canvas shelter. A portable lodge, made of canvas or other material, stretched and sustained by poles and pins, used for shelter. To pitch tents upon.

Tent Fly. The outer canvas of a tent with a double top.

Tentage. A supply of tents.

Tented. Covered with, sheltered by, or provided with tents.

Terminal Angle. The angle formed by the tangent to the trajectory with the horizontal plane at the point of impact.

Terminal or Final Velocity. The velocity of a projectile at the point of fall or on impact. Striking velocity.

Terminology. The technical or special words or terms used in any business, art, science, or the like.

Terrain. The ground. The surface of the earth. An area of ground considered with reference to some special purpose.

Terrain Appreciation. Understanding of the adaptability of a terrain or terrain features to various military purposes.

Terrain Compartment or **Corridor.** See Compartment of Terrain.

Terrain Exercise. A tactical exercise in which a military situation, the troops being imaginary or outlined, is laid out and solved on the ground, the solution generally being in writing.

Terrain Flying. Directing the flight of an aircraft by actual view of the terrain and the routes and landmarks thereon, as distinguished from instrument or blind flying. The landmarks may be previously known, or recognized from a map, or both.

Terrain Nomenclature. The names of the various features, especially the natural features of a terrain.

Terrain, Open, Close. Open terrain or country is free or relatively free from obstruction to passage or view. Close terrain or country is generally broken and wooded.

Terrain Sensing. A positive sensing based upon the position of a shot on the terrain and not sensed with reference to the target proper.

Terreplein. The horizontal surface in rear of the parapet of a fortification, on which guns may be mounted.

Terrestrial Observation. Observation from the ground or from any natural or artificial structure on the ground.

Territorial. Designating or pertaining to forces organized primarily for territorial defense. A member of a territorial force. Geographical.

Territorial Army. An army or force, organized in time of peace, composed of reserve troops. Armed forces maintained in territorial posessions other than the homeland.

Testing Tank. A large tank in which submarine mines are tested.

Tete-de-pont. A work constructed at the end of a bridge to cover and protect communication across it. A bridgehead.

Tetryl, Tetryl Cap. A high explosive used as a boosting charge, especially for TNT. A blasting or electric cap using tetryl as a booster.

Theater of Operations. Any part of a theater of war in which actual military operations are, or are to be conducted; there may be several such theaters, as in the World War. An area within which all military operations are under the control of a commander of the field forces.

Theater of War. The entire area of land and water which may become involved in the operations of a war. It includes the territory of both belligerents and their allies, and may include the territory of neutrals violated by a belligerent. In the case of maritime nations it also includes the seas.

Thermit, Thermite. A mixture of powdered aluminum and metallic oxide, capable of producing great heat (3,000°C) when ignited. It is used in welding, and as an incendiary in shells and bombs.

Thin Lines. Widely deployed lines. Lines deployed at wide intervals; often used for advance under heavy fire.

Thirty-seven mm Gun. A light, quick-firing gun of 37mm caliber, which can be readily carried forward by advancing infantry.

Thousand-inch Range. A miniature target range, measuring 1000 inches from firing point to targets; used for machine gun and rifle practice, sometimes in connection with landscape targets.

Three-point Landing. The landing of an airplane in which contact is made with the surface with three points at once; in case of a landplane with two wheels and tail skid touching the ground at the same time.

Throttling Bar, Groove, Rod, Valve. The various devices by which the flow of oil in hydraulic cylinders is regulated.

Throttling Pipe. A pipe connecting the ends of two recoil cylinders. See also Connecting, Equalizing Pipe.

Thrust. An attack on a limited front. A hostile attack with any pointed weapon. Extending a rifle, with bayonet attached, suddenly to the full length of the arms, without movement of the feet. The force exerted in prolongation of its shaft by the propeller of a vessel or aircraft, due to the reaction of the water or air against its flukes or vanes; which serves to propel the craft.

Thumb Nut, Thumb Screw. A nut or screw designed to be turned by hand. A wing nut.

Thumbing the Vent. In old guns having open vents for the insertion of friction primers, it was the custom to close the vent by thumbing as a precaution against premature explosion from burning fragments in the bore.

Tilt. An inclination to the vertical or horizontal, as of an airplane or aerial camera. A thrust with a rapier. A mounted combat between knights in an arena. A joust. To throw out of level or plumb.

Time and Space. The distances over which elements must march and the time required for such movements, considered with reference to their effect upon tactical plans.

Time Element. The amount of time that is available for the purpose under consideration.

Time Fuze. A fuze which can be set to function at any desired time after the projectile leaves the muzzle of the gun. A tube filled with a relatively slow burning compound, used to explode charges in demolition and blasting; also called safety fuze.

Time Interval. The interval of time between two successive observations made on a moving target during tracking.

Time Interval Clock. A clock to indicate time intervals in tracking, by the ringing of a bell.

Time of Attack. The hour designated for the forward movement from the line of departure to begin. H-hour.

Time of Flight. The time in seconds required for the projectile to travel from the muzzle of the piece to the point of impact or burst.

Time Train. The powder of a time fuze by means of which the time of explosion is fixed.

Title. The military or naval rank of an officer or enlisted man, used as a form of address and in official designation.

Title of an Order. The official designation of the command by which issued; it may, where circumstances require, be shown by a code name.

To Arms. A summons to war. An alarm call, being the signal for soldiers to fall in on their company parades, under arms, as quickly as possible.

To Horse. The signal for mounted men to assemble at a designated place, mounted and under arms, as quickly as possible.

To Lie. To be in quarters, in camp, in cantonment, in ambush, under cover, in wait, etc.

To the Color, Standard. A bugle call sounded when the color (standard) is saluted at ceremonies.

To the Rear. A movement executed by each man turning to the right about in marching and stepping off in the new direction.

TOG. The triangle formed by the target, observer and gun.

Tompion. A plug fitted into the muzzle of a cannon, when not in use, to protect the bore from injury.

Top Carriage. The upper assembly of a gun carriage, which immediately supports the piece (or cradle), and moves in recoil. See Cradle; Sleigh.

Topographical Crest. The top or highest part of a ridge.

Topographical Reconnaissance. A reconnaissance of an area of terrain, usually including a sketch of the ground, showing in detail all the military features, and accompanied by a written report.

Topography. The ground forms and other physical features of a terrain considered as a whole. Map making, map reading, and sketching.

Torpedo. A metal case containing explosives, so arranged as to explode on contact or when fired electrically; it may be stationary, capable of being projected, or self-propelled.

Torpedo Plane. An airplane designed to carry and discharge torpedoes or bombs. A bombing plane.

Torpedo, Submarine. A self-propelled, cigar-shaped projectile containing a large charge of high explosive used to attack vessels below the water line. Torpedoes are classed as simple automobile and dirigible torpedoes; the latter being under control by means of an electrical connection to the shore.

Touch Signals. Signals conveyed by touch, as those of a tank commander to his driver.

Tour of Duty. Any duty performed successively or by order. Duty in succession. The length of time covered by a prescribed duty.

Tourniquet. A tight bandage or ligature used to stop hemorrhage in case an artery is severed.

Towed Target. A mobile target for practice firing, representing a vessel, towed by a tug. A sleeve towed by an aircraft. A sled or vehicle towed by a truck or tractor.

Tower. A projection from a wall for purposes of defense. A citadel. A fortress. A tall, movable structure of wood, used in attacking the walls of fortified places. A pier of a suspension bridge.

Town Major. An officer in charge of the billeting of troops, etc., in a town.

Toxic. Having or producing a powerful deleterious physiological action.

Toxic Chemical Agent. See Chemical Agent, Toxic.

Trace. The outline of a fieldwork in plan. The horizontal projection on the terrain of the interior crest of a work or the forward edge of a trench. To lay out the trace of a defensive work.

Tracer. Any device, as one producing a train of smoke or fire, attached to a projectile to enable the firer and others to observe its flight.

Tracer Bullet. A bullet containing a combustible pellet (as magnesium) in the base; the pellet being ignited from the propelling charge and burning during a part of the bullet's flight, emits a bright light or smoke by means of which the flight of the bullet and the shape of the trajectory can be seen and followed.

Tracer Control. Control of fire with the aid of tracer bullets.

Tracer Limit. The range within which a tracer pellet burns and is normally visible.

Tracer Shell. High explosive or other projectiles which are provided with tracers.

Tracing. A map or drawing on translucent paper or cloth, from which an actinic print may be made. Laying out on the ground the lines of trenches or other works of fortification.

Tracing Party. A detail of men to trace out the lines of intrenchments.

Tracing Tape. Tape used in tracing trenches on the ground.

Track. The actual path of an aircraft over the surface of the earth. The path of a vessel. A flexible belt, usually of steel, which forms the bearing surface for a vehicle of the track-laying or caterpillar tractor type. Marks on the ground made by the feet of men or animals, especially if visible to hostile aerial observation or on a photograph. A footpath or trail. Any sign left by the enemy of his presence, movements, etc. To follow a moving target with the gun sights. To plot the course of a moving target.

Track Plate. One of the links or sections of which the track of a track-laying (caterpillar tractor) vehicle is composed.

Tracking. The act of determining the expected course of a moving target over a limited time by establishing its position periodically and prolonging the line thus formed. The process of making successive observations on a moving target for the purpose of plotting its course, and predicting its future positions. Following a moving target with the sight so that the piece is continually pointed at the target.

Track-laying Vehicle. See Caterpillar Tractor.

Traction. The act of drawing an object, specifically, a vehicle, by animal or mechanical power. Propulsion of an airplane by means of a propeller or propellers mounted in front of the supporting surfaces.

Tractor Airplane. An airplane in which the propeller or propellers are forward of the main supporting surfaces.

Tractor-drawn Artillery. See Artillery, Tractor-drawn.

Traffic. The movement of vehicles, animals, troops, on a route of travel. The carriers and things carried in transportation.

Traffic Circulation. The movement and direction of traffic on the roads in an area.

Traffic Control. The control of the movement along roads, or other channels of traffic, of persons, animals and vehicles. It includes all measures taken by a command to insure the uninterrupted and systematic flow of traffic in the area under consideration.

Traffic Control Posts. Places, usually important intersections, at which one or more men are stationed for the purpose of controlling traffic.

Traffic Patrols. One or more men who constantly move along roads for the purpose of controlling traffic between traffic control posts.

Trail. The part of the stock of a gun carriage which rests upon the ground when the piece is unlimbered. A foot path; a route for pack animals.

Trail Ferry. See Rope Ferry.

Trail Handspike. A lever or handle inserted in the end of the trail of a gun, by means of which it is moved in azimuth.

Trail Spade. A metal spur, or prong, on the under side of a gun trail at its end, which is driven into the ground by the recoil of the first shot, and thereafter prevents movement of the carriage during firing.

Trailer. A vehicle having no motive power of its own, which is attached to the rear of and towed by another vehicle.

Trailer Mount. A trailer used as a mount for a weapon. A weapon mounted on a trailer.

Trailing Edge. The rearmost edge of an airfoil or of a propeller blade.

Train. That portion of a unit's transportation, including personnel and animals, operating under the immediate orders of the unit commander primarily for supply, evacuation and maintenance. To instruct, especially by practice. To aim a gun.

Trainee. One who is undergoing training.

Training. Instruction of any kind in any branch of the military art, especially practical instruction.

Training Area. Any place or area for the training of troops, which provides the terrain and other facilities required.

Training Center. A large, organized training area, or a group of such areas.

Training Guide. A detailed explanation of each training subject required by the training program and schedule; it includes the following subheads, viz: references; additional data; standards required; and method of inspection.

Training Management. The planning and direction of training so as to make the most effective use of the means and time available. It is a function of command, and therefore the responsibility of every unit commander.

Training Marches. Marches intended to harden the troops and keep them in condition, and to instruct them in the duties incident to campaign; practice marches.

Training Manuals. A series of pamphlets published in elaboration of Training Regulations and Technical Regulations and containing instructions on methods of procedure to be followed in the performance of or instruction in certain training.

Training Mission. The ultimate objective or goal of the training efforts of a unit, to be attained during a complete cycle of training.

Training Objective. Every phase of military training has a well-defined objective, contributory to the ultimate objective. Each phase is marked by certain well-defined steps which provide systematic progression from phase to phase, until the ultimate objective is approached or attained.

Training Order. An order which announces general training provisions of a permanent or semi-permanent nature; it prescribes the general doctrines, policies, scope, and character of subjects, as well as the general methods of procedure applicable in principle to all arms and branches.

Training Program. An outline or general statement of the objects to be attained in training, the subjects to be covered, the standards of proficiency expected, methods of inspection, and, when desirable, the season of the year or the relative order in which the subjects are to be taken up.

Training Regulations. A series of pamphlets containing only those regulations which govern the training of units, but may incidentally deal wholly or in part with so much of the training of the individual and the technique of the arm or service as is considered essential to secure the proper degree of unit or team training.

Training Schedule. A detailed statement, often in tabular form, showing the date, hour, and duration of the periods of instruction or the subdivisions thereof, the place and character of instruction, special equipment or uniform required, etc., governing instruction during a stated period.

Training Standards. Standards of proficiency for individuals and organizations in the various grades and categories, established by the War Department.

Training Stick. A stick having a knot at one end and a ring at the other, used in bayonet training.

Training Supervision. A function of command which should be exercised, when practicable, by the unit commander in person.

Training Unit. A company, battalion, or other unit, organized and maintained for training purposes.

Trajectory. The curve described by the center of gravity of a projectile in flight. The path of an aircraft or other object through the air.

Trajectory Chart. A graphic range table. A chart used in antiaircraft firing which shows graphically the trajectories for various slant ranges, together with times of flight, fuze settings, elevations, super-elevations, angles of site, altitudes and horizontal ranges.

Trajectory, Rigidity of. See Rigidity of Trajectory.

Transfer. To change from one organization to another. To shift fire from one target to another, or from a transfer point to a target. See Fire, Transfer of.

Transfer Limits. Limits of distance within which transfers of fire are reasonably accurate.

Transfer Point. A point at which control over railway trains, motor convoys, or reinforcements passes from one commander to another. A check or registration point in artillery fire. See Fire, Transfer of.

Transformer. A device used to change the voltage of an alternating current of electricity by induction from one circuit to another; or to change alternating to direct current or conversely. An apparatus which projects the oblique photographs of a multiple-lens aerial camera into the plane of the central photograph, and makes the prints of same (wing prints).

Transient or Transitory Target. A target that affords a very limited time for adjustment and fire for effect. A fleeting target. A target of opportunity.

Translating Crank. The crank by means of which a breechblock is withdrawn from or returned to the breech recess.

Translating Roller. A long, threaded roller in the breechlock tray of certain cannon, by which the breechblock is withdrawn from and returned into the breech.

Translating Roller Breech Mechanism. A mechanism used in seacoast cannon, in which on opening the breech the breechlock is first withdrawn straight to the rear onto a tray, and then swung clear of the breech.

Transmitted. Forwarded through proper channels of communication. Sent or transferred from one person or place to another.

Transom. A horizontal plate, beam or frame across an opening as between the cheeks of a top carriage, the side frames of a chassis, the trusses of a bridge, etc.

Transport. A ship, train or other means used to move troops and their equipment and supplies. That part of the personnel and equipment of an organization that is concerned with transportation. All equipment used for purposes of transportation. Trains in general. To move personnel or materiel from one place to another.

Transport Corporal. A noncommissioned officer in charge of the transport of a machine gun or other small weapon.

Transport Vehicles. Motorized vehicles employed for transport. They include: (1) Cross-country; (a) cargo, (b) special (such as communication, motor-servicing, and salvage vehicles); (2) Road; (a) automobiles, (b) cargo trucks, (c) tank carriers, (d) miscellaneous special.

Transportation. The act or means of transporting. The movement of individuals, troops, and supplies by land, air, and water. All vehicles and equipment used in transportation.

Transports. Government vessels for the conveyance of troops.

Traps. Devices or contrivances used to catch an enemy unawares. Snares or stratagems.

Travel Pay. An allowance for traveling expenses. See also Mileage.

Travel Ration. The ration prescribed for troops traveling otherwise than by marching, and separated from cooking facilities.

Traveling Position. The position in which a gun or any apparatus is placed for transport, as distinguished from its firing or working position.

Traverse. A bank of earth constructed to afford cover against enfilade fire and to localize the effect of shell bursts. The movement of a piece about a pivot or along an axle in a horizontal direction. An irregular line, from a known point to another point, established by successive directions and distances, used for control and location in mapping and sketching, for establishing gun positions, etc. A journey. A transom. To move the muzzle of a gun to the right or left in pointing. To travel across. To run a traverse. To determine and stake out on the ground, or plot on the map, the lengths and azimuths of a series of straight lines, usually from a known point to an unknown point.

Traverse Wheels. Wheels supporting the rear of the chassis in certain types of fixed mounts, and rolling on a base ring or traverse circle, to permit movement of the piece in azimuth.

Traverse Circle. A base ring on which the traversing rollers bear in certain types of gun carriages.

Traversed Trench. A trench consisting of a series of alternating fire bays and traverses.

Traversing Arc, Rack, Segment. A geared circular rack attached to a gun or carriage by means of which the piece is moved in azimuth.

Traversing Handwheel. A handwheel by which a gun or gun carriage is revolved in azimuth or direction.

Traversing Mechanism. The mechanism for moving a gun in direction.

Traversing Rollers. A set of live rollers on which the racer of a gun carriage revolves in azimuth.

Traversing Stops. Devices used to prevent traversing beyond certain limits.

Travois. An improvised means for transporting sick and wounded. It consists of two long poles, forming shafts for a horse, the free ends dragging on the ground

Tray, Breechblock. See Breechblock Tray.

Tray, Loading. See Loading Tray.

Tray Latch. A latch on the face of the breech of a cannon which secures the breechblock tray in the firing position.

Tread. The berm in a fortification on which a soldier stands while firing over the parapet. The fire step. The distance between the wheels of a vehicle, on the same axle. The bearing surface of a railroad rail.

Treaty. Any formal agreement between two or more independent nations.

Trebuchet. A ballistic machine which threw great stones, used in ancient and medieval warfare. It was developed from the Roman onager and consisted of a frame resting on the ground, to the front end of which a heavy vertical frame was rigidly attached; through the vertical frame ran an axle which had a single stout spoke. At the extremity of the spoke was a cup to hold the projectile. When firing, the spoke was forced back and down against a compensating weight, a torsion of twisted ropes or springs, by a windlass, and then suddenly released. The spoke then kicked the crosspiece of the vertical frame and shot the projectile forward. A dropping weight was also used as the propulsive power. See also Balista; Ballistic Machines; Catapult.

Trench. An excavated ditch with an embankment (usually formed of the excavated earth) on one or both sides for the protection of men while firing or moving from one part of the position to another. See Trenches, Classification of.

Trench Accessory. Any special work or facility designed to increase the usefulness of an intrenched position or to provide for the safety, health, or convenience of the troops.

Trench Artillery. See Artillery, Trench.

Trench Boards. See Duck Boards.

Trench Cavalier. A work consisting of a parapet raised on a mound of earth to give it command over the covered way of the besieged work.

Trench Equipment. Special equipment designed for use in trench warfare.

Trench Foot. An affection similar to chilblains, marked by discoloration of the feet and, in severe cases, by gangrene, caused by the combined effect of cold and wet upon the feet.

Trench Gun or Mortar. A small portable cannon or mortar used at short ranges for dropping bombs or similar projectiles upon the enemy.

Trench Knife. A small knife or dagger designed for use in the trenches.

Trench Standards. The prescribed dimensions for the traces and profiles of all types of trenches.

Trench System. Comprises all field works included in a defensive zone.

Trench Trace. See Trenches, Classification of.

Trench Warfare. A kind of warfare in which both the opposing armies have lost the initiative and power of maneuver and, for the time being, have constructed and occupied intrenchments for protection against the enemy. Stabilized warfare.

Trenches, Classification of. Trenches are classified: (1) According to the direction in which they run, as parallels, approaches, and switch trenches; (2) According to their traces, as traversed, octagonal, wavy, zig-zag, and echelon; (3) According to their profiles, as fire and communication trenches; (4) According to the situation in which employed, as hasty and deliberate; (5) According to the position of the firer, as prone (shelter), kneeling, and standing. There are other special classifications. See type in question.

Trenches, Nomenclature of. The nomenclature of the trench includes the following items whose definitions are given elsewhere: trace, profile, exterior slope, superior slope, parapet, fire crest, elbow rest, fire step, revetment, interior front slope, interior rear slope, berms, A frames, trench boards, drain, parados, height, depth, cover, relief.

Triage. A place where casualties are sorted and classified for further disposition. The process of sorting casualties. See also Clearing Station.

Trial Elevation. The bracketing elevation of the fork obtained in trial fire. The elevation or correction arrived at as a result of trial fire.

Trial Judge Advocate. See Judge Advocate.

Trial Range. In precision fire, the range to the center of a bracket or a range giving a target hit. The trial elevation is the quadrant elevation corresponding to the trial range.

Trial Shot. A shot fired during trial fire.

Trial Shot Correction. An adjustment correction resulting from trial fire, used in entering fire for effect.

Trial Shot Point. A point in the field of fire, visible from the observing stations, at which trial shots are fired. A point in space at which trial fire is directed.

Trial-shot Method. In this method, four shots are fired, singly or by salvo, with the same data, at a fixed point in the water, and full correction is made for the deviation of the center of impact from the trial shot point.

Triangular System of Organization. An organization in which a unit includes three subordinate units, as a three-platoon company, three-regiment brigade or division.

Trigger. An exterior lug on a firearm which, when pressed, releases the hammer or firing pin and causes the explosion of the cartridge or cap.

Trigger Guard. A metal loop protecting a trigger.

Trigger Squeeze Exercises. Preparatory exercises intended to teach a man how to squeeze the trigger of a rifle properly.

Trinitrotoluene, Triton, Trotol, T.N.T. A high explosive, well adapted to military uses, as for charging shells and for demolition.

Trip Wires. Wires placed low in front of a defensive position for the purpose of tripping individuals so as to betray their presence or to discharge bombs or mines.

Triplane. An airplane having three main supporting surfaces (wings) placed one above the other.

Tripod. A three-legged stand for a gun, observing instrument, sketching case, etc. A form of field derrick, consisting of three timbers lashed together at one end.

Tripping. The act of releasing the counterweights of a disappearing carriage and thus causing the piece to go into its firing position. Releasing a latch or detent.

Tromblon. A firearm which was formerly fired from a rest and discharged several balls or slugs. An ancient wall-piece. A rifle grenade discharger.

Troop. A mounted unit composed of a headquarters and two or more platoons, constituting the basic unit of its type.

Troop Leading. The art of leading troops in maneuver and battle. Minor tactics.

Troop Leading Problem. See Problem of Execution.

Troop Movements. The movements of troops from one place to another by marching, motor transportation, by rail, or by water.

Troop Parade. A parade held in the afternoon. See Company Parade.

Troop School. A school for newly commissioned officers under the direct control of post commanders, wherein are taught the basic principles for officers of the combatant branches.

Troop Training. The preparation of individuals and units for battle by practice and application on the drill ground and in tactical exercises and maneuvers, supplemented by instruction in schools.

Trooper. A cavalry soldier.

Troops. Soldiers in general or collectively. An armed force.

Trous-de-loup. An obstacle of military pits in the form of inverted cones or pyramids, usually having a pointed stake at the bottom of each.

Trowel Bayonet. A bayonet shaped like a trowel, designed to serve both as a bayonet and as an intrenching tool.

Truckhead. A supply point where loads are transferred from truck transportation from the rear. See also Railhead.

Truce. A suspension of hostile operations for a specified time for the purpose of parleying, burying the dead, or any other purpose agreed upon.

True Azimuth. Azimuth measured from the true meridian.

True Center of Impact. The theoretical center of impact of an infinite number of shots.

True Copy. A copy of an order or other document, certified as true by an officer, over his signature.

True Distance. The actual distance to the target as measured along the line of position.

True Error. The divergence of a point of impact (or burst) from the true center of impact (or burst).

True North. The direction at any point to the true north pole.

Trumpet. A wind instrument, without keys, similar to but somewhat larger than the bugle.

Trumpet Call. A service call sounded on the trumpet.

Trunnion. One of the two pivots supporting a cannon and forming the horizontal axis about which it revolves in elevation.

Trunnion Band. The hoop to which the trunnions of a gun are attached or of which they form a part.

Trunnion Bearings. The bearings in which the trunnions of a gun or gun lever arms turn.

Trunnion Cap. A removable latch at the top of a trunnion bearing. A cap square.

Trunnion Carriage. That part of a gun carriage in which the trunnions rest; a top carriage.

Tube Sight. A sight formed by a small tube placed on top of a cannon at the muzzle.

Tune In. To adjust the receiver of a radio set to receive messages at a given wave length.

Turn and Bank Indicator. A gyroscopic instrument combining in one case a turn indicator and lateral inclinometer; for use in aircraft.

Turn Indicator. An instrument for indicating the existence and approximate magnitude of angular velocity about the normal (vertical) axis of an aircraft.

Turning Movement. A form of attack in which the command is separated into two forces which operate beyond supporting distance of each other, one force designated to hold the enemy in place, while the other, usually the stronger, makes a detour to strike him on the flank.

Turret. A cylindrical, hemispherical, or cylindro-spherical structure, heavily armored and sometimes revolving or portable, containing one or more cannon. A high wooden tower on wheels or skids, which was pushed toward a fortification to enable attackers to surmount its walls. Part of the superstructure of a tank, mounting one or more guns.

Twenty-five Per Cent Zone. That portion of the dispersion diagram the dimensions of which are two probable errors in range by two probable errors in deflection, and the center of which is at the center of impact. It is that area which is common to the 50 per cent zones in range and direction and in which 25 per cent of the shots are expected to fall.

Twist. The inclination of the spirals formed by the grooves to the axis of the bore in a rifled piece. Twist may be uniform or increasing, and is usually expressed as one complete turn in a length of so many calibers.

Twisted Pair. Two insulated conducting wires twisted together, used to form a complete metallic circuit.

Two-lined, Three-lined Crater, etc. A crater produced by a subterranean mine, whose diameter is two, three, etc., times the line of least resistance. See Mine, Common; Mine, Undercharged.

Two-track Road. A road wide enough for two columns of traffic throughout its entire length.

Two-way Controlled Traffic. A system of traffic control in which two-way traffic is maintained on one-track roads, or narrow sections of roads, by permitting the traffic to flow alternately in the two directions.

Two-way Road. A road wide enough to permit traffic to move in both directions at the same time, and which is so designated.

U

Ultimatum. The final conditions or terms offered by one government or commander to another government or commander for the settlement of differences. A final demand.

Umpire. An officer who assists in the conduct of a field exercise or maneuver.

Unattached. Not attached to any organization. Unassigned.

Uncase. To take out of the covering, or spread to view, as a flag or colors.

Unconditional Surrender. A surrender not limited by terms or stipulations.

Uncoordinated Attack. An attack, usually resulting from a rencontre or advance guard action, in which the attacking elements are committed to action piecemeal, as they come up.

Uncover. To move from in front of. To divest of head covering.

Uncovered Approach March. An approach march in which there is no adequate covering force to provide security against a ground attack. See Approach March, Covered Approach March.

Under. Subject to the orders of. Below, in a subordinate position. In subjection to.

Under Arms. Armed and equipped and in readiness for action, or duty. In the rendering of salutes and courtesies, under arms means with arms in hand, or having attached to the person a hand arm or the equipment pertaining directly to the arm, such as cartridge belt, pistol holster, automatic rifle belt. Exceptions: officers wearing officer's belt without arms attached.

Under Canvas. In tents.

Under Command. To be subject to orders for any duty.

Under Cover. Protected from the fire or observation of the enemy by any means.

Under Fire. Exposed to and enduring the fire of the enemy.

Under Orders. Having received orders, and awaiting the designated time to execute them.

Under Sentence. Liable to punishment as the result of a sentence of a court-martial.

Undercarriage. See Landing Gear.

Undercharge. To use too small a charge.

Underground Warfare. Fighting carried on from positions below the natural surface of the terrain. Subterranean or mine warfare.

Undisciplined. Not trained to regularity or obedience to orders.

Unditching Gear. Devices, such as small logs and ties, fastened to the track plates of a track-laying vehicle to aid in moving it when it is stuck.

Unfavorable Target. A target unsuitable to the characteristics of a particular weapon.

Unfurl. To unroll and display.

Uniform. The costume of a soldier. The prescribed habiliments and equipment for various arms, services, grades and occasions. Regular and systematic. Conforming to rule. To place in uniform.

Unilateral Adjustment. An adaptation of the bracketing method of adjustment by unilateral observation.

Unilateral Observation or Spotting. Lateral observation of fire by a single observer. Lateral observation of fire from one station only, when the target angle is greater than 100 mils and less than 1300 mils.

Union. The white stars on a blue field in the flag of the United States. A confederation, coalition or league. The United States or Federal Union of North America. The act or condition of being united or leagued. A form of joint for pipes or rods that is easily disconnected.

Union Jack. The British military flag, a combination of the flags of England, Scotland and Ireland.

Unit. A military force having a prescribed organization. Units vary in size from a squad to a division. Higher units (corps, army) have no fixed organization.

Unit Assemblage. The equipment and material necessary for the performance of a specific functional duty by an organization, assembled at and received from a single source, as a complete evacuation hospital, etc. See also Group of Parts.

Unit Defensive Area. See Defensive Areas.

Unit Engineer, Quartermaster, Signal Officer, Medical Officer, etc. The senior engineer, quartermaster, signal officer or medical officer, etc., belonging to a division or higher unit, who commands any troops of his branch which are a part of such unit, and serves also as a member of the technical and administrative staff of the unit commander.

Unit Gas Officers. Officers corresponding to chemical officers, on the staffs of brigades and smaller units. They are detailed from the officer personnel of the unit concerned.

Unit Mobilization Plan. A plan, in minute detail, of all steps to be taken in mobilizing any organization.

Unit of Equipment. A container and its contents, authorized for issue to an organization for a special purpose, such as a shoe repair outfit or an armorer's chest, each with its complement of tools. A set of tools for any purpose, as a carpenter's set.

Unit of Fire. The average quantity, in rounds or tons of ammunition, bombs, grenades, and pyrotechnics, which a particular organization or weapon may be expected to expend in one day of combat. See Day of Fire.

Unit Plan. A plan for the mobilization, organization, and methods of operation of the office or unit concerned.

Unit School. A school intended primarily to prepare the personnel of any organization to carry out successfully the current program of training. A school organized in any unit.

Unit Staff. The staff pertaining to any unit.

United Services. The armed services as a whole; the army, navy, and marine corps.

United States Carbine, Caliber .30, M1. The Winchester light carbine, adopted (1942) to replace about 80% of the automatic pistols (Caliber .45) formerly carried. It weighs 4.63 pounds, has an overall length of 36 inches, and an effective range of 300 yards. It fires a light, pistol type projectile, its magazine having a capacity of 15 cartridges.

United States Military Academy. A school, founded in 1802, located at West Point, N. Y., for the education and training of young men for appointment as officers of the army.

United States Rifle, Model 1903. The authorized rifle for general issue to the military forces of the United States; it is a breech-loading magazine rifle of the bolt type and has a caliber of .30 inch; the magazine holds 5 cartridges, and the rifle, without bayonet, weighs 8.69 pounds. It is also known as the Springfield Rifle, as it was first manufactured at the Springfield, (Mass.) Armory. Used also in the Cuban and Philippine Armies. It will ultimately be superseded by the M1 (Garand semiautomatic) rifle.

Units. This term, as used in the War Department Mobilization Plan, refers to divisions, brigades, regiments, battalions, and companies of infantry, or corresponding organizations in other arms and services.

Unity of Command. Signifies that all the forces engaged in the accomplishment of a common end are operating under the will of a single commander; generally considered in connection with the forces of two or more nations. Undivided authority or singleness of control in a military force of any size.

Unlimber. To disconnect the limber from the gun or caisson.

Unload. To remove all cartridges from the chamber and magazine of a rifle; to remove the projectile and powder charge from the bore of a cannon without firing.

Unmilitary. Contrary to the rules of discipline or military procedure. Unworthy of a soldier. Lacking in soldierly appearance or attributes.

Unofficial. Not official. Not clothed with authority.

Unorganized Militia. Comprises all persons, not included in the organized land or naval forces, who have been or may be declared by Congress to be liable to perform military duty in the service of the United States.

Unprofitable Target. A target of such a nature, or so situated, that the results to be expected from firing upon it do not justify the necessary expenditure of ammunition, or the danger of loss of airplanes in attacking it.

Unqualified. Not qualified. Those who in the last practice season failed to qualify as marksman or better, or those who did not fire the course and are not otherwise qualified.

Unserviceable. A term applied to all stores which are no longer of use, being either worn out, damaged, or obsolete.

Unsheathe. To draw from the sheath or scabbard.

Unsupported Attacks. Surprise attacks or raids made by light forces for the accomplishment of a minor mission.

Unwritten Rules. Rules which are not promulgated as written regulations, but which through long observance and custom have become recognized as rules of conduct or warfare.

Uprising. An insurrection or revolt. A rebellion. A mutiny.

Upstream Anchors. The anchors of a ponton bridge upstream from the structure.

Usages of War. Certain customs or practices in time of war which have come to be generally recognized and accepted by all civilized nations.

Utilities. Buildings, structures, plants, and systems; and facilities necessary in connection with their maintenance, repair, and protection.

Utility Railways. Railway facilities, standard gauge or otherwise, at posts, camps, construction projects and other establishments, not in the theater of operations.

Utilization of the Terrain. Taking advantage of all natural and artificial features of the terrain to strengthen one's position, secure concealment or cover, or facilitate any tactical operation.

V

Vacancy. An office or commission, or place or post unfilled.

Van. The front of an army or an advancing body of troops. The troops marching in the front of an army. The leading unit.

Vanguard. That part of a force which marches in front. An advance guard.

Variable Recoil. Recoil automatically varying in length according to the elevations of the piece. For small angles of elevation recoil is long to prevent **overturning**; for large angles of elevation it is short because of the reduced space available.

Variable Velocity. Velocity which changes from instant to instant.

Variations of the Compass. Changes in the pointing of the needle of a compass due to many causes.

Vedette. Vidette. A mounted sentry of an outpost.

Vedette Post. An outguard of two or more vedettes, one of whom is constantly on the alert.

Velocity. The speed of a projectile, aircraft or other moving object, or of the wind; expressed in miles per hour for low speeds, and in feet per second for high speeds, as of projectiles.

Velocity Adjustment. A change in muzzle velocity from that of the range tables due to condition of piece or powder, or weight of charge.

Velocity Error. A value, found from registration fire, whose effect, when combined with the effects for known conditions affecting range, makes the computed range agree with the adjusted range of the registration.

Velocity Error Transfer. A transfer of fire wherein the initial range is computed with the use of a velocity error; for this type of transfer a metro message is required. Commonly called a VE transfer.

Velocity of Recoil. The rearward velocity imparted to a gun and carriage by the force of discharge.

Vent. A passage through the breech or breechblock of a cannon using separate loading ammunition, whereby the flame from the primer is communicated to the propelling charge. An axial vent is one parallel to the axis of the bore. A radial vent is a hole drilled through the piece from the top, at right angles to the axis of the bore; used in muzzle loading cannon. A small outlet of any kind.

Vent Cover. A safety device on certain breechloading cannon, consisting of a radial arm which covers the vent and prevents the insertion of a primer until the breechlock is securely locked in its closed position.

Verbal Order. An oral order usually communicated by the leader in person. More properly, an oral order.

Vernier Sight. A rear sight with a vernier scale attached.

Vertical Base Position-finding System. A system of position finding for moving targets, using one station equipped with a depression position finder.

Vertical Cover. The elevation of the crest of the parapet of a trench or fortification above the bottom of the trench or level of the terrain inside the fortification. Head cover.

Vertical Dead-time Angle. The vertical angle between the line of position at observation and the line of present position.

Vertical Deflection. In antiaircraft fire, the vertical angle between the plane of site at the instant of firing, and the line the axis of the bore of the piece would occupy were there no superelevation.

Vertical Deflection Angle. The angle which must be added to the present angular height to allow for the vertical angular effects due to travel of the target and pointing corrections.

Vertical Deflection Setting. The setting on the vertical deflection scale of the sight corresponding to the vertical deflection angle.

Vertical Interval. The difference in elevation between adjacent contours on any map. Contour interval.

Vertical Jump. The vertical component of the angle between the line of departure and the line of elevation. See Jump.

Vertical Photograph. An aerial photograph taken with the axis of the lens vertical or nearly so; the terrain is shown approximately to scale.

Vertical Pointing Correction. That part of the vertical deflection angle due to causes other than the travel of the target.

Vertical Probable Error. On a vertical target, the range probable error, multiplied by the tangent of the angle of fall.

Vertical Shot Group. A shot group on a vertical target, placed at right angles to the direction of fire.

Vertical Spotting. A method of determining approximate deviations in range by measuring apparent vertical heights of impacts of observed from an elevated station.

Vertical Stabilizer. See Fin.

Vertical Velocity. The vertical component of the velocity at any point of the trajectory.

Very Pistol. A pistol that discharges flares, used for signalling.

Vesicant. A chemical agent whose principal physiological effect is to burn or blister the skin or other tissues.

Veteran. A long-trained and experienced soldier, especially one who has taken part in a war. A person who has served in the armed forces of a nation in time of war.

Veterans' Administration. An independent bureau of the federal government, charged with the control and direction of all agencies concerned with benefits to war veterans, including pensions, insurance, hospitalization, vocational training, domiciliary care, etc.

Veterans of Foreign Wars. A society of persons who have served in the armed forces of the United States during a foreign war; founded in 1899.

Veterinary, Veterinarian. A practitioner of veterinary medicine and surgery. Pertaining to diseases and injuries of domestic animals, specifically horses and mules, and their treatment.

Veterinary Aid Station. A station which relieves organizations of disabled animals, and renders first aid.

Veterinary Collecting Station. A collecting station for animal casualties where they are given primary treatment and sorted for further disposition.

Veterinary Company. A part of the medical regiment whose function is to collect, render first aid to and evacuate animal casualties.

Veterinary Corps. A branch of the medical department charged with the care and medical treatment of animals.

Veterinary Hospital. A hospital for the care and treatment of sick and wounded animals.

Vickers Guns. Machine guns and quick-firing cannon of various types and calibers, manufactured at the Vickers-Armstrong Works in England; used in many armies.

Victoria Cross. A decoration instituted on the termination of the Crimean campaign in 1856, by Queen Victoria of England, and granted to a soldier of any rank for a single act of valor.

Victory Medal. A bronze medal distributed to all persons who saw service in the World War; it is similar for all allied nations.

Vise. To examine and endorse as to correctness.

Visibility. The determination as to whether a point, a route of travel, or an area, is visible from a given point. The state or condition of the weather as to clearness. The distance to which one can see in a haze, wood, or poor light.

Visibility Chart. A chart or map showing the areas visible or invisible from an observation post.

Visiting Patrols. Patrols employed in maintaining communication with the pickets, outguards, supports, and reserves of an outpost, as well as with adjacent units.

Visor or Vizor. That part of armor which covers the face, often hinged to permit uncovering. A projection of a military cap which shades the eyes.

Visual Signals. Signals conveyed through the eye; they include flags, lamps, panels, heliograph, pyrotechnics, etc.

Voice Radio. A small two-way, portable radio-telephone for field use. See Radio.

Volley. The simultaneous discharge of a number of firearms. A flight of missiles. To fire by volley.

Volley Gun. A gun having several barrels for simultaneous discharge.

Volume of Dispersion. See Antiaircraft Volume of Dispersion.

Volume of Fire. The amount of fire that can be or is being delivered.

Volunteer. One who enters the military service of his own free will. To enter into service, or offer one's service, without compulsion.

Volunteer Army. An army recruited by volunteer service.

Vomiting Gas. Chloropicrin, classed as a lacrimator.

Voucher. A paper which serves to attest to the truth of accounts, or to establish facts, usually in connection with payments or issues.

W

Wagon Convoy. A train of wagons carrying supplies, including its escort.

Wagon Cover. A canvas cover for use on wagons.

Wagon Master. A person in charge of a number of wagons.

Wagon Train. A number of wagons used in transporting supplies. A convoy.

Wagoner. A wagon driver.

Waiting Guard. Men held in readiness to replace sick or disabled members.

Walk. The maneuvering of a balloon or airship by the handling crew walking from place to place. A gait of a horse.

Walk Through. To search an area with artillery fire delivered with successive changes in range or elevation.

Walking Wounded. Wounded who are able to walk without assistance.

Wall. A work or structure of stone, brick, or other materials, raised to some height, and intended for defense or security. A fortification. A rampart. The body of a projectile surrounding an interior cavity or core.

Wall Tent. A tent having upright side walls.

Wall Tower. A tower built into or against a wall, common in medieval times.

Walled. Furnished or inclosed with a wall or walls. Fortified.

Wall-piece. A very small cannon, or earlier, a harquebus, mounted on a swivel on the wall of a fortress for short range fire.

War. An armed conflict between states or nations. The profession or art of war. To carry on war. See also Civil War.

War Bread. Bread made from various grains and combinations of grains, used as a measure of economy during war.

War Bride. A woman who marries, or has recently married, a soldier ordered into active service during war.

War Council. An advisory body to assist in formulating war policies.

War Department. The executive department of the government which has charge of all matters relating to the army and land warfare.

War Department General Staff. A body of selected officers charged with developing and recommending plans and policies, and, after their approval by proper authority, with supervising and coordinating their execution.

War Department Intelligence. The military intelligence produced under the direction of the War Department General Staff in time of peace and war, largely strategic in nature.

War Department Overhead. The agencies and individuals, collectively considered, which pertain to the War Department itself and to certain subordinate activities under its direct control.

War Diary. A record of events, kept by an organization in campaign. See Military Diary.

War Establishment. The complete military establishment to meet the requirements of war. The war footing of an army.

War Game. The game of war played on a map by officers to familiarize and perfect them in the art of handling troops in war. A form of tactical exercise.

War Industries Board. A board appointed to coordinate the large industrial problems arising in connection with the prosecution of war and its aftermath.

War of Movement. A war of rapid maneuver. Mobile warfare.

War of Position. Trench or siege warfare. Stabilized warfare.

War Plans. A group or series of separate plans prepared by or under the direction of the War and Navy Departments, and designated to meet a given situation involving a declaration of war and the employment of both the Army and Navy; classed as Joint Plans, Army Plans, and Navy Plans. A plan made to meet a specific situation requiring the use of all or part of the Army of the United States; as a rule it consists of a strategical and logistics plan.

War Plans Division. A division of the War Department General Staff charged with the formulation of plans for the employment of the military forces in various possible situations, either separately or in cooperation with naval forces.

War Reserves. All stocks of raw materials, finished articles, and components held in storage or in the hands of troops in accordance with law and War Department policies, for the purpose of meeting war requirements prior to the establishment of adequate production and distribution.

War Risk Insurance. Life insurance at low rates with many privileges, provided by the government for those serving in the armed forces during the World War; intended as a substitute for pensions.

War Service Chevrons. Gold chevrons worn on the lower half of the left sleeve, one for each six months service in the theater of operations in the World War.

War Vessels, Classification of. War vessels include capital ships, non-capital ships, and auxiliaries. Capital ships include battleships and battle cruisers. Non-capital ships include cruisers, light cruisers, destroyers, gunboats, submarines, aircraft carriers, and mine planters.

War Whoop. The battle cry of the North American Indians.

Warfare. War. State of war. Armed conflict.

Warming-up Effect. The effect on the functioning of any apparatus, as a gun, of becoming warm.

Warning Call. The first call for a formation or duty.

Warning Order. An order issued as a preliminary to a field order, especially for a movement, which is to follow; it may be a message or a field order, and may be either written or oral. The purpose is to give advance information so that commanders may make necessary arrangements to facilitate the execution of the subsequent field order.

Warp. To change the form of an airplane wing by twisting it. Warping was formerly used to perform the function now performed by the ailerons.

War-paint. Paint put on the face by savages when going to war.

Warrant. A certificate of rank or appointment given to one below the grade of commissioned officer. An authority for the payment or receipt of money. Proper authority. A legal writ of arrest, search or seizure. To authorize or guarantee.

Warrant Officer. The next grade below that of commissioned officer.

Watch. A sentinel. A guard. A lying in wait. An ambush. The time during which a guard does duty. Vigilance. To keep guard. To act as a guard. To tend.

Watch-and-sun Method. A method of determining the direction of north. When the hour hand of a watch is pointed at the sun the meridian lies midway between the hour hand and XII. The method is only approximate.

Watchword. A word of parole communicated only to officers of the guard and not to other members. A countersign. A password.

Water Bag. A large canvas bag, mounted on a tripod and provided with faucets, for the supply of potable water to organizations in the field, and the sterilization (chlorination) of water when necessary.

Water Battery. A seacoast battery located very nearly on a level with and close to the water.

Water Box or Chest. A box designed to contain and for the easy carrying of water. A chest containing cooling water for an automatic weapon.

Water Discipline. The observance and enforcement of all regulations and orders governing the use and care of water.

Water Jacket. A space surrounding the cylinders of an internal combustion engine, or the barrel of a gun, in which water for cooling circulates or stands.

Water Point. A place where water is supplied to organizations.

Wattling. The woven part of brushwork. Brush used in weaving gabions and hurdles.

Wave. One of a succession of lines of tanks, skirmishers, foragers, or small columns often employed in attack.

Wave Formation. A formation in successive waves or thin lines for combat.

Waver. To become unsteady. To hesitate or be undetermined.

Wavy Trace. A trench trace which follows a wavy line, and affords protection from enfilade fire.

Way Planks. Longitudinal planks laid under the wheels of vehicles over the deck of a bridge.

Weapon Carrier. Any vehicle that transports a weapon that is not a part of the vehicle. Specifically, the term is applied to a light, wheeled truck, of great cross-country ability, which carries the organic supporting weapons (machine guns and light mortars) of the infantry.

Weapons. In a broad sense the term includes all implements, machines or devices used in war or combat, either for offense or defense. This definition thus includes purely defensive devices such as body armor, shields, steel helmets and gas masks. In a more restricted and more correct military sense the term means any implement, machine or device designed and intended to be used to inflict injury or damage upon enemy personnel or materiel.. The latter definition excludes protective devices, and classes all weapons as offensive, inasmuch as any known weapon may be used either by the attacker or the defender.

Wear and Tear. Ordinary usage in the military service.

Weather Effects. The effects on a standard trajectory of weather or meteorological conditions not standard. See Error of the Moment.

Weatherproof Shelter. A shelter against weather but not necessarily against fire.

Wedge Formation. A body of troops drawn up in a wedge-shaped formation. A defensive flying formation of airplanes, in the general shape of a wedge.

Wedge Type of Breechblock. See Sliding Wedge.

Weight of Metal. The total weight of the projectiles that are or can be fired from a single gun or an assemblage of guns in a given time or in a single discharge.

Welfare. All measures designed to improve the physical, mental and spiritual condition of personnel, and to promote contentment and harmony; an important element of good discipline and morale.

West Point. The seat of the United States Military Academy, about 50 miles north of New York City on the Hudson River.

Wheel. A movement in which troops in line change direction while preserving their alignment; commonly used for change of direction by columns with narrow fronts, or in changing from line to column of squads or the converse.

Wheel Shoes. Tread blocks attached to wheels to facilitate passage over soft or yielding ground. Caterpillar shoes.

Wheel Team. The rear pair of a 4-, 6-, or 8-line team of draft animals.

Wheellock. An improvement on the matchlock, invented at Nuremburg in 1515. It consisted of a small wheel, which revolved at the side of the barrel; the hammer of the gun held a piece of pyrite which, on being rubbed against the rough edge of the rotating wheel, emitted sparks and ignited the priming mixture. The cost of the mechanism and the uncertainity of its action prevented its general adoption. A small arm having a lock of this type. See also Flintlock; Matchlock.

Whippet Tank. A small British World War tank capable of a speed of about 8¼ miles per hour.

Whistle Signals. Signals blown on a whistle.

White Flag. A plain white flag recognized as a flag of truce, and as a token of surrender when displayed over a place, position, or body of troops.

Who is There? The night challenge of a sentry on post.

Wide Sheaf. A battery sheaf of fire, wider than a parallel sheaf.

Wig-wag. A method of visual signaling by means of one flag, waved from side to side and downwards. To signal with a flag.

Wild Shot. A shot of which the armament error is greater than four developed probable armament errors, and is also greater than six firing table errors, either in range or in direction. A shot which falls outside the limits of the area of dispersion or so-called 100% zone.

Will Clear. A term meaning that all elements of the designated organizations shall have passed the indicated point by the designated hour.

Will Follow. A term meaning that a rear element will regulate its march on an element in front.

Will Precede. A term meaning that the leading element of a force in the march will regulate its march on a rear element.

Winchester Rifle. An American repeating rifle which first appeared about 1867 and was extensively used. It was a development of the Colt, Henry and Spencer rifles.

Wind a Crossbow. To draw back the string of a crossbow and set it for firing, by means of a small attached windlass or lever.

Wind and Parallax Computer. An instrument used to determine wind and parallax corrections in antiaircraft fire.

Wind Component Indicator. A device for resolving the ballistic wind into its components.

Wind Components. The component velocities of the wind parallel and perpendicular to the line of fire.

Wind Deflection. The deflection of a projectile caused by the action of the wind, which is allowed for by corrections made on the rear sight.

Wind Drift and Refraction. The effect of constant and variable winds on the apparent direction of sound.

Wind Gauge. A graduated attachment on the rear sight of a gun by which allowances may be made in aiming for the effect of the wind upon the projectile during flight.

Wind Gauge Rule. A rule for moving the strike of a rifle bullet a desired distance right or left, by means of the wind gauge on the sight.

Wind Rule. A rule for setting the wind gauge on the rear sight of a rifle to compensate for the effect of the wind.

Wind Vane Sight. A sight used in aerial gunnery to compensate for the motion of the gunner's plane.

Wind Velocity. The velocity of the ballistic wind. The actual velocity of the wind.

Windage. The influence of the wind in deflecting a projectile from the point at which it is aimed. The amount of change on the wind gauge necessary to counteract the effect of the wind.

Windshield. A cap of light metal placed over the point of an armor piercing projectile to give a streamlined ogive, and increase ballistic efficiency.

Wing. That part of a deployed military force to the right or left of the center, or the center unit. An ornament worn on the shoulders. A small epaulet. The side or flank of a crown-work, horn-work, etc., connecting it with the main work. A general term applied to the whole, or a portion of one of the main supporting surfaces of an airplane, usually designated as right wing, left wing, upper wing, etc.

Wing, Aviation. Two to four groups, which may be of different types of aviation.

Wing Prints. The peripheral, trapezoidal prints from the peripheral lenses of a multiple-lens aerial camera, which have been transformed to the plane of the central print.

Wing Signal. A visual signal displayed on the wing of an airplane, as by lights, movement of ailerons, etc.

Wing Skid. A skid placed near the wing tip of an airplane to prevent damage to the wing in case of contact with the ground.

Wingheavy, Right or Left. The condition of an airplane whose right or left wing tends to sink when the lateral control is released in any given attitude of normal flight.

Wing-over. A maneuver in which the airplane is put into a climbing turn until nearly stalled, at which point the nose is allowed to fall while continuing the turn, then returned to normal flight from the ensuing dive or glide in a direction approximately 180 degrees from that at the start of the evolution.

Wire. A thin filament of metal, used for many purposes. For military uses wire is classified as smooth, barbed, conducting or electric, insulated, weatherproof, twisted pair, outpost, field, firing or lead, mesh, etc., which see. A telegram.

Wire Cart. A small cart, provided with a reel, for carrying and laying field signal wire. A reel cart.

Wire Cutter. A device for cutting strands of wire. Wire-cutting pliers.

Wire Entanglement. An obstacle of barbed wire, erected in place on pickets, or constructed in rear of the site and brought up and placed in position.

Wire Entanglement Drill. The erection of a wire entanglement, which takes the form of a drill.

Wire Entanglements, Types of. There is a great number of types of entanglements. In general they are classified as fixed, including high wire, spider wire, double apron fence, low wire, etc.; and portable, including concertina and ribard spirals, plain spirals, hedgehog, gooseberry, wire cheval-de-frise, etc.

Wire, Field. See Field Wire.

Wire Line. A line consisting of two or more circuits of the same type of construction along the same route, forming an integral part of a wire signal system.

Wire Mesh. A woven mesh, known commercially as chicken or rabbit wire, extensively used for superficial revetments, to increase the bearing power of roads in sandy soil (wire mesh road), and for other purposes.

Wire Mesh Road. A road in sandy soil made by laying wire mesh on the surface.

Wire Net. The telephone stations of a unit and its next subordinate units and of attached and supporting units, considered as a whole. It is designated by the name of the superior unit, as brigade net.

Wire Party. A detail of men to erect or repair barbed wire entanglements.

Wire Patrol. A patrol sent out to inspect wire entanglements for necessary repairs.

Wire Signals. Signals conveyed through electric circuits, such as telegraph, telephone, buzzer, etc.

Wire Systems. All means of signal communication utilizing wire lines. Wire circuits are classified as: (1) Trunk circuits, which connect telephone centrals; (2) Local circuits, which connect telephones to centrals or to other telephones. A further classification is: (1) Metallic circuits, in which two wires form a complete path for the electric current; (2) Ground return circuits, in which one wire is used, with the earth taking the place of the second wire. Additional special circuits are: (1) Simplex circuit, in which a single metallic circuit provides for both telegraph and telephone communication without interference; (2) Phantom circuit, in which two metallic circuits may provide a third telegraph or telephone circuit without interference from another by placing a repeating coil at both ends of each of the two wire circuits and connecting the center points of these coils.

Wire Tapping. The act of attaching devices to signal wires in order to read any messages being sent over them. Cutting in on a wire line for the purpose of sending a message.

Wire Trench. A trench carrying electric wires.

Wire-wound Gun. A built-up gun in which some of the reinforcing rings or hoops are replaced by wire wrapped under tension around the inner tube.

Withdraw. The act of withdrawing. To draw back. To retire or retreat. To quit a position.

Withdrawal. A planned retrograde movement by a part or all of a deployed force engaged with the enemy, made for the purpose of improving the tactical situation. A breaking of contact with the enemy in battle as a preliminary to retreat or retirement.

Witness Target. A visible point on which fire is adjusted immediately after adjustment on a target which may become obscured.

Women's Army Auxiliary Corps. A voluntary organization of women for non-combatant service in war.

Woods Fighting. Fighting conducted in woods which are sufficiently dense to prohibit the use of usual methods and formations.

Word of Command. A prescribed word or phrase used in directing the movements of soldiers. The orders of authority.

Working Party. A detail, preferably of one or more complete units, which engages in any class of work, especially the construction of works of fortification. For fortification, working parties are usually divided into reliefs so that work may be continuous during daylight and, if necessary, also at night.

Works. The fortifications about a place. The several lines of trenches occupied by an army. All work done to fortify a position, or for the attack or defense of a fortified place.

Wound Chevron. A gold chevron authorized to be worn by each person in the military or naval service who shall have been wounded in action with an enemy of the United States.

X

X-line. An east and west line of a grid, perpendicular to the Y-lines.

X-X Line. A limiting line in the direction of depth defining the primary responsibility of corps and division artillery for certain classes of fire missions.

Y

Y-Azimuth. An azimuth measured from grid or Y-north.

Yaw. The angle at any instant between the longitudinal axis of a projectile and the tangent to the trajectory at the center of gravity of the projectile. An angular displacement about an axis parallel to the normal axis of an aircraft. To move unsteadily or irregularly. To move irregularly or change direction violently, as a projectile, vessel or aircraft. To steer wildly.

Y-line. In the standard grid system, one of the north and south lines, parallel to the central true meridian of the zone. In an improvised grid, an assumed north and south line. A grid meridian.

Y-Y Line. A limiting line in the diection of depth defining the primary responsibility in the handling of certain fire missons by subdivisions of the corps artillery.

Y-north. The direction of north as established by the Y-lines. Grid north.

Z

Zeppelin Airship. A large dirigible airship, of German manufacture, capable of extended journeys.

Zero Hour. The hour designated for a movement or attack to begin. H-hour.

Zero Lines. The parallel lines from which the direction of fire of guns in battery is determined; each gun has its zero line, which is parallel to the zero lines of the other guns of the same battery.

Zero of a Rifle. The corrected sight settings for a rifle, determined from calibration firing.

Zero Plane. The horizontal plane through a gun position.

Zero Stake. A stake put out to mark a zero line.

Zigzag Trace. A form of trench trace which follows a zigzag line.

Zone. A term of wide application. An area bounded by lines approximately parallel or approximately perpendicular to the front, being one of those into which a theater of operations, combat zone, defensive position, etc., is divided for a particular purpose. An area in which the fire of certain units or a certain class of artillery is delivered under certain circumstances. An area, bounded by concentric circles, in which artillery projectiles fall when a fixed propelling charge is used and the elevation varies from the minimum to the maximum. An area over which projectiles fired with the same data are dispersed, etc.

Zone Bag. An increment of the propelling charge added for each zone of mortar fire.

Zone Defense. A form of defense which includes an outpost area and several successive battle positions, more or less completely organized.

Zone, Objective. See Objective Zone.

Zone of Action. A lane or belt of terrain, usually normal to the front, assigned to a unit in attack, or retirement. See also Sector, Defensive.

Zone of Departure. The belt of terrain, generally parallel to the front, in which troops are massed prior to attack, and from which the attack is launched.

Zone of Dispersion. The area covered by the points of impact of a number of projectiles fired with the same data and under the same conditions; on the ground it is the same as the beaten zone. The zone which would include all impacts of an infinite number of shots fired from a gun using the same firing data for each shot.

Zone of Investment. The belt of ground immediately outside of the line of investment, occupied by the investing forces.

Zone of Operations. The territory which contains the lines of operations, or lines on which an army advances, between its base and its objective.

Zone of the Interior. That part of the territory of or controlled by a belligerent which is not included in the theater of operations.

Zoom. To climb for a short time at an angle greater than the normal climbing angle. Colloquially, to turn on the full power of an airplane engine after having shut off power, as in climbing after stunt flying, especially near the ground. The noise made by an airplane in zooming.

Z-Z Line. A limiting line in the direction of depth defining the primary responsibility of corps and army artillery for certain classes of fire.

ADDENDA. MODERN SLANG

A

Ace Pilot. A good pilot.
Ack Ack. Antiaircraft fire.
A.G. Adjutant General.
Air Hog. One who flies at every opportunity.
Air Log. An altimeter.
Air-blind. Nervous and watching the instruments constantly.
Amex. American soldiers in the first World War.
Angel's Whisper. Bugle call.
Anzac. An Australian or New Zealand soldier.
Archie. Antiaircraft artillery or shell.
Armored Cow. Canned milk.
Army Bible. Regulations.
Army Strawberries. Prunes.
Awkward Squad. Men who require extra instruction at drill.
Axle Grease. Butter.

B

Babe. The youngest man in his class.
Baby Carriage. Machine gun cart.
B-Ache or Belly-Ache. To complain.
Back-biter. A tattle-tale.
Bad Time. Time spent in the guard house.
Badgy. One who enlists under age.
Bail Out. To jump from an aircraft in flight with a view to landing by parachute.
Baldie. A haircut.
Battery Acid, Blackstrap, Bootleg. Coffee.
Bean Gun. A rolling kitchen.
Beans. A mess sergeant or the commissary.
Bean-shooter. Officer in charge of the commissary.
Bearded Lady. A searchlight with diffused rays.
Beast. A new cadet while receiving preliminary training.
Beef. To complain.
Behavior Report. A reply to a love letter.
Belly Robber. A mess sergeant.
Bend. To damage an airplane.
Bertha, Big Bertha. Applied to various German guns of large caliber, from Frau Bertha Krupp.
B. G. A Brigadier General.
Big John. A recruit.
Big Willie. Pilot model of the Mark 1 British tank, the original tank; also known as Mother.
Big-boat Pilot. A good sport.
Blab Off. To talk out of turn.
Black Strap. Liquid coffee.
Blanket Drill. Sleep or bunk fatigue.
Blighty. England, or the mother country.
Blind. A forfeiture of pay without confinement by sentence of a court-martial.
Blitz. Bombardment.
Blitzes. Air patrols.
Blob-stick. A bayonet training stick.
Blood. Ketchup.
Blooied. Relieved from duty for reclassification.
Blue Devils. French Alpine Chasseurs.
Blue Ticket. A discharge for disability, confinement by civil authorities after sentence by a civil court, or the like. The discharge paper itself.
Bobtail. Discharge without honor.
Boche. A German soldier.
Bolo Squad. Men who show inaptitude in marksmanship.
Bone. To study.
Bone Bootlick. To cultivate; to curry favor.
Booked It. To make a mess of anything.
Bootleg. Coffee.
Bootlick. To flatter.
Boudoir. A squad tent.
Bowlegs. A cavalryman.
Brace. An exaggerated position of attention.
Brass Hats. High military officials.
Brassed or Browned Off. To be very bored.
Break. To reduce a noncommissioned officer to the grade of private.

Brown Bombers. C. C. pills.
Bubble Dancing. Washing dishes while on kitchen police.
Buck Private. A private without specialist rating, the lowest military rank.
Bucket Carriers. Ration parties.
Bucking for Orderly. To give special attention to uniform and equipment before going on guard with a view to competing for orderly for the commanding officer.
Buddy. A companion or pal.
Bug. Any solid in soups; a defect in a machine.
Bugle. To avoid reciting in class.
Build a Wooden Horse. To make an imperfect landing.
Bull. Bull Durham smoking tobacco; a line of chatter.
Bull Pen. Guard house or place of confinement.
Bully Beef. Canned meat.
Bunk. One's bed.
Bunk Fatigue. Lying on one's bed when off duty.
Bunk Flying. Aviation talk in barracks.
Bunkie. A tentmate or pal.
Bus, Crate, Buggy, Wagon, Wreck. An airplane.
Bust. To reduce a noncommissioned officer to the grade of private.
Busted. Reduced to the grade of private.
Busy Bertha. A German 42 centimeter shell.
Butcher. The company barber.
Butt. Balance of enlistment period.
Buttoned Up. Orders carried out.
Buzzard. A discharge paper.
Buzzard Meat. Chicken or turkey.
Buzzing a Town. To fly over it.

C

Cabbage. Money.
Cadet Widow. A young lady who has known and been popular with several classes of cadets.
Campaigner. An old or veteran soldier.
Canned Horse, Canned Willie. Canned beef.
Cannon Fodder. A derisive name applied by pacifists to men called to the colors in defense of their country.
Caraburger. A steak or hash, etc. of carabao meat.
Carry On. To continue or to resume a situation or occupation.
Caterpillar Club. A mythical organization among flying men to which only those who have jumped for life from a disabled aircraft are eligible.
Cauliflower. A trench mortar shell with wings to destroy wire entanglements.
Chasing Prisoners. Guarding prisoners at work.
Chat. A louse.
Chicago Atomizer. An automatic rifle.
Chinese Attack. An attack simulated or threatened by the use of dummies.
Chinese Landing, Chinese Ace. To land with one wing low.
Chinese Three Point Landing. A crash.
Chips. Regimental pioneer sergeant; a carpenter.
Chow. Food; meals.
Chow-hound. A man fond of eating.
Chucking a Dummy. Fainting at any formation.
Chute. A parachute.
Circus Water. Iced drinks with meals.
Cit. A civilian.
Cits, Civies. Civilian clothing.
Clara. The all-clear air raid signal.
Climb. To scold or berate.
C.O. The commanding officer.
Coal Box. A German high-explosive shell which emits a heavy black smoke.
Coffee Cooler. One who seeks and secures a detail more congenial than his regular duties; one who is always looking for an easy job.

Cold Feet. Lack of courage; fear.

Com. The commandant.

Comb It. Make a detailed inspection.

Come-Alongs. Loops of barbed wire thrown over the heads of prisoners to force them to come along.

Come and Get It. The call of the cook when mess is ready.

Commence Firing. To request one to begin talking.

Commissaries. Groceries.

Completely Cheesed. Badly bored.

Cootie. A louse or flea.

Corn Willie. Corned beef.

Cosmolines. Coast artillery.

Coughing Clara. A British aircraft gun with a hoarse report.

Couple of Jugs. Several beers.

Crab. To find fault.

Crash. An omelet.

Crate. An airplane.

Crawl. To admonish; to berate or abuse.

Creep. A bicycle.

Crump. A German shell.

Crump Hole. A crater made by an explosion.

Cuckoos. Dive bombers.

Cum-Shaw. Graft or bribe.

Cushy. Soft and easy.

D

D.A. Delayed action bomb.

Dad. Oldest member of a class, group, society, etc.

Daddy. A medium caliber German gun; Grandpa was the largest, Grandma the next, Daddy the next, and so-on down to Emme-Gee, the machine gun.

D.D. Dishonorable discharge.

Dead Soldier, Dead Dog. An empty beer or liquor bottle.

Devil's Piano. A machine gun.

Dizzies. High mounds of earth where communicating trenches widen or change direction.

Dizzy Pilot. One who is always taking chances.

Dock. Forfeiture of pay by sentence of a court-marital; the hospital.

Dodo. A ground school student; a new flying cadet.

Dog and Bystander. A standing and a prone silhouette target set up side by side.

Dog Face. An old soldier.

Dog Fight. A general melee in the air.

Dog Leg. A chevron.

Dog Show. Foot inspection.

Dog Tags. Identification discs.

Dog Tent. A shelter tent.

Doodle Bug. A tank.

Dopey Crate. A slow plane.

Doughboy. An infantry soldier.

Dough-Puncher. A baker.

Down Wind. In a predicament.

Downhill. On the last half of an enlistment period.

Draftee. One who has been drafted for military service.

Drag. Influence with one's superiors; to take a lady to a hop or other entertainment; the lady so escorted.

Draped. Drunk.

Draw the Long Bow. To indulge in exaggeration.

Driving the Train. Leading two or more air squadrons into battle.

Dry Run. A practice run on the target range in which firing is simulated.

Duck. An amphibian airplane.

Duckboards. Boards used in the bottom of wet and muddy trenches.

Dud. A shell or bomb that fails to explode; a failure.

Duff. Any pastry, dumpling or pudding.

Dustbin. Enemy rear gunner's lower position in an airplane.

E

Eagle. A flying student; a bomb.

Elsie. A fine imposed by an inferior court-martial.

Emma-Gees. Machine guns.

Empennage. Human buttocks.

Ether. Radio telephone.

Erks. Airplane mechanics.

E-Boats. Enemy torpedo boats.

Eye Wash. Prettying up barracks and grounds for inspection. Anything done or added for the sake of effect only.

F

Fag. A cigarette.

Fanny. A plane's empennage.

Fat Friends. Balloons.

Fem. A woman, or young lady.

File. A number on the promotion list; a male person.

Fin. The vertical stabilizer of an airplane.

Fin Out. To turn the palms to the front

Find Out the Score. Get the correct information.

Fireworks. Effect of antiaircraft shells.

Fishtail. The motion made when the tail of an airplane is swung from side to side to reduce speed in approaching ground for a landing.

Flap. An alarm; a row; excitement.

Flat Spin. To act in a befuddled manner.

Flat Tire. A failure.

Fleas. Small Fords and Austins

Flippers. The elevators of an airplane.

Flower Pot. The power turret of a plane.

Fly by the Seat of his Pants. To fly by instinct.

Flying Elephant. A balloon.

Flying Pig. A large German aerial torpedo.

Flying Shotgun. Shrapnel.

Flying Streamers. Effect when a plane swirls downward.

Flying the Iron Beam. Flying along a railroad.

Flying the Wet. Flying along a river.

Fogy, Fogie. A periodical increase in pay for additional years of service; an old soldier.

Foot. The infantry; an infantry soldier.

Football. An exercise, raising and lowering the legs while lying on the back.

For It. In trouble; to be tried.

Found. To be deficient or wanting in anything, as in an examination.

Fox Hole. An individual rifle pit.

400W. Maple syrup.

French Leave. Unauthorized absence; A.W.O.L.; desertion.

Fritz. A German soldier, shell, airplane, or the like.

Frog. A French soldier.

Frog-Sticker. An infantry sword.

Funeral Glide. A plane out of control.

Fuselage. One's body.

G

Gas House. A beer garden or saloon.

Gasoline Cowboy. One in the armored force.

Geese. Enemy bomber formation.

General's Car. A wheelbarrow.

Get a Pasting. To be badly bombed.

Get Off the Ground. To make the grade.

G.I. Anything carried by the Quartermaster.

G.I. Can. Galvanized iron garbage can.

Gigolo. One who receives a phone call from a girl.

Gimper, Goofer. A skilled airman.

Give It the Gun. To step on the gas.

Glasshouse or **Conservatory.** A power-operated turret.

G-Man. Garbage man.

Go Into a Tailspin. To get mad.

Goat. The lowest ranking man in a class, post, regiment, etc.

Goaty. Awkward or deficient in knowledge.

Gold Brick. A lazy person; an unattractive girl.

Gold Fish. Canned salmon.

Gone West. Dead or fatally wounded.

Goof Off. To make a mistake at drill.

Grandpa. The largest caliber German gun.

Grapevine. Rumor or unconfirmed gossip.

Grass-cutter. A small airplane bomb used against personnel.

Gravel Crushers. The infantry.
Grayback. A German soldier; a louse.
Grinders. Static in radio.
Gringoes. Mexican name for American soldiers, or Americans in general.
Gripe. To grumble.
Groundhog. A nonflying member of the air service.
Grousing. Grumbling or complaining.
G-2. Inquisitiveness.
Guardhouse Lawyer. One having a slight knowledge of military law and regulations, and who is liberal with advice to men in trouble.
Gugus. Philippine natives.
Gun. The throttle of an airplane engine.

H

Hand Grenade. A hamburger.
Hanger. A drinking establishment.
Hard-boiled. Strict or severe and unyielding in discipline.
Hash Mark. A service stripe.
Hay Burners. Cavalrymen.
H.E. High explosive.
Heap, Job. An airplane.
Hedge Hopping. Flying close to the ground.
Heinie. A German soldier, airplane, shell or gun.
Hell Buggy. A tank.
Hickboo. Signal that hostile aircraft are approaching.
Highbrow. An officer regarded as a military pedant; a member of the general staff.
Higher than a Georgia Pine. Unduly excited. cited.
Hike. To march; a march; the act of hiking.
Hit the Deck. To land a plane.
Hit the Hay. Go to bed.
Hitch. An enlistment period.
Hive. To catch on or discover.
Hobo. The provost sergeant.
Holy Joe. A chaplain.
Hop. To rise from the ground in flight; a flight in an aircraft; a dance.
Horizontal Flying. Talking shop while lying down.
Hornet. A bomber.
Horse. Cavalry; a mounted soldier.
Horsing er' Down. Flying close to the ground and scaring people.
Hot Cat. The French Chauchat automatic rifle, used in the First World War.
Hot Crate. A fast plane.
Hot Landing. A landing at too high speed.
Housewife. A sewing kit.
How. The customary army salutation before taking a drink, meaning "My best regards," etc.
Huffed. Killed.
Huns. Germans.

I

I.C. Inspected and condemned for further military use by an inspector, after which these letters are placed on any article so acted upon.
In a Spin. In an unsettled state of mind.
In Stir. In confinement as a punishment.
Ink or Java. Coffee.
Irish Grapes. Potatoes.
Iron Ration. The emergency ration.
Iron Horses. Tanks.

J

Jake. All right, in good order.
Jam Pots. Small bombs made of jam or other tin cans.
Java and Side Arms. Coffee with cream and sugar.
Jawbone. Credit; to buy without money; to fire a weapon over a qualification course when it does not count for record.
Jawbone Corporal or Sergeant. An acting corporal or sergeant.
Jazzing It. Diving close to the ground with an airplane.
Jeep or Jitterbug. A reconnaissance car.
Jeepy. Screwy.

Jeeter. A lieutenant.
Jerry. A German soldier, plane, shell or bomb; a steel shrapnel helmet.
Jinking. Dodging antiaircraft fire.
Jobs. Girls, classified as "neat jobs," "fast jobs," etc.
Joe-Emma. A trench mortar.
John. A recruit or newly inducted man.
Josephine. The 75mm gun.
Joy Ride. A short flight for pleasure only; the initial flight given a student aviator.
Joy Stick. The Joyce stick, named for the inventor of the airplane control system; the control stick of an airplane.
Joy Hop. A flight for pleasure only.
Jump. To admonish; to censure severely.
Jungle Money. Canteen checks.
Junk. Salt Meat.

K

Kamerad. Comrade; a German appeal for quarter.
Keen. A joke or witty saying.
Kennel Ration. Hash or meat loaf.
Khaki. Cotton uniform.
Kick. A dishonorable discharge.
Kicking 'er Around. Manuevering dangerously in an airplane.
Kicks. Shoes.
Kid. A bomber co-pilot.
Kiltie. A Scottish Highlander in kilts.
Kip. Bed.
Kiwi, Kewie. A groundhog or air service man who does not fly; a timid pilot.
K.O. The commanding officer.
K.P. Kitchen Police.

L

Ladies from Hell. German name for Scotch soldiers in kilts.
Lame Duck. A damaged plane.
Lance Jack. A lance corporal.
Lance Sergeant. An acting sergeant.
Landing Gear, Running Gear. One's legs.
Latrine Rumor. An unsubstantiated tale.
Latrine Sergeant. An enlisted man who issues unauthorized orders.
Leatherneck. A marine.
Leviathan, Mary Ann, Vampire. A tank.
Lingo. Language.
Little Poison. The 37mm gun.
Long-Faced Chum. One's horse.
Loggy Ship. A plane that handles sluggishly.
L.P. A leg-puller; a girl who cannot dance.
Loot. A lieutenant.

M

Mad Minute. Firing 15 rounds from one's rifle in one minute.
Made. To be promoted to noncommissioned officer grade, or from corporal to sergeant, etc.
Make. To promote a soldier; a soldier so promoted.
Mae West. An aviator's life belt or jacket.
Mainten. Overhaul base.
Major. The sergeant major.
Make a Forced Landing. To slip and fall down.
Makings. Cigarette papers and tobacco.
Manchued. Relieved from a duty in compliance with the "Manchu Law" passed in 1912, the year the Manchus were dethroned in China.
M and D. Medicine and duty for one feigning sickness.
Mark Time. To spend time in any rank or grade, place or position.
Meat Hound. One who always thinks of eating.
Meat Wagon. An ambulance.
Mex Rank. Temporary promotion to a higher grade, as in war time.
Mickey Mouse. The lever that releases the bombs in an airplane.
Mill. The guardhouse or other place of confinement; an airplane motor.
Minnehaha. A minenwerfer or German trench mortar shell.

Minnies. German trench mortars or their shells.

Misery Pipe. A bugle.

Missouri National. A well known tune that induces rain.

Mitt Flopper. A chronic handshaker.

Mona. An air raid signal.

Monkey. A company clerk.

Mopping Up. To crush hostile units passed over by the assault echelon.

Mother. An English gun of 9.2 inch caliber; the pilot model of the M1 tank, also known as Big Willie.

Mother McCrea, Mother Machree. A sob story; a sad story.

Mud-Crusher. An infantryman.

M.P. Military police.

Mufti. Civilian clothing.

Mule Skinner. A packer or teamster.

Mustard. A smart pilot.

N

Napoo. No more.

Night Bomber. A pilot who sleeps in the daytime and visits night clubs at night.

No-Man's-Land. The ground lying between hostile trenches or lines.

Noncom. A noncommissioned officer.

No Savvy. I do not understand.

Nose-Over. The accidental turning over of an airplane on its nose on landing.

O

O.D. Olive drab woolen uniform; the officer of the day.

Office. The cockpit of an airplane.

O.G. The officer of the guard.

Old Issue, Old File, Old Settled. An old soldier.

Old Glory. An affectionate name for the flag of the United States or the Stars and Stripes.

Old Man. A company or regimental commander; sometimes applied to any commander.

On Official Terms. Means that two people avoid relations except on official matters.

On the Carpet. To be called before the commanding officer for admonition or discipline.

Onion. A dope or wet blanket.

Onions. Antiaircraft shells.

Openers. Cathartic pills.

Orderly Bucker. One who makes every effort to be selected as orderly for the commanding officer.

Outfit. One's organization or unit.

Outside. Civil life or outside the service.

Over the Hill. To desert or in desertion.

Over the Top. The jump-off of an attack.

Over There. The European battlefields.

Overslawed. To be jumped by one's juniors in promotion.

P

Pacifico. A peaceful person; a pacifist.

Packed Up. Dead.

Padre. A chaplain.

Pancake. To level a plane off a very short distance above ground in landing.

Pass the Buck. To shift responsibility to another.

Paul Pry. A searchlight.

Pay Hop. A flight to qualify for flying pay.

Pearl Diver. A man on kitchen police.

Peashooter. A pursuit plane.

Peel Off. To curve away from other aircraft.

Peep. A bantam car.

Penguin. A ground school student; a non-flying member of the air force.

Peppy Crate. A plane that flies well.

Pie Wagon. An inclosed tool wagon.

Piece. One's rifle or gun.

Pig Boat. A submarine.

Pill Box. A concrete emplacement for machine guns.

Pill Rollers. Medical department soldiers.

Pills. A medical department sergeant.

Pineapple. A hand grenade.

Pipped. Wounded.

Plain Butch. A medical officer.

Plebe. A new cadet. A first year man at the U.S.M.A.

P.M.E. Practical Military Engineering, as taught at the U.S.M.A.

Police, Police Up. To clean up.

Poilu. A French soldier.

Popsickle. A motorcycle.

Pounding His Ear. Sleeping.

Pretty Perch. A good landing.

Prop. The propeller of an airplane.

Prop Wash. An expression of disbelief.

Punk. Light bread; very poor.

Pulpit, Office. The cockpit of an airplane.

Pup Tent. A shelter tent.

Purple. Warning that enemy aircraft are approaching.

Pushing up the Daisies. To be dead and buried.

Put her on Hot. To make a fast landing.

Put on the Feed Bag. Go to mess.

Put up a Black. An unsuccessful effort to carry out an order.

P.X. The post exchange.

Q

Q Company. A squad for special training for awkward soldiers; the awkward squad; recruit receiving center.

Quirk. A flying student.

Q.M. The quartermaster or supply officer.

R

Rain Room. A bathhouse.

Ranked Out. To be superseded or compelled to vacate by a senior, as quarters.

Rat Trap. A balloon barrage.

Red. Next warning state after PURPLE.

Real Gen. Inside information.

Red Tape. Routine procedure, rigidly adhered to.

Red Leg. An artilleryman.

Regimental Monkey. The drum major.

Returns. Casualties sent to the rear.

Re-up. To reenlist.

Right Drill. Something done correctly.

Roller Skates. Tanks.

Rookie. A recruit.

Ropey. Rotten weather.

Rosalie. Cold steel or the bayonet.

Rum Hound. One who likes his liquor.

S

Salivate. To knock out.

Sammy. An American soldier in the first World War; black molasses.

Sand Rat. One on duty in the target butts.

Santa Claus in the Pits. A good target record.

Sarge. A sergeant.

Sausage. A non-rigid airship; an observation balloon shaped like a sausage.

Saw-bones. A doctor.

Scag. A cigarette.

S.C.D. Discharge on surgeon's certificate of disability.

Scrounger. One who is resourceful in getting what he wants.

Seconds. Second helpings at mess.

See the Chaplain. Shut up!

Sep. A cadet who enters the Military Academy about September 1.

Serum. Liquor.

Set her Down. To land an airplane.

Sewer Trout. White fish.

Shanghaied. Transferred without one's requestor contrary to one's wishes.

Shavetail. A newly commissioned second lieutenant.

Shoestring Corporal. A lance corporal.

Shoot. Go ahead and talk.

Shoot Down in Flames. To give someone a bawling out.

Shot. An inoculation.

Show. An action in the air.

Shut Eye. Sleep.

Shutter. Camphor or opium pills.

Sideslip. Bread and butter.

Silent Susan. A German shell of high velocity.

Six Months and a Kick. A sentence of six months in prison and a dishonorable discharge.

Skipper. Commander.

Skirt Patrol. A man in search of a girl.

Sky Pilot, Sky Scout. A chaplain.

Slum Burner. A cook.

Slumgullion, Slum. Meat and vegetable stew; hash.

S.O.L. Sure out of luck.

Sojer, Soldier. To shirk; to make a pretense of working.

Soldiers' One Per Cent. An agreement by which a soldier borrows a dollar and pays back two dollars on pay day.

Sound Off. Speak up.

Soup. Thick clouds or fog, wet weather, etc., for flying.

Soup Job. An extra fast plane.

Sow-belly. Bacon or salt meat.

Sparks. A radio operator.

Spuds. Potatoes.

Started to Spoil. Under the influence of liquor.

Stars and Stripes. Baked beans.

Step. A grade in promotion.

Stick. The control column of an airplane. See Joy Stick.

Stir. Confinement, in a guardhouse or prison.

Stone Crushers. The infantry.

Stovepipe. A trench mortar.

Street Monkeys. Bandsmen.

Striker. A soldier who works for an officer.

Stripes. Chevrons.

Stunt. To perform aerial acrobatics.

Sugar Report. A letter from one's best girl.

Suicide Club. Machine gunners, grenade carriers, and flame throwers.

Sweat. To expect.

Sweet Pilot. A good pilot.

Swinging the Lead. Telling the tale.

T

Tac. A tactical officer.

Tail-end Charlie. The rear gunner in a bomber.

Tail Skid. A lamb chop.

Take an Outside Loop. Give someone a runaround.

Take another Blanket. To reenlist.

Take Off. To bawl out.

Take On. To reenlist.

Taxpayers' Pudding. An elaborate bread pudding enriched with fruits and sauce.

Tear Off a Strip. Give someone a bawling out.

Tell that to the Marines. I don't believe what you say.

Three-Point Landing. Ham and eggs.

Ticket. Certificate of discharge.

Tiger Meat. Canned meat.

Tin Can. A destroyer.

Tin Drawers, Pants, Plus Fours. Wheel cowlings.

Tin Hat. A steel helmet; a combat soldier.

Tin Soldier. One who is or has been a cadet at a military school.

To Herd. To pilot an airplane.

To Porpoise. To land in a series of bounds.

Toe Parade. Inspection of feet.

Tommies, Tommy Atkins. British regular soldiers.

Tommy Gun. A Thompson submachine gun.

Top Sergeant, Top Kick, Top Knocker. A first sergeant.

Total Loss. A failure.

Tub. A scout car.

Tuck-Emmas. Trench mortars.

Turned In. Reported for delinquencies.

Turtle. A very destructive German hand grenade in the form of a turtle.

Typewriter. A machine gun or automatic rifle.

V

Valley Forge. A temporary camp during a cold spell.

Visiting Firemen. Visiting officers to whom courtesies must be paid.

Vulture. A half-trained student pilot.

W

WAAC. A member of the women's Army Auxiliary Corps.

Wagon Soldier. A field artilleryman.

Walkie-Talkie. A portable radio.

Waldorf. The mess hall.

Ward Mamma. A hospital nurse.

Wash-out. The complete wreckage of an airplane; to fail in any test.

When the Balloon Goes Up. When things begin to happen.

Whiz Bang. A projectile of great velocity, so-called because the sound of its flight and explosion come close together.

Wife. One's particular pal, a roommate or tentmate.

Wig-Wag. To send a message by means of signal flags.

Wind Jammer. A trumpeter or bandsman.

Wood Butcher. A company artificer.

Wooden Willie. A tiresome exercise with the rifle in the position of aim.

Y

Yard Bird. A raw recruit.

Yellow Leg. A cavalryman.

Yellow Ticket. A dishonorable discharge, made out on yellow paper.

Z

Zep. A Zeppelin airship.

Zeppelins In a Fog. Sausage and mashed potatoes.

Zero. Anyone ignorant of military matters.

Zoom. To turn on full power of an airplane engine after having shut off its power.